NUREG/CR-7156
PNNL-19222

I0482622

Fitness for Duty in the Nuclear Power Industry: An Update of Technical Issues on Drugs of Abuse Testing and Fatigue Management

Office of Nuclear Regulatory Research

United States Nuclear Regulatory Commission

Protecting People and the Environment

NUREG/CR-7156
PNNL-19222

Fitness for Duty in the Nuclear Power Industry: An Update of Technical Issues on Drugs of Abuse Testing and Fatigue Management

Manuscript Completed: August 2011
Date Published: June 2013

Prepared by:
Kristi Branch, Kathryn Baker

Pacific Northwest National Laboratory
P.O. Box 999
Richland, WA 99352

Marina Skumanich, Nancy Durbin

Nancy E Durbin Consulting

DaBin Ki, NRC Project Manager

Office of Nuclear Regulatory Research

ABSTRACT

This report is part of a series of updates of technical issues concerning fitness for duty in the nuclear power industry. It discusses technologies relevant to the detection and management of two key elements of a fitness-for-duty program: drug and alcohol testing and fatigue management. On drug and alcohol testing, the report provides an introduction to the pharmacokinetics of drugs of abuse in different bodily fluids and substances (matrices), a review of the technologies used to separate, identify, and quantify drugs in workplace drug testing programs, and a description of emerging research in developing and validating the technology systems capable of testing alternative matrices as well as newly appearing drugs of abuse, both in the laboratory and at the point of collection. On fatigue management, the report reviews recent research on sleep and fatigue, describes efforts under way to develop and deploy technologies to aid fatigue assessment and management, reviews the status of fatigue management in industries and governmental sectors where fatigue is a significant safety concern, and discusses implications for the nuclear power industry. Finally each chapter includes an extensive bibliography of documents to support further, more in-depth reviews.

TABLE OF CONTENTS

LIST OF FIGURES

LIST OF TABLES

EXECUTIVE SUMMARY

The U.S. Nuclear Regulatory Commission (NRC), in 10 CFR Part 26 (73 FR 16966; March 31, 2008), requires nuclear power plant (NPP) licensees and other entities regulated by the NRC to implement fitness-for-duty programs. Part 26 specifies a number of elements that must be included in a licensee's fitness-for-duty program. Two program elements are the focus of this report: testing for drugs of abuse and fatigue management. The science and technologies associated with drug testing and fatigue management are rapidly evolving. The NRC has an ongoing interest in developments in these areas. Therefore, this report provides an update on (1) innovations in technologies pertinent to drug testing and (2) advances in fatigue science and the tools and technologies available to manage fatigue and its consequences.

Drug Testing: Since the NRC first required licensees to implement drug and alcohol testing programs in 1989, workplace testing for drugs of abuse has expanded greatly among both Federally regulated and private-sector employers. In 1987, Congress mandated the U.S. Department of Health and Human Services (HHS) to establish requirements for drug testing programs in the Federal workplace (Section 503 of Public Law 100-71 also known as The Drug Testing Workplace Act of 1988). The HHS published the first Mandatory Guidelines for Federal Workplace Drug Testing Programs (53 FR 11970) on April 11, 1988 and has continued to update them in response to research findings, implementation experience, and technological advances. The NRC relies on the HHS Guidelines for the majority of the drug testing requirements that are incorporated into Part 26.

In 2004, the HHS published proposed revisions to the HHS Guidelines (69 FR 19673) which would have expanded the Federal drug testing program to (1) permit testing of specimens other than urine to include hair, oral fluid, and sweat, and (2) allow the use of point-of-collection testing (POCT) for urine and oral fluid specimens. Point-of-collection testing uses devices that assess a specimen's validity and perform initial immunoassay drug tests without the need for a laboratory and associated instrumentation. The final 2008 amendment to the HHS Guidelines (73 FR 71858) did not permit testing of alternative specimens or the use of POCT devices. As a result, the NRC also does not permit the use of alternative specimens or POCT devices for drug testing under Part 26, other than in a few circumstances.

Since 2008, however, the tools available to identify, and measure substances, such as drugs of abuse and their metabolites, are rapidly becoming increasingly sophisticated, more widely available, and, in some cases, easier, less costly and less intrusive to use. Because workplace drug testing programs require very well-validated technologies and place a priority on convenience and cost, they often adopt or adapt innovations developed and tested elsewhere. Basic science, medicine, pharmaceuticals, toxicology, forensic science, and the national security sectors are all active participants in research and development that is generating innovations in many scientific measurement tools, including those essential to drug testing: chromatography, mass spectrometry, and immunoassay.

A major development in chromatography and mass spectrometry technologies has been the ability to link them together in various combinations. These advances permit the identification and measurement of substances at very low concentrations, such as a few molecules. Along with standardized specimen preparation procedures, these technologies are now capable of providing reliable test results for specimens containing drugs or drug metabolites at the low concentrations found in oral fluids, hair, and sweat. Concentrations in these specimens are generally much lower than those found in urine. The equipment and procedures necessary for this type of testing are moving from research laboratories into commercial testing laboratories.

Several institutional and policy questions are raised by the potential use of alternative specimens that have not yet been resolved, however. These questions include, for example, how (and whether) to require comparability across test results for testing that is conducted on different types of specimens.

Immunoassay technologies also are advancing rapidly. The advances are enabling ever smaller and more dilute samples to be tested reliably and accurately by a widening array of specialized immunoassay tools. Immunoassays are central in many scientific fields: biology, medicine, pharmaceuticals, toxicology, and forensics. Innovative immunoassay techniques are being developed not only for use in laboratories, but increasingly, in POCT devices, which are becoming more sophisticated, capable, easier to use, and in many cases, less expensive. Micro- and nanotechnology innovations are contributing to the development of ever smaller and more capable "labs on a chip." Some POCT devices on the commercial drug testing market include tests for multiple specimen attributes to confirm specimen validity and for multiple drugs of abuse. Some devices provide test results in minutes rather than hours or days. Numerous evaluation studies have compared the performance of different immunoassay methods and specific collection and testing devices for different drugs in the different types of biological specimens. However, questions do remain about the consistency of results from POCT immunoassay testing devices and a number of implementation issues with onsite drug testing using POTCs remain unresolved and are currently being reviewed by HHS.

The amount of practical information available about the use of alternative specimens and new testing technologies is growing, however. Some of the innovative collection techniques, such as sweat patches, have been pilot tested and adopted by criminal justice, addiction recovery, and clinical medicine programs. Pilot tests of devices to collect and test oral fluids at the roadside have identified, and contributed to the resolution of, issues with the collection devices. For example, the roadside pilot tests demonstrated that differences in collection procedures can significantly affect test results. A growing number of private-sector workplace testing programs are also successfully using alternative specimens and POCT devices for drug testing.

Fatigue Management: The NRC's fatigue management requirements in Part 26 were, in part, based on an extensive review of the available scientific literature and consideration of the technologies available to support the prediction, detection, and response to fatigue in the workplace. In the decade after the scientific and technical bases for the fatigue management requirements in Part 26 were developed, research related to fatigue management has advanced.

Scientists are continuing to conduct research on the neurophysiological processes that regulate sleep, wakefulness, and circadian cycles. The same advancements in technologies to identify and measure extremely low concentrations of drugs and drug metabolites have also improved scientists' ability to identify and measure the biochemical substances that are associated with sleep, wakefulness and circadian cycles. This research is providing a better understanding of the mechanisms that create the empirically observed patterns of wakefulness, desire to sleep, and sleep. The most useful outcome from this research for fitness-for-duty programs would be biochemical tests to aid in assessing fatigue.
Continuing research is also addressing the underlying neurophysiological and genetic bases for individual differences in susceptibility and resilience to fatigue and individuals' different patterns

of sleep and wakefulness. The extent of individual differences has made it difficult to develop reliable methods for predicting individual responses to short-term sleep loss or prolonged periods of restricted sleep. More research is needed, but the significant advances made during the past decade provide the groundwork for clarification of some of these fundamental mechanisms and traits in the relatively near future. The results of this research will enable further refinement of the biomathematical models that are increasingly used to design work schedules that manage fatigue effectively.

Research on the underlying mechanisms of sleep regulation is being combined with a growing body of information about the physiological indicators associated with different stages in the sleep-wake cycle and states of fatigue. This information is needed to incorporate individual differences in biomathematical models and to develop more precise tools to predict and monitor impairment from fatigue and circadian cycles. Although tools are now available to measure most of the identified physiological indicators in the laboratory (i.e., core body temperature, melatonin levels, brain wave patterns and brain activity, and pupil reactivity), it has been challenging to adapt these tools to workplace settings.

More success has been achieved in adapting laboratory methods to measure and track physical and behavioral indicators of fatigue to workplace settings. The successes have been enabled largely by the extraordinary advances in sensors, data processing and storage, and wireless communication that has occurred in the last several decades. Portable, durable, and reliable actigraphs (equipment that measures body movement and can accurately determine the duration and quality of an individual's sleep) have allowed this technology to move from sleep laboratories to home and field use. Similarly, laboratory-based technologies to measure eye gaze, eye movement, and eyelid closure have been incorporated into eyeglasses and workplace monitors that are small and robust. Efforts are under way to integrate the two technologies into tools that are acceptable to the users and proven effective in enhancing safety.

A key focus of fatigue research has been to develop instruments to measure impairment from fatigue and adapt the instruments for field deployment. The approach is to integrate physiological and physical/behavioral data with measures of cognitive and psychomotor impairment, such as slower psychomotor and cognitive speed, degraded alertness/attention, and decrements in memory. A substantial database has been established linking degree of sleep restriction to performance on specific psychomotor vigilance tests in various contexts. Less attention has been given to relating impairment to actual job performance, except in transportation and aerospace. Substantial work remains to adapt, field-test, and validate the benefit of fatigue monitoring and prediction tools to other work environments.

Conclusions from this review of potential interest to the NRC and its licensees include:

- Tools to evaluate work schedules in terms of their impact on fatigue are now commercially available and have been shown in other sectors to be useful for informing workforce and crew scheduling approaches and strategies.
- Technologies, such as actigraphs, have been tested and used in a sufficient number of contexts that they can be considered potential resources to address particular fatigue management issues, including verification of the effectiveness of fatigue management practices.

- Analytic tools now available could be used to assess and enhance fatigue management program effectiveness by helping to identify where and why any fatigue hazards may be occurring.

ACKNOWLEDGMENTS

Preparation of this report has benefited from the assistance and support of many individuals. We gratefully acknowledge the assistance provided by DaBin Ki, project manager, and Valerie Barnes, senior level advisor. We would also like to thank Jodi Amaya for assistance with the patent search, Jessica Sandusky for her assistance obtaining and reviewing information about the various drugs of abuse testing technologies, Sadie Johnson and Burk Dowell for assistance with report preparation, and Cornelia Brim and Susan Tackett for technical editing. We would like to acknowledge Thomas Sanquist, Mary Zalesny, Oksana Elkhamri, Jon Olson, Stephanie Morrow, and Mike Boggi for their technical input and attentive technical reviews. In addition, we would like to thank the many individuals who either in-person, by phone, or e-mail helped us find and interpret information about the research and technologies discussed in this report.

The accuracy of the information and the views presented in this report are the responsibility of the authors and do not necessarily represent the opinion of the NRC or of any particular individuals or licensees.

ACRONYMS

CHAPTER 2: DRUGS OF ABUSE TESTING TECHNOLOGIES

6-AM:	6-acetylmorphine
AAFS:	American Academy of Forensic Sciences
AAMRO:	American Association of Medical Review Officers
ABFT:	American Board of Forensic Toxicology
AMA:	American Medical Association
BAC:	Blood Alcohol Concentration
BZE:	Benzoylecgonine, a metabolite of cocaine
CAP:	College of American Pathologists
CE:	capillary electrophoresis
CEDIA:	cloned enzyme donor immunoassay
CLIA:	chemiluminescent immunoassay
DATIA:	Drug and Alcohol Testing Industry Association
DEA:	U.S. Department of Justice Drug Enforcement Administration
DOD:	U.S. Department of Defense
DOT:	U.S. Department of Transportation
DRE:	Drug Recognition Experts
DUI:	driving under the influnce
DUID:	driving under the influence of drugs
EBT:	evidential breath testing device
EIA:	enzyme-immunoassay
ELISA:	enzyme-linked immunosorbent assay
EMIT:	enzyme multiplied immunoassay technique
EtG:	Ethyl Glucuronide, a metabolite of ethyl alcohol
FDA:	U.S. Food and Drug Administration, part of the U.S. Department of Health and Human Services
FFD:	Fitness-for-Duty
FPIA:	fluorencence polarization immunoassay
GC/MS:	gas chromatography/mass spectrometry
HHS:	U.S. Department of Health and Human Services
HPLC:	high performance liquid chromatography or high pressure liquid chromatography
IA:	immunoassay
IAFT:	International Association of Forensic Toxicologists
IEC:	International Electrotechnical Commission
IR:	infrared light
ISO:	International Organization for Standardization
KIMS:	kinetic interaction of microparticles in solution
LC/MS:	liquid chromatography/mass spectrometry (also known as, LC-MS)
LC/MS/MS:	liquid chromatography/tandem mass spectrometry (i.e., serially linked mass spectrometers) (also known as, LC-MS/MS)
LLE:	liquid-liquid solvent extraction
LOD:	limit of detection
MASK:	Multiple Adulterant Strip Kemistry
MDA:	methylenedioxyamphetamine
MDEA:	methylenedioxyethylamphetamine
MDL:	mapping description language

MDMA: methylenedioxymethamphetamine, also known as Ecstasy
MRO: Medical Review Officer
MS/MS: tandem mass spectrometry (i.e., mass spectrometry/mass spectrometry)
NIDA: National Institute on Drug Abuse
NLCP: National Laboratory Certification Program
NPT: near patient technologies
NTSB: National Transportation Safety Board
PCR: polymerase chain reaction
QMS: quadropole mass spectrometer
POCT: point-of-collection testing
QA/QC: quality assurance/quality control
QC: quality control
ROSITA: The Roadside Testing Assessment
RIA: radioimmunoassay
SAMHSA: Substance Abuse and Mental Health Services Administration
SAPAA: Substance Abuse Program Administrators Association
SOFT: The Society of Forensic Toxicologists
SOHT: Society of Hair Testing
SOP: Standard Operating Procedure
SPE: solid-phase extraction
SPR: surface plasmon resonance
SVT: Specimen validity test
THC: tetrahydrocannabinol (an active ingredient in marijuana)
TLC: thin layer chromatography
UPLC: ultraperformance liquid chromatography

CHAPTER 3: FATIGUE MANAGEMENT

AECS: average eye closure speed
ART90:   Act-React-Test system 90
ANS: autonomic nervous system
ASRS: Aviation Safety Reporting System (NASA)
AVCLOS: average length of eyelid closure
BAC: blood alcohol concentration
CDC: Centers for Disease Control and Prevention
CFR: Code of Federal Regulations
DDDS: Drowsy Driver Detection System
DDMS   Driver Drowsiness Monitoring System
DOT: U.S. Department of Transportation
EDS: excessive daytime sleepiness
EEG: electroencephalography
EKG or ECG: electrocardiograph
EMG: electromyography
EOG: electrooculogram
EPVT: enhanced psychometer vigilance task
ESS: Epworth Sleepiness Scale
FAA: Federal Aviation Administration
FACTS: Fatigue Accident Causation Testing System
FAID: Fatigue Audit InterDyne
FAST: Fatigue Avoidance Scheduling Tool

FFD:	fitness for duty
FIRM:	Fatigue Index Risk Module
FMCSA:	Federal Motor Carrier Safety Administration
FRA:	Federal Railway Administration
FRI:	Fatigue Risk Index
FRMS:	Fatigue Risk Management System
GINA:	Genetic Information Non-Discrimination Act
HEV:	heading error variability
HOS:	hours of service
Hz:	hertz (wave frequency in waves per second)
IR:	infrared
KSS:	Karolinska Sleepiness Scale
LDV:	lane deviation variability
MRO:	Medical Review Officer
MSE:	micro-sleep events
NASA:	National Aeronautics and Space Administration
NHTSA:	National Highway Transportation Safety Administration
NPP:	nuclear power plant
NRC:	U.S. Nuclear Regulatory Commission
NREM:	non-rapid eye movement
NTSB:	National Transportation Safety Board
OSA:	obstructive sleep apnea
PDA:	personal digital assistant
PERCLOS:	percentage of eye closure
POC:	point of collection
PTC:	positive train control
PVT:	performance vigilance test (sometimes referred to as psychomotor vigilance test)
QHPT:	Queensland Hazard Perception Test
REM:	rapid eye movement
RNA:	ribonucleic acid
RTP:	readiness to perform
SAFE:	System of Aircrew Fatigue Evaluation Model
SAFTE:	Sleep, Activity, Fatigue, and Task Effectiveness Model
SATS:	Shiftwork Adaptation Testing System
SMS:	safety management system
SOFI:	Swedish Occupational Fatigue Inventory
SRRT:	stimulus response reaction test
SSS:	Stanford Sleepiness Scale
SWA:	slow wave activity
TPMA:	Three Process Model of Alertness
ULR:	ultra-long range

1.0   INTRODUCTION

The U.S. Nuclear Regulatory Commission's (NRC) regulation 10 CFR Part 26 requires nuclear power plant (NPP) licensees and other entities to implement fitness-for-duty programs that meet the five specific performance objectives shown in Table 1.1.

Table 1.1.   NRC FFD Performance Objectives (10 CFR § 26.23)[a]

26.23 Performance objectives
Fitness-for-duty programs must--
(a) Provide reasonable assurance that individuals are trustworthy and reliable as demonstrated by the avoidance of substance abuse;
(b) Provide reasonable assurance that individuals are not under the influence of any substance, legal or illegal, or mentally or physically impaired from any cause, which in any way adversely affects their ability to safely and competently perform their duties;
(c) Provide reasonable measures for the early detection of individuals who are not fit to perform the duties that require them to be subject to the FFD program;
(d) Provide reasonable assurance that the workplaces subject to this part are free from the presence and effects of illegal drugs and alcohol; and
(e) Provide reasonable assurance that the effects of fatigue and degraded alertness on individuals' abilities to safely and competently perform their duties are managed commensurate with maintaining public health and safety.

[a] Source: U.S. NRC 2008. 10 CFR § 26.23.

10 CFR Part 26 specifies a number of elements that must be included in a licensee's fitness-for-duty program. Among them are two key program elements that are the focus of this report: drug and alcohol testing and fatigue management.

Workplace drug and alcohol testing programs identify individuals who have used/abused drugs or violated the licensee's required drug and alcohol policy. Since June 7, 1989, the NRC has required drug and alcohol testing of NRC licensees and other entities (54 FR 24468). Part 26 establishes the parameters of a drug and alcohol testing program to ensure that it is fair, accurate, valid, effective, and efficient in detecting the consumption of prohibited drugs or the illicit use of alcohol. Drug and alcohol testing programs are now operating within a well-established institutional infrastructure, and the technologies used to separate, identify, and quantify target drugs and alcohol are proven and mature. Nevertheless, interest in testing specimens other than urine (e.g., oral fluids, sweat, and hair) for drugs as well as steady improvements in existing technologies suggest the potential for changes to the established protocols and approaches currently used in workplace testing for drugs of abuse.

Fatigue management seeks to reduce the hazard to safety created by fatigued workers by addressing the causes of fatigue, identifying workers who are impaired by fatigue, and preventing fatigued workers from performing tasks that require alertness and vigilance to protect public safety and security. In 2008, the NRC amended Part 26 to impose requirements related to the management of worker fatigue among NPP licensees. In developing the revised rule, however, the NRC recognized the importance of monitoring the research on sleep, wakefulness, and fatigue; the technologies and practices that are emerging for measuring, assessing, and managing fatigue; and the approaches being taken in other industries to integrate some of the emerging science and technology into regulation and best practice.

This report seeks to address both of these areas of innovation and research. It focuses on emerging technologies and practices as well as the institutional context in which they are (or might be) deployed. The report represents an update to the previous NUREG/CR-6470 (Durbin et al. 1996). Given the breadth of research occurring in the areas of drug and alcohol testing, and fatigue management, it is intended to be the first in a regular series of updates on these issues.

The report is organized as follows: Chapter 2 focuses on Drugs of Abuse Testing Technologies. Chapter 3 addresses Fatigue Management Technologies. Appendix A includes descriptions of available technologies for assessing fatigue and performance. Appendix B presents a method for investigating the potential role of fatigue in events.

2.0 DRUGS OF ABUSE TESTING TECHNOLOGIES

2.1 Introduction

The NRC's regulation, 10 CFR Part 26 (hereafter referred to as Part 26 in this report), requires nuclear power plant (NPP) licensees and other entities to implement comprehensive fitness-for-duty (FFD) programs. Part 26 specifies a number of elements that must be included in a licensee's FFD program. Among them is a drug and alcohol (D&A) testing program to, in part, deter and detect the use of prohibited drugs and illicit use of alcohol.

Workplace drug and alcohol testing programs identify individuals who have used and/or abused drugs or violated the licensee's required D&A policy; they do not measure impairment. Part 26 requires FFD programs to include other elements to detect or identify impairment, for example, behavioral observation. The rule establishes the parameters of a D&A program that ensure its fairness, accuracy, validity, effectiveness, and efficiency in detecting and measuring evidence of certain drugs or the consumption of alcohol. These programs are now operating within a well-established institutional infrastructure. Existing technologies are capable of meeting the requirements established for Federally-mandated testing in the United States by the U.S. Department of Health and Human Services (HHS) and the U.S. Department of Transportation (DOT) in terms of specificity, reliability, and interpretability when applied to urine specimens when testing for drugs, and breath or oral fluids specimens when testing for alcohol.

This chapter provides an update on technologies with the potential to affect workplace testing for drugs of abuse. It is based on a review of the extensive and rapidly expanding literature and discussions with a select set of experts. The main impacts on workplace D&A testing programs are expected to result from the intersection between (1) continued improvements in technologies to collect, screen, confirm, and interpret results for different biological specimens and, (2) concern about workplace impacts of an ever-wider range of drugs of abuse. Many of the innovations discussed in this chapter are already being implemented in private-sector workplace testing programs. However, the development and issuance of drug testing guidelines using alternative specimens by HHS in Federally-mandated programs will raise many policy and implementation issues. This chapter is intended to provide a survey of those issues as they relate to specific technology choices, as well as present preliminary considerations as to how they might be resolved.

This chapter is organized as follows. Following the introduction, Section 2.2 provides a brief overview of the institutional infrastructure of which Federally-mandated workplace testing. Section 2.3 provides an introduction to alternative specimens, pharmacokinetics, the factors that influence the presence of drugs and their metabolites in different bodily fluids and substances (also referred to as matrices), and an overview of the benefits and challenges, and remaining issues about each of the alternative specimens. The characteristics of the drugs being tested and the specimens being used affect the characteristics of the methods and technologies needed to separate, identify, and measure them effectively and efficiently. Section 2.4 provides a description of the current technologies used to separate, identify, and quantify the drugs and their metabolites that are of greatest interest for workplace drug testing programs, and describes emerging research in the development and validation of technology systems capable of testing alternative specimens and additional drugs in the laboratory, and at the point of collection. It concludes with a brief review of innovations that may affect workplace testing in the future. Section 2.5 provides a brief summary of the key areas of innovation and the journals, conferences, and professional associations of interest specific to the NRC. Finally, Section 2.6 provides a bibliography of documents reviewed for this report. It is included both to provide full references for sources cited in the chapter and to illustrate the nature and scale of the work being conducted in this arena. For ease of reference, a glossary of terms and a list of acronyms used in this chapter are included separately in the frontpiece materials of this report.

2.2 Drug and Alcohol Testing Institutional Infrastructure

Since the inception of widespread workplace drug testing in the late 1970s and early 1980s, an extensive research, regulatory, and industrial infrastructure for drug testing has developed in the U.S. and worldwide. This infrastructure extends well beyond the workplace testing arena to include biological and pharmaceutical research; medical research and clinical practice; law enforcement, criminal justice, and homeland security forensics and forensic toxicology; athletic performance enhancement testing; and addiction treatment and recovery follow-up and compliance monitoring. It influences the methods and technologies that are developed, validated, field-tested, commercialized, certified/approved, and institutionalized through guidelines, standards, mandates, and use. This infrastructure, therefore not only includes the development of innovative methods and technologies, but also their availability in laboratories and other service-providing organizations in addition to the administrative burden and cost of their use. This section provides an overview of the evolution of this infrastructure and discusses the role of different sectors in the development and institutionalization of methods and technologies applicable to workplace D&A testing programs.

2.2.1 Workplace Testing: A Well-Established Infrastructure Concerned Primarily with Effectiveness, Efficiency, and Defensibility

In 1970, the U.S. Congress passed the Comprehensive Drug Abuse Prevention and Control Act (PL 91-513) that consolidated regulations of "controlled substances" with the exception of alcoholic beverages and tobacco. Commonly known as the Controlled Substances Act, this legislation established the five categories, or "schedules," of controlled substances, and reinforced the distinction between "legal" and "illegal" drugs. Concern over the use of illegal drugs during the 1970s and 1980s, exemplified by the "war on drugs" initiated by President Nixon in 1971, provided an impetus for workplace drug testing.[1] As directed by President Nixon in Executive Order 11599, the U.S. Department of Defense (DoD) implemented a drug testing program in 1971 to identify service members in need of rehabilitation as they returned from Vietnam. In 1974, with the issuance of DoD Instruction 1010.1, the DoD drug testing program expanded to a random testing program, still focused on identifying individuals in need of treatment. High rates of positive test results for illegal drug use and a serious accident in 1981 on the USS Nimitz in which drugs were identified as a contributing factor led the military to refocus its drug testing program on deterrence, including the imposition of severe sanctions on individuals who tested positive. However, a 1983 review of the U.S. Army and Air Force drug testing procedures concluded that the program did not meet forensic standards and that the results were not legally defensible (U.S. DoD 2009; Caplan and Huestis 2007:732; Langman and Kapur 2006:504).

Although initially required only for military employees, private-sector employers started voluntarily instituting workplace drug testing programs. The early programs faced significant

---

[1] Catlin et al. (1992) point out that this interest in preventing illegal drug use and sale has led to two distinct models for drug testing: (1) a penalty model; and (2) a medical model.

legal challenges and employee opposition, due in part to a lack of established standards and consistency within and between programs. As Caplan and Huestis (2007:732) noted:

> During the 1983-1986 time frame, many companies in the oil, chemical, transportation, and nuclear industries voluntarily implemented drug-testing programs. Without standards and recognized procedures, almost every action incurred controversy. Lawsuits and arbitration caseloads mounted rapidly. Reports of laboratory errors in the massive military program raised concerns that the application of this state-of-the-art technology might be premature. Allegations of employees stripped naked and forced to provide specimens in view of other employees were often repeated and added justification for regulations.

A regulatory framework to address these problems began to emerge shortly thereafter. In 1983, the National Transportation Safety Board (NTSB), in collaboration with the National Institute on Drug Abuse (NIDA), started work on a drug regulation for the DOT (Caplan and Huestis 2007:732; Jenkins 2003:31). In 1984, DoD issued Directive 1010.1 defining drug testing requirements for military personnel and responsibilities for program administration (U.S. DoD 1984, referenced in U.S. DoD 2009).

In 1986, President Reagan issued Executive Order 12564 (51 FR 32889). His Commission on Organized Crime issued its final report and Congress passed the Drug Free Workplace Act of 1986 (PL 99-570) (also known as the Anti-Drug Abuse Act of 1986) creating an additional impetus for workplace drug testing, with a strong focus on deterring the use of illicit drugs. To address the many issues associated with such testing, NIDA convened a conference that led to consensus on some of the principles that have become foundations of workplace drug testing programs, including a requirement for positive results from an initial screening method to be confirmed by a second, alternative method (Caplan and Huestis 2007:733).

In 1987, Section 503 of Public Law 100-71 mandated HHS to specify the general requirements for drug testing programs within the Federal workplace. In 1988, HHS established the first *Mandatory Guidelines for Federal Workplace Drug Testing Programs* (53 FR 11970) and implemented the National Laboratory Certification Program (NLCP) to standardize procedures and quality assurance practices in the laboratories authorized to conduct the required drug tests (Bush 2007; Caplan and Huestis 2007:732).[2] In 1989, both the NRC (10 CFR Part 26 at 54 FR 24468)[3] and the DOT (interim final rule DOT 49 CFR Part 40 at 54 FR 49854) published regulations that required drug testing of private-sector employees. Both regulations incorporated many aspects of the HHS Guidelines and specified that all laboratory tests conducted under the regulations be performed by HHS-certified laboratories. In 1991, The Omnibus Transportation Employee Testing Act required the DOT to conduct drug and alcohol testing of transportation employees performing safety-sensitive jobs in aviation, trucking, railroads, mass transit, pipelines, and other transportation industries. The Omnibus Act required DOT to incorporate the HHS scientific and technical guidelines relating to laboratory standards and procedures (U.S. DOT 2010), further establishing HHS and NIDA as the lead Federal

---

[2] HHS subsequently published a proposed revision to the Mandatory Guidelines that included specifications for the use of alternative specimens and point of collection devices in 2004 (U.S. HHS 2004) and a Final Revision of the Mandatory Guidelines in 2008 that retained urine as the sole authorized specimen and did not authorize use of point of collection devices in 2008 (U.S. HHS 2008).

[3] The NRC has independent regulatory authority to establish requirements for drug and alcohol testing for its licensees: The HHS Guidelines do not apply to NRC-regulated entities. However, one of the NRC's specific goals in making its 2008 revisions to Part 26 was to increase consistency with the HHS and DOT requirements.

agencies responsible for the scientific basis, methods, and technologies for workplace drug testing.

Because of its responsibility for regulating large numbers of safety-sensitive workers in the transportation area, the DOT has established a leadership role in researching and evaluating new testing devices and practices pertinent to the field requirements of its component entities, and for methods and technologies to test for alcohol, which are not covered by the HHS Guidelines. Through the National Highway Traffic Safety Administration (NHTSA), DOT establishes specifications for devices to measure breath alcohol (69 CFR 42237); certifies breathalyzers, and publishes and updates a "conforming products list" for alcohol testing;[4] and establishes training requirements for breathalyzer technicians. NHTSA has also undertaken a wide variety of research projects to evaluate, test, and pilot drug- and alcohol-testing methodologies (e.g., saliva testing for drugs and alcohol in drivers) (Jones et al. 2003; Kadehjian 2005:16).[5] In addition, DOT conducts studies and sponsors workshops to improve the technical basis of testing standards.[6]

The guidelines and regulations issued by HHS, DoD, DOT, NRC as well as other Federal agencies have codified the requirements and procedures for testing programs and have applied drug (and in some cases alcohol) testing to Federal employees and other personnel working in regulated environments. The result has been an expanding demand for workplace drug testing capacity. In 2009, The Substance Abuse and Mental Health Services Administration's Drug Testing Advisory Board (U.S. HHS 2009) reported that the DoD was testing an estimated 4.5 million specimens per year from military personnel, the NRC-regulated entities were conducting about 140,000 tests a year, and DOT regulations covered an estimated 10 million employees, overall.[7]

During this same period, workplace D&A testing became increasingly common in the private sector. Workplace D&A testing is now widespread across workplaces in the U.S. and the infrastructure that supports it is well established.[8] By 2007, Caplan and Huestis (2007:732) estimated that almost half of the American workforce was subject to testing for illegal drugs. Reynolds (2005:7) reported that 67 percent of all major U.S. corporations have drug-testing policies. Many states have enacted regulations governing workplace testing, and in the mid-1980s, the College of American Pathologists established the Forensic Urine Drug Testing Program to provide guidelines for testing programs not governed by the HHS Guidelines or regulations (Cone 2001).

Although slower to adopt workplace and military D&A testing, the European Union and individual countries have become active participants in the development of guidelines and procedures, and the examination and validation of methods, devices, and technologies (Kintz and Agius 2009; Lillsunde et al. 2008; de la Torre et al. 2004). They have been particularly

---

[4] NHTSA issued the first "qualified products" list of evidential breath measurement devices in November 1974.

[5] For example, NHTSA conducted performance evaluations of non-evidential alcohol screening devices for saliva for use in the DOT testing program and in 2001, approved on-site oral fluid (saliva) testing for alcohol in transportation workers ("Conforming Products List of Screening Devices to Measure Alcohol in Bodily Fluids" issued May 4, 2001 National Highway Traffic Safety Administration [Docket No. NHTSA-2001-9324]).

[6] For example, DOT conducted a "Water Loading Study" to determine whether it was possible for an individual to drink a sufficient quantity of water to reduce creatinine levels to below 5 mg/mL and to provide additional information about the range and distribution of creatinine levels in human urine and, with the Federal Aviation Administration (FAA), sponsored a colloquium on "Workplace Urine Specimen Validity" in 2003.

[7] In 2009, the Federal Transit Association (FTA) testing program covered 3,264 employers with 280,731 safety-sensitive employees and conducted over 100,000 random drug tests (U.S. DOT FTA 2010:3).

[8] Many private sector drug testing programs also require the use of an HHS-certified laboratory (U.S. HHS 2004).

active in evaluating technologies to enable roadside testing for drugged drivers (including those who are also using alcohol). De la Torre et al. (2004) reported that the Home Office of the United Kingdom was already doing up to 250,000 oral fluids tests a year in the early-mid 2000s and that up to 60 percent of German police forces were routinely using sweat testing at the roadside.

An infrastructure of private-sector laboratories, service providers, research organizations, technology developers, and professional and industry associations evolved to support the growing demand for expertise, materials, and facilities. An important component of this private sector infrastructure is the HHS-certified laboratories capable of serving the Federally-regulated employers. In 2010, 37 laboratories were certified by HHS to conduct Federally-mandated drug tests on urine (U.S. HHS 2010). The Forensic Toxicology Council (2010) estimates that U.S. laboratories collect and test approximately 6.5 million Federally-regulated workplace samples and 50 million non-Federally-regulated (e.g., private sector or state/local government) workplace samples annually. In addition to urine tests, which continue to be the only type of drug tests authorized by the HHS Guidelines, these laboratories report testing a growing number of oral fluids, hair, and sweat samples (The Forensic Toxicology Council 2010; Quest Diagnostics, Inc. 2011).

As workplace testing expanded, so did a "drug testing subversion" industry. Subversion of the testing process has the potential to jeopardize the benefits of workplace drug testing programs. As noted in NUREG/CR-6470 (Durbin and Grant 1996:2-9), the effectiveness of a drug testing program to identify users of prohibited substances:

> …hinges on valid, accurate results and fairness. Effectiveness is lost if some drug users evade detection and compromise the integrity of the FFD program. When subversion of FFD testing programs occurs, the program fails in its mission to identify and remove drug abuse and its consequent effects. In addition, allowing subversion implies to workers that the program is not taken seriously by management and undermines support for the program. It is also not fair to workers who must put up with a testing regimen but do not have the benefit of removal of those abusing drugs and alcohol from the workplace.

Subversion can take a number of forms, including:

- avoidance of a test (either refusing the test directly or subverting the selection process to avoid testing);
- providing a surrogate specimen for testing (substitution);
- diluting a specimen either in-vitro or in-vivo (dilution); and
- adding an adulterate to a specimen (adulteration).

The industry providing products to subvert the testing process has become larger and increasingly sophisticated.[9] Currently, all Federally-mandated drug testing requirements specify urine as the specimen for drugs and require the specimen to be tested for validity.[10] The DOT and NRC allow oral fluids and breath for initial testing for alcohol, but specify breath

---

[9] A SAMHSA representative noted (Bush 2008b:114): "Marketing products to 'beat the drug test' continues to proliferate. In September 2002, a Google search on the phase 'beat a drug test' yielded 158,000 hits in 0.4s; in May 2005, the same search yielded 1,210,000 hits in 0.6 s."

[10] The 1989 NRC regulation allowed, but did not require, licensees to test urine samples for validity.

as the specimen for confirmatory alcohol testing.[11] The main types of products available to potentially subvert urine testing include: (1) dilution and cleansing products (e.g., teas that, along with consumption of large amounts of water, are intended to "cleanse" the urine of drugs or dilute urine in-vivo); (2) adulteration additives (e.g., chemicals added in-vitro after it is provided in the collection cup); and (3) substitute urine products (e.g., actual urine from a "clean" donor or a dried product advertised to mimic urine when added to water) (Dasgupta 2005). Although many of these products are not effective, "some are effective and detectable, while others are not yet detectable or disappear on their own." (Bush 2008b:115.) During the period from May 2004 to April 2005 HHS-certified laboratories tested 6.8 million specimens under the Federally-regulated program. Of these about 2.1 percent were positive for drugs and 0.15 percent were identified as adulterated, substituted, or invalid[12] (Bush 2008b:116).

As new counter-subversion procedures are developed, new subversion techniques quickly emerge to overcome them (Dasgupta 2010). For example, although Federal drug testing programs (and the HHS Guidelines) do not allow the use of hair, oral fluids, or sweat as alternative specimens, non-regulated workplace programs are using them in increasing numbers. Products such as shampoos for hair and mouthwashes for oral fluids claiming to help subvert these tests are already being heavily marketed. Counter-subversion measures generally lag well behind new subversion techniques due to the time it takes to develop a mitigating strategy, validate it, and eventually incorporate into a regulation. Regulators began considering validity testing in the early 1990s as a way to thwart subversion of urine testing. Validity testing was included in the NRC Part 26 rule in 1998 and in the HHS Guidelines in 2004.

The expansion of drug testing programs has created a market for methods and technologies/devices that can meet the evolving testing specifications faster, less expensively, and with high reliability, in part through less susceptibility to subversion. The Food and Drug Administration (FDA) plays an important role by certifying methods and technologies for screening tests for drugs of abuse. It has established industry guidance for validation of analytical procedures and methods, and reviews test results for new assays, applications for tests of new equipment, and tests using alternative specimens. Based on these reviews, the FDA grants 510(k) approvals and clearances-for-marketing, which allow the commercial sale of cleared products (Reynolds 2005:4). The Clinical Chemistry and Clinical Toxicology Panel of FDA's Medical Devices Advisory Committee makes recommendations for these reviews. The HHS Guidelines require FDA clearance-for-marketing for all products used in Federally-mandated testing programs, including immunoassays. To strengthen the basis for evaluating and approving drug testing methods, NIDA sponsored a national research program on the science, basic techniques, and clinical applicability of drug testing methods, with a particular focus on establishing sensitivity, specificity, reproducibility, and reliability parameters for tests of hair, oral fluids, urine, and sweat (Schultz 1997). The FDA has also initiated the development of standards for the validation of on-site test devices, and guidelines for validation and evaluation of point-of-collection testing (POCT) devices (Reynolds 2005:4-5).

---

[11] A number of recent studies reflect a growing interest in testing for chronic excessive alcohol use in urine, oral fluids, sweat, and hair. The target metabolites are ethyl glucuronide (EtG) and fatty acid ethyl esters (FAEE) (Kintz 2010). Because ethanol is present in so many commonly consumed products, SAMHSA has issued an advisory stating that tests for EtG alone are insufficiently sensitive to be used as a stand-alone indication of prolonged alcohol consumption (U.S. HHS 2006).

[12] However, "there are an unknown number of successfully adulterated or substituted specimens submitted to testing and not identified as adulterated or substituted." (Bush 2008b:116)

A number of professional and industrial associations provide expertise, disseminate information, and participate in the ongoing development of guidance, standards, and good practices related to workplace testing programs. These include the American Board of Forensic Toxicology (ABFT), the Society of Forensic Toxicologists (SOFT), the American Academy of Forensic Sciences (AAFS), the International Association of Forensic Toxicologists (IAFT), the College of American Pathologists (CAP), the Drug and Alcohol Testing Industry Association (DATIA), and the Substance Abuse Program Administrators Association (SAPAA) (The Forensic Toxicology Council 2010). In 2009, the ABFT initiated a process to align its laboratory accreditation program with the ISO (International Organization for Standardization) and IEC (International Electrotechnical Commission) standards, particularly ISO/IEC 17025 (Testing and Calibration Laboratories) and ISO/IEC 15189 (Clinical Laboratories). These standards have done much to promote consistency across countries and industries (Penders and Verstraete 2006).

As a result, workforce D&A testing programs now operate within a well-established institutional framework that influences the development and adoption of methods and technologies. This framework includes specification of the conditions under which drug tests are conducted, the processes and procedures followed in workplace testing programs, the substances included in testing panels, the types of specimens (matrices) collected, the types and performance parameters of the technologies used to determine results, and the concentrations of substances (cutoff levels) that distinguish positive from negative results. This process has established that workplace testing programs must include the following elements:

- Policy and administrative procedures. These include (1) specification of prohibited behavior, administrative roles and responsibilities (including training and qualifications), and conditions for testing; (2) procedures for selecting, notifying individuals to report for testing, and tracing compliance; (3) substances to be included in test panel; and (4) cutoff levels, etc.;
- Specimen collection. This includes collection protocols, conditions of the collection site, authorized containers/devices, initial inspection, labeling;
- Specimen packaging and storage, chain of custody maintenance, transport, and ascension procedures;
- Initial specimen validity testing (i.e., to determine whether the specimen is altered, dilute, substituted, or invalid);
- Initial screening/testing of the specimen for the presence of prohibited drugs and their metabolites to differentiate "negative valid" specimens from all others – only specimens that are *not* "negative valid" require further testing;
- Confirmatory testing of those specimens determined by initial or screening tests to be not valid or to be valid but not negative; confirmatory testing includes protocols for specimen preparation, separation, and measurement;
- Recordkeeping and reporting; and
- Medical Review Officer (MRO) review of results.

Similarly, a set of conditions under which employees may be tested has also become well established and are incorporated in the 2008 HHS Guidelines and the NRC and DOT regulations. These are:

- applicant/pre-employment;
- random, especially for personnel in safety-sensitive industries or positions;
- reasonable suspicion/for-cause;

- post-accident/event;
- return to duty;
- follow-up.[13]

Each of these conditions imposes different constraints and considerations for the selection and notification of the individual to be tested, the conditions under which the specimen is collected, and the salience of assessing the individual's current state of fitness/impairment. As discussed in Sections 2.3 and 2.4, this creates the potential that different devices, methods, technologies, and matrices might be better suited for or applicable to different testing conditions. Each innovation has the potential to affect other elements of the overall testing program.

The requirement to protect individuals from false positive results by subjecting non-negative specimens to a second, confirmatory test using a different method affects the technologies used in workplace drug testing. This strategy of initial and then confirmatory testing creates a demand for low-cost, fast, and reliable initial "screening" technologies that are sufficiently sensitive to accurately identify non-negative specimens, and for "confirmatory" testing technologies that use a different chemical principle and are both highly sensitive and highly specific. Additionally, it creates a demand for experts capable of interpreting the results of these tests and determining whether they provide a basis for concluding that the individual providing the specimen has violated program policy.

The strategy of testing only for a limited, pre-established drug test panel is a third feature of workplace testing that has become well institutionalized. This convention affects the methods, instruments, and technologies used in workplace testing programs by removing the need for a technology system capable of identifying unknown substances. For many years, the test panel specified in the HHS Guidelines included only the "NIDA 5" now known as the "SAMHSA-5" or "HHS-5" drugs (SAMHSA 2006):

- cannabinoids (THC);
- cocaine;
- opiates;
- phencyclidine (PCP); and
- amphetamines and methamphetamines.

The NRC specified this same test panel in the 1989 Part 26 regulation. However, Part 26 also provided a mechanism for licensees to add additional drugs to address variability in use patterns at different sites and worker populations. DoD has revised its test panel several times since its initial specification in the 1960s. The 2008 HHS Guidelines added 6-acetylmorphine (6-AM), a metabolite of heroin, and additional "designer drug" variants of amphetamine – MDMA (methylenedioxymethamphetamine, commonly known as Ecstasy), MDA (methylenedioxyamphetamine), and MDEA (methylenedioxyethylamphetamine) to its required test panel. In its recent rule revisions, the DOT adopted these changes in the composition of its drug panel. Although the NRC has not yet revised its drug panel to include MDMA, it continues to allow (and specifies the procedures for) adding additional drugs of abuse to the test panel. Both DOT and NRC also require testing for alcohol.

---

[13] These conditions are slightly adapted in the 2008 10 CFR Part 26.31(c) requirements: Pre-access, for cause, post-event, follow-up, and random.

In addition to establishing the drug panel, the HHS Guidelines and other regulating agencies specify a "cutoff level" for each drug – the concentration of a drug and/or metabolite in the specimen that differentiates a negative from a positive result. Cutoff levels are specified separately for each drug for each of the two stages of testing: initial (screening) and confirmatory. In general, cutoff levels are designed to be low enough to capitalize on the best ability of the technology to detect and quantify the drug/metabolite but high enough to exclude drug/metabolite levels that may be caused by legitimate behavior. Consequently, cutoff levels have evolved over the course of institutionalizing workplace testing, reflecting advancements in technology and accumulation of empirical data.

2.2.2 Other Institutional Sectors that Develop, Test, or Create a Market for Technologies Pertinent to Workplace Testing

Workplace testing programs are not alone in creating a demand for improved technologies to collect, store, clean, and prepare biological specimens and to separate, identify, and measure drugs and related substances (Walsh 2008; Kraemer and Paul 2007). Indeed, because workplace testing requires technologies that are well established, with highly accurate performance and low cost, the workplace testing sector is generally a technology follower rather than a technology developer or leader. It draws primarily on technologies used and developments made in the other sectors shown in Table 2.1, and the scientific advancements being made in the disciplines of forensic toxicology, analytical and clinical chemistry, analytical toxicology, pharmacology, separation science, biology, and medicine.

Table 2.1. Sectors Contributing to the Development and Validation of Methods and Technologies Pertinent to Workplace Testing

Analytical Application	Goals	Priority Technology Attributes	Methods and Technologies Being Used/Developed; Substances Tested
Workplace testing, including the military	Detection and deterrence, with special focus on safety-sensitive workers	Efficient, sensitive, reliable, established/ standardized, defensible, inexpensive	Urine specimen collection protocols to prevent subversion Urine specimen validity testing protocols LC-MS/MS[14] confirmation allowed DOT-certified breathalyzers Specified panel of drugs of abuse
Safety and law enforcement, especially driving under the influence of drugs	Detection and deterrence	Efficient, sensitive, reliable, established/ standardized, defensible, inexpensive	Urine, breath, and oral fluids specimen collection, with protocols to prevent subversion Urine specimen validity testing protocols LC-MS/MS confirmation allowed DOT-certified breathalyzers and oral fluids tests for alcohol Specified panel of drugs
Forensic/ criminal investigations	Identification, matching, and interpretation; toxicology	Efficient, sensitive, reliable, defensible	Wide variety of specimens/ matrices used, some collected without subject participation/knowledge Less concern about preventing or testing for subversion Wide variety of analytical methods used Screening→confirmation sequence not always required Frequently testing to identify unknown substances
Follow-up, monitoring, and recovery/ treatment support, including criminal justice programs	Detection, monitoring, and reinforcement	Efficient, reliable, defensible	Apply collection protocols to prevent subversion, including specimen validity testing Increasing use of hair and sweat as matrices to extend window of detection and increase effectiveness of monitoring function Detection and identification at drug family level often adequate (i.e., less need for highly sensitive methods) Drugs of abuse, including alcohol
Athletic doping tests	Detection and prevention	Efficient, sensitive, reliable, established/ standardized, defensible	Focus on performance enhancing substances, including increased levels/concentrations of naturally-occurring substances

---

[14] Liquid chromatography w/ tandem mass spectrometry.

Analytical Application	Goals	Priority Technology Attributes	Methods and Technologies Being Used/Developed; Substances Tested
Medical/ clinical	Diagnosis and treatment, therapeutic drug monitoring	Reliable, quick. For treating overdoses, may need to know only the family of drug, not the specific substance	Less concern about specimen subversion Increasing use of oral fluids as specimen to monitor dose (usually for therapeutic drugs) Need capability to test for wide range of drugs and metabolites, both drugs of abuse and therapeutic drugs Testing often used to identify unknown substances, but detection and identification often needed only at drug family level (adequate to determine treatment). Screening→confirmation sequence often not required Drugs of abuse, pharmaceuticals, and poisons
Biological and pharmaco-logical research	Understanding biological processes and interactions	Sensitive	Developers of state-of-the-art methods and technologies to separate, identify, and measure a very wide range of substances in all variety of matrices
Homeland Security	Detection of drugs and other dangerous substances	Portable, sensitive	Investing in sensitive and portable technologies, especially miniature mass spectrometers Major focus on testing environment or physical substances, secondary focus on biological fluids Interest in technologies capable of detecting concealed substances and detecting and identifying unknown substances

The methods and technologies that will affect workplace testing are likely to emerge from these sectors through the process illustrated in Figure 2.1, and studies demonstrating their performance are likely to be presented in the journals and conferences of these professions.

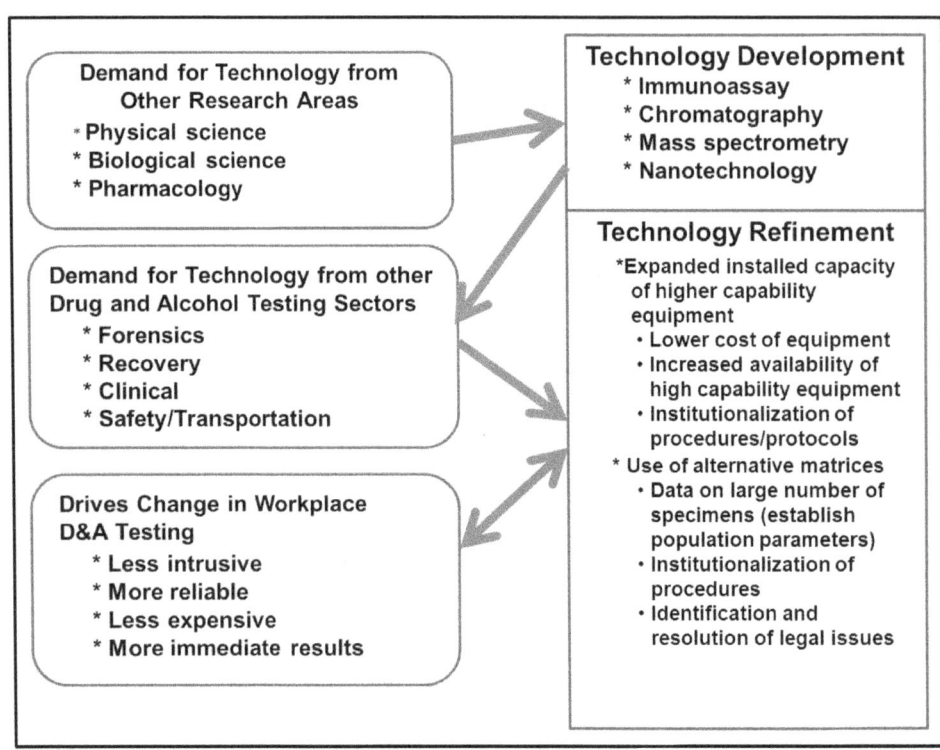

Figure 2.1.   Dynamics of Technology Development Benefitting Workplace Drug and Alcohol Testing

As shown in Figure 2.1, the more basic science sectors are leading the development of improvements in separation and measurement technologies, including mass spectrometers, chromatography, and associated data acquisition and analytical software. Advances in micro- and nano-technology are introducing new separation and measurement options. Advances in pharmacology and immunology are contributing to the development of new types of immunoassays and an expanding inventory of analyte-specific immunoassays and efforts to improve therapeutic drug monitoring and management are contributing to the development of point-of-collection devices and procedures (Langman and Kapur 2006; Dams et al. 2007). This new equipment is enabling analytical chemists and analytical toxicologists to examine the behavior of drugs of abuse in alternative matrices and in this new equipment. It is also providing the tools necessary for researchers to examine impairment resulting from drug use, which has been of particular interest to those addressing the drugged driver problem (Ramaekers et al. 2006; Verstraete 2005a,b). Drawing upon and contributing to these advances in technology, and responding to demand from these sectors, the private sector is developing devices for the collection and analysis of alternative specimens in a variety of settings. This has generated demand for improved preparation protocols and automation of the separation and measurement processes.

2.2.3 Adoption of New Technologies into Workplace Testing

Although the inertia of current practice imposes a significant barrier to the introduction of new technologies in workplace testing, the large market represented by workplace testing creates considerable technology pull. For example, the demand created by drug abuse screening provided an important stimulus for the adaptation of immunoassays to drugs of abuse and "probably provided the major impetus for the development and refinement of non-isotopic immunoassays" (Jatlow 1988:108). However, technologies and processes developed in other sectors require tailored adaptation and extensive validation before they are adopted into workplace testing. In addition, adoption of innovative technologies requires assessment and modification of policies, protocols, and procedures and confirmation that laboratories and other supporting infrastructure (e.g., suppliers of devices, forms, and laboratory services) have the equipment and know-how to implement the innovation cost effectively (Catlin et al. 1992; Penders and Verstraete 2006). Working through these considerations slows the diffusion of technology innovations, especially when regulations must be revised before an innovation can be adopted (Caplan and Huestis 2007:734).

Within this procedural and performance framework, the primary drivers for updated technologies in workplace testing are a desire for reduced costs and increased efficiency, reliability, and acceptability through technologies that:

- are more effective in preventing subversion;
- reduce inconvenience and embarrassment in specimen collection;
- reduce time, skill, and personnel required for specimen collection and are safe for the subject, collector, and others in the testing process;
- are stable under established, readily available preparation and storage conditions;
- reduce time between specimen collection and the availability of confirmed results;
- further reduce false positives and false negatives in initial testing;
- use small quantities of inexpensive materials and reagents;
- can be automated;
- are thoroughly validated and documented for an appropriate range of field conditions, substances, and populations;
- have an adequate base of reliable and capable providers; and
- maintain consistency across Federally-mandated testing requirements.[15,16]

To meet these objectives, employers implementing workplace testing programs not subject to Federal regulations have begun to explore new technologies, in particular, the use of alternative matrices (oral fluids, hair, and sweat) for specimen; collection devices and procedures specific to these alternative matrices; and point-of-collection testing devices to perform initial validity and screening tests (Quest Diagnostics 2011). To provide a basis for evaluating emerging technologies, the HHS Drug Testing Advisory Board and the FDA collaborated to articulate the principles and procedures to be used in validating the performance of bioanalytical methods. Following its decision not to authorize either the use of specimens other than urine or point-of-collection devices in its 2008 update of the HHS Guidelines, HHS has stated that it is actively evaluating the information available on these technological and procedural innovations. Their evaluation is focused on determining whether the innovative technologies and procedures have been adequately validated and meet performance requirements and implementation considerations (HHS 2009). The Federal agencies regulating workplace testing place a priority on coordination to maintain consistency to the extent appropriate, given differences in purpose

---

[15] This list is ordered to roughly reflect the testing process, from collection through confirmation, rather than by priority or importance.

[16] In undertaking its 2008 revision of Part 26, the NRC was specific about its goal of enhancing consistency with the HHS Mandatory Guidelines (and DOT alcohol testing requirements) (73 FR 16965 to 17235; March 31, 2008): (Goal) "Update and enhance the consistency of 10 CFR Part 26 with advances in other relevant Federal rules and guidelines, including the HHS Guidelines and other Federal drug and alcohol testing programs (e.g., those required by the U.S. Department of Transportation [DOT]) that impose similar requirements on the private sector...").

and field conditions. Thus, when considering adoption of new technologies, these agencies must consider the implications of change and variability in requirements on the overall drug testing system in addition to the purely technical aspects of an innovation.[17]

Meeting the validation and documentation requirements is expensive and time-consuming, as demonstrated by the elapsed time and amount of work undertaken between the introduction of an innovative technology, its approval by the FDA, and its wide-spread adoption within the workplace testing industry revealed in the literature included in Section 2.6. The employer and laboratory experience from the non-Federally-regulated programs is contributing valuable implementation experience and data for the validation of these technologies and identification of issues associated with their application in workplace testing.

As discussed below, the innovations with the greatest potential for adoption by or adaptation to Federally-regulated workplace testing include:

- Alternative specimens. Oral fluids, hair, and sweat, in addition to urine;
- Immunoassays. Continued advancements in immunoassay sensitivity and specificity, and devices for conducting immunoassays of drugs of abuse and their metabolites in a variety of bodily fluids and tissues, including the development of high-speed, fully automated systems with which laboratories can test and report results within 1-2 hours of specimen receipt (Chyka 2009:48);[18]
- Point-of-collection devices. Collection devices that include testing capabilities (e.g., specimen validity screening and initial drug screening tests);
- Protocols for specimen storage, preparation, separation, and measurement. Continued advancements in specimen preparation and separation methods and technologies and the development of standard protocols; and
- Chromatography and mass spectrometry systems. Continued advancements in mass spectrometry, including "hyphenated" instruments that link mass spectrometers together and with other analytical tools to increase sensitivity and decrease the volume of specimen needed for confirmatory testing.

Further in the future, advanced separation, identification, and quantification technologies based on advances in micro- and nanotechnology and miniaturization have potential for application in workplace testing programs. These technologies are likely to enable particularly significant advances in on-site testing devices and processes.

The very extensive literature reporting laboratory and field trials of various methods, devices, and technologies as applied to alternative specimens, a subset of which is referenced in Section 2.6, reflects the interest and effort being expended on these issues in recent years. As is frequently the case, the development of methods, devices, and equipment capable of analyzing drugs of abuse in alternative specimens and interest in their use have been synergistic. Advancements in technology enable measurements with the degree of specificity and sensitivity needed for the small volumes and low concentrations of analytes in alternative specimens, and their potential use drives further refinements in those technologies.

---

[17] HHS updated guidelines for validating the expanded confirmatory testing technologies include linearity, limit of detection, limit of quantitation, accuracy, and precision at cutoff and 40 percent of cutoff, analytical specificity, and carryover (HHS-SAMSA 2009).
[18] Even hospital clinical laboratories now are capable of conducting these automated tests.

2.3 Alternative Specimens and the Pharmacokinetics of Drugs of Abuse

There has been longstanding interest in alternative specimens for forensic toxicology, therapeutic drug monitoring, and testing for drugs of abuse. Workplace testing programs are interested for the following reasons:

- Restricting drug testing to only a single specimen increases the opportunity for subversion by enabling those being tested to know in advance which specimen will be collected;
- Some specimens require less invasive and/or less embarrassing collection procedures than urine does, facilitating collection in a range of field conditions;
- Specimen collections either by the collector or under direct observation are less subject to subversion than urine (or other specimens whose collection is not directly observed);
- Different specimens have different "windows of detection"[19] for drugs of abuse. Depending upon the condition and purpose of testing, a longer or more immediate window of detection may be most informative and useful;
- Some specimens contain the parent drug rather than (or in addition to) its metabolites, which allows more definitive identification of the drug that was consumed and avoids interferences that complicate interpretation of results; and
- Some specimens provide a better indication of the current level of parent drug in the system than others do, which is pertinent for decisions concerning treatment and impairment.

However, an extensive base of information about the pharmacokinetics of the drugs of interest in each alternative specimen is needed to establish its appropriateness for workplace testing programs. The pharmacokinetics of a drug in a specimen determines the performance requirements of testing technologies and establishes the basis for interpreting test results. Attributes of and the cost, performance, and availability of the set of technologies needed to implement the entire testing program are key factors in determining the feasibility and desirability of a particular specimen. Data validating dose-response characteristics and technology performance under field conditions are also needed. Every combination of the "drug - specimen - collection device - storage, transport, and preparation protocol - initial screening device or test - confirmatory test" system has to be studied. Generating this information and building scientific consensus about the findings is a long process: Human subject protection considerations make studies involving drugs of abuse difficult.[20] However, as discussed below, the body of research that has been assembled and reviewed has begun to yield pharmacokinetic information about a range of drugs of abuse in different specimens and a consensus on "best practice" protocols for some aspects of the testing process for the alternative such as of oral fluids, hair, and sweat.

2.3.1 Pharmacokinetics of Drugs of Abuse

Pharmacokinetics describes how the body acts on drugs, with an emphasis on developing mathematical models to describe what substances are present in which fluids and tissues over what period of time after the administration of a drug. Pharmacokinetic studies attempt to

---

[19] Sometimes also referred to as "surveillance window."
[20] Consequently, much of the data is epidemiological or based on known drug users.

describe the time course of the processes (absorption, distribution, metabolism, and excretion) a drug undergoes in the body and to establish the quantitative relationship between administered doses of a drug and the observed concentration of the drug and its metabolites in body tissues and fluids. Pharmacokinetic studies also seek to clarify how the drug's physicochemical properties and factors such as mode of administration, use patterns, user demographic characteristics, drug-drug interactions, and variations in physiological condition affect these processes.

Figure 2.2 illustrates the processes at work within the body that influence the disposition of a drug. These are dynamic processes that involve complex, interactive physiological systems. Where a drug goes after entering the body, what levels it reaches, and how long it stays there depend upon a number of factors, including:

- dose;
- mode of administration;
- blood flow patterns;
- extent of protein binding;
- lipid solubility of the drug;
- acid/base character of the drug;
- the pH of the tissues and fluids and the pH gradient between them;
- metabolic processes and the chemical characteristics of the metabolites; and
- pathways and rates of elimination.

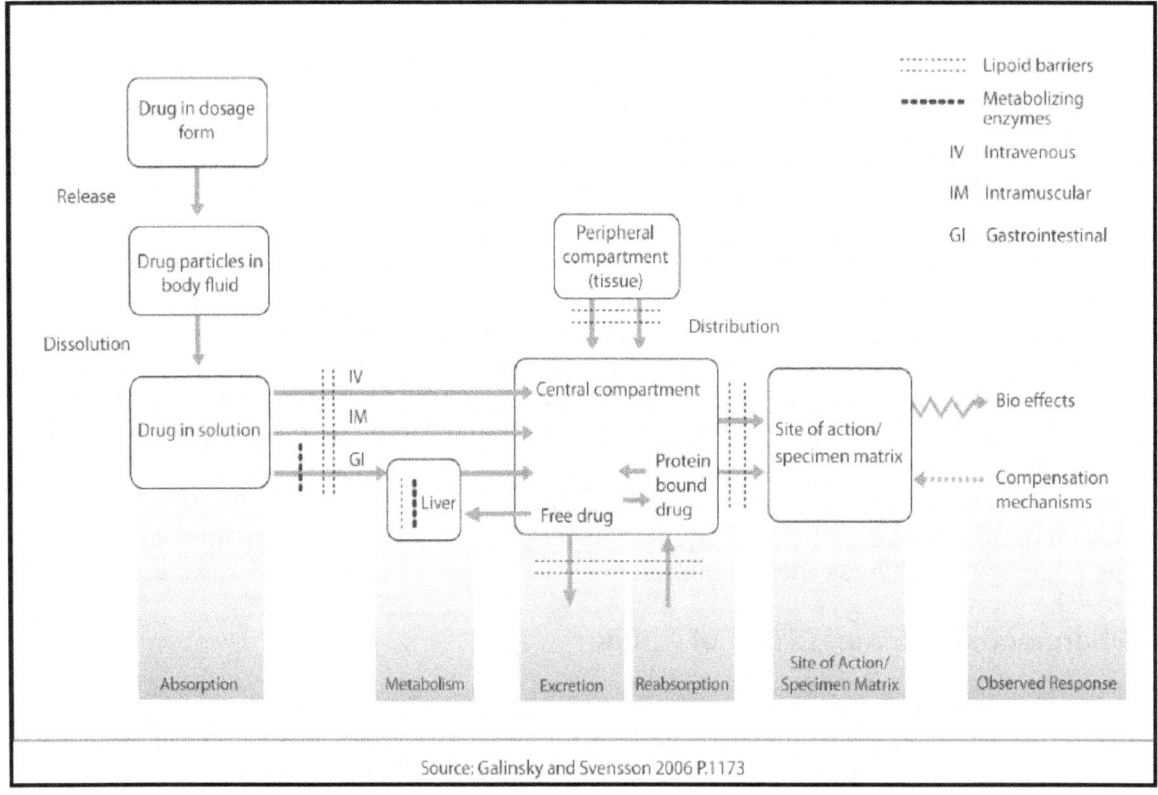

Source: Galinsky and Svensson 2006 P.1173

Figure 2.2.   The Pharmacokinetic Processes

Detailed studies are needed to explicate the parameters of these processes for each drug of interest and to determine how they are affected by variables such as:

- consumption patterns (e.g., chronic or occasional);
- mode of administration (e.g., smoking, insufflation (snorting through nose)), or intravenous);
- dose size; and
- consumption of other substances (e.g., other drugs of abuse and/or substances consumed in an effort to speed up excretion or reduce drug/metabolite concentrations).

Consequently, establishing the timing and concentration of a drug and its metabolites in a specimen requires studies in which controlled doses of the drug are administered, followed by sampling and testing the specimen of interest at specific time intervals. Because studies involving drugs of abuse are difficult to conduct, information about these processes for many drugs is still incomplete (Huestis and Smith 2006; Langman and Kapur 2006; Milman et al. 2010). However, as seen in Section 2.6, the number of controlled drug administration studies is growing (see for example, Barnes et al. 2009; Cone 1993; Cone et al. 1997; Drummer 2005; Jenkins et al. 1995; Jufer et al. 2000; Kacinko et al. 2005; Kato et al. 1993; Kim et al. 2002; Navarro et al. 2001; Schepers et al. 2003). Additionally, many of these studies involve very small sample sizes.

Figure 2.3 illustrates the information pharmacokinetics seeks to establish about the time course of a drug in a body and for each tissue or fluid of interest. Of special importance is delineating the mechanisms underlying the metabolism of a drug, which exerts a major influence on the concentration of the drug in the body, as well as the nature and concentration of resulting metabolites (Spiehler and Levine 2003; Yamada et al. 2005).

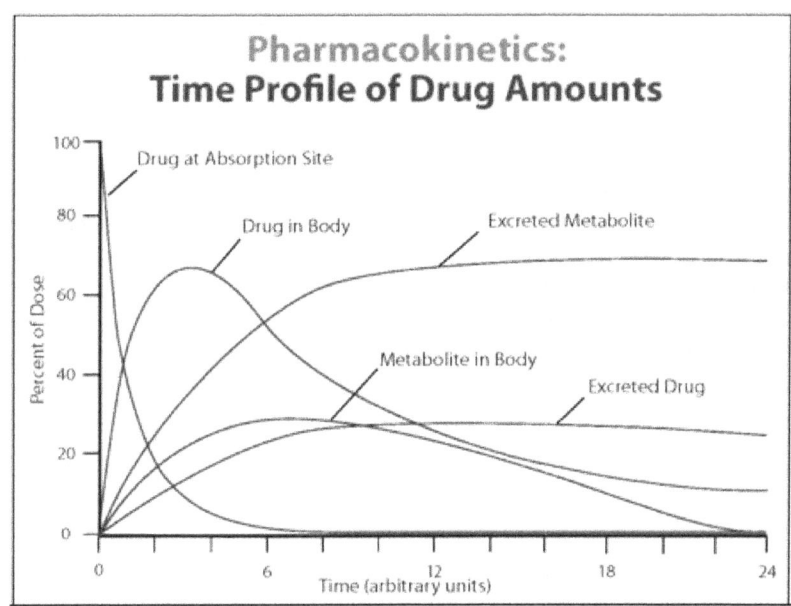

Figure 2.3.   Time Profile of Drugs in the Body

The same factors that influence the disposition of a drug in the body influence the concentration and time course of the drug and its metabolites in different bodily fluids and tissues. All of the specimens identified as candidates for workplace testing (blood/plasma, urine, oral fluids, sweat, and hair) have rich blood flow patterns.[21] This enables drugs that have entered the blood stream to be delivered to various matrices.

Similarly, a number of factors influence how the drug is distributed from the blood into another matrix. Protein-binding is one of these factors. Drug molecules that are bound to plasma proteins (1) lose pharmacologic activity, (2) are not able to diffuse or transfer across cell membranes, (3) are not metabolized by the liver and other tissues, and (4) are not excreted from the body. Consequently, only the unbound drug is in equilibrium throughout the body and only the unbound drug is metabolized (transformed into a different compound).[22] For these reasons, the portion of a drug that is protein-bound affects the drug's distribution into the other matrices and also its metabolism rate.[23] Drugs that are unbound and lipid soluble generally diffuse across cell membranes more easily than water soluble drugs and metabolites (which are frequently water soluble).

Other physicochemical properties of a drug that interact with the properties of the matrix to influence concentration are its molecular size, solubility, and the relative lipid solubility of its ionized and non-ionized forms.[24] The degree of ionization is affected by the tendency of the drug to dissociate (i.e., its pK_a[25]) and the pH of the fluid/tissue. These properties affect the unbound drug's ability to move across cell membranes (Jenkins 2007a). In general, non-ionized forms of an unbound drug are lipid soluble and pass readily across cell membranes. Because the portion of the drug that is non-ionized is affected by pH, changes in pH can affect drug concentration.[26] This is one reason the concentration of unbound drugs that are weak bases (e.g., cocaine, opiates, benzodiazepines, nicotine) is higher in saliva, which is slightly more acidic (pH = 6.2 - 7.4), than in plasma (pH = 7.4), which is slightly more basic (Cone 1993; Kidwell et al. 1998; Navarro et al. 2001). In general, drugs with high lipid solubility, a high non-ionized fraction, and low protein binding in plasma are distributed widely in the body.

The rates and processes by which the body metabolizes drugs of abuse influence the time course of a drug, and the nature and extent of its transformation into other substances, some of which may also be pharmacologically active. Some drugs of abuse are metabolized quickly, rapidly reducing the amount of "parent" drug in the system, but increasing the amount of metabolite present. Metabolism often produces multiple metabolites. Drugs that belong to the same family (e.g., opiates) often produce the same metabolites.[27] Metabolism is a precursor to excretion for many drugs of abuse, transforming them from lipid soluble to water soluble substances. As metabolism and excretion proceed, the concentration of the parent drug in the system declines. At any time, the ratio of parent drug to metabolite may be quite different in different body fluids and tissues. Urine, which is a primary route for drug excretion, generally contains primarily drug metabolites rather than the parent drug itself. The concentration of unbound drug in plasma has become the standard basis for determining and monitoring the

---

[21] In the case of blood/plasma, the specimen is the blood itself.

[22] The protein bound drug is in equilibrium with the unbound drug, and acts as a reservoir of the unbound drug as the concentration of unbound drug is reduced by metabolism and excretion.

[23] The concentration of the drug, the drug's affinity for protein, and the amount of protein available determine the fraction of drug that is protein-bound. Many drugs are extensively metabolized by the liver (Jenkins 2007a).

[24] These properties also affect metabolism. Lipid-soluble substances are biotransformed into water soluble substances before they are eliminated from the body by the kidneys.

[25] The pK_a is the negative logarithm of the acid dissocation constant, K_a; the lower the pK_a (low pH), the stronger the acid and vice versa.

[26] As discussed below, this attribute affects the collection protocol for oral fluids because stimulating saliva flow lowers its pH and alters the drug concentration in the oral fluids (Crouch 2005; Drummer 2008).

[27] This adds complexity to drug testing when it is necessary to determine the identity of the parent drug.

therapeutic or toxic effects of a drug. Consequently, those interested in therapeutic drug monitoring or impairment from drug use have undertaken studies to establish the ratio of drug concentration in plasma and other specimens, particularly oral fluids, for different drugs and under different conditions (Gjerde and Verstraete 2010; Laloup et al. 2005; Ramaekers et al. 2006; Verstraete 2005b).[28] Despite extensive research, the mechanisms by which drugs are distributed into and removed from the different body fluids and tissues are still often not well understood (Jenkins 2003; Sachs 2000).

Table 2.2 illustrates some of the physicochemical properties of drugs commonly included in workplace test panels.

Table 2.2.   Physicochemical Properties of Drugs of Abuse

Drug Property	Protein-Binding	Lipid Solubility	Acid/Base	pKa	Metabolism[29] Half-Life in Plasma
Amphetamine/ Methamphetamine[a, b]	15 - 40% (Low)	High	Base	9.9	Renal excretion; significant portion unaltered ½ life = 12 - 13 hrs
Marijuana (Cannabinoids)	~95 - 99% (High)	High	Acid	9.5	Slow, Multi-step ½ life is variable depending on use patterns
Opiates[d] (e.g., Morphine) Heroin	~30%, but variable	Variable	Base	~6.5 - 8.7	½ life = 2.5 - 3 hrs Heroin metabolized to 6-AM and morphine; ½ life = 6 - 25 min
Cocaine[e]	91% (High)	Variable	Base	8.6 - 8.7	Very rapid ½ life = 1 hr
Phencyclidine[f]	65% (Medium)	High	Base	8.5	Extensive to inactive; ½ life = 7 - 46 hrs

[a] Sources: Wikipedia "Amphetamine" at http://en.wikipedia.org/wiki/Amphetamine; [b] De La Torre et all. 2004; [c] Huestis 2007; Crouch et al. 2004;[d,e,f] Couper and Logan 2004

Figure 2.4 illustrates the variety of information that is used to interpret test results regarding drugs of abuse. Knowledge of the pharmacokinetics of the drugs included in the test panel provides much of the foundation for informed interpretation, which is essential for fair and effective testing programs.

---

[28] To date, alcohol/oral fluids is the only drug/specimen combination for which the relationship between testing concentration and impairment has been well enough established to be relied upon in workplace testing.

[29] Half-lives are influenced by a variety of factors and the half-life of the parent drug and each of its metabolites may be very different (often very different) (Jatow1988).

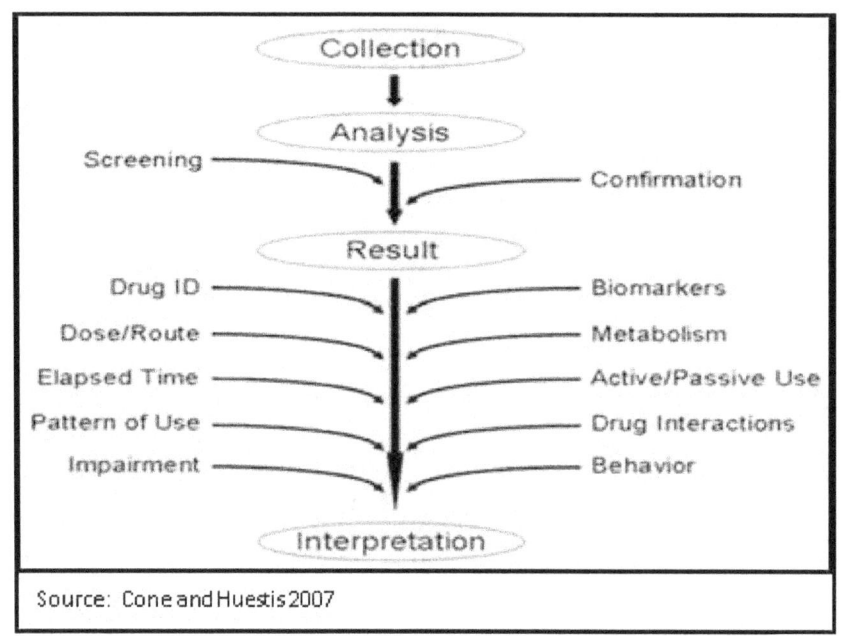

Figure 2.4.   Information Used in Interpreting Drug Testing Results

2.3.2 Characteristics of Biological Specimens

Although a wide variety of bodily fluids and tissues can serve as a specimen to provide information about an individual's drug use, the specimens of greatest interest for workplace testing programs are urine, oral fluids, hair, and sweat. Blood and plasma, no longer widely used in workplace testing, are still widely used in therapeutic and clinical testing, and continue to play a role in workplace testing as the specimen against which other specimens are compared. Because the drug level in plasma is considered the best indicator of the amount of drug reaching the brain or other target organs, plasma testing plays a central role in efforts to establish the impairing effects of the various drugs of abuse. However, because blood and plasma are not commonly used in workplace testing, they are not discussed in detail in this report.

Urine is by far the most commonly used specimen for workplace drug testing. Quest Diagnostics, Inc., one of the largest workplace testing laboratories in the U.S. (and an HHS-certified laboratory), conducts tests for both Federally-mandated and non-Federally-mandated testing programs. Quest Diagnostics reported conducting over 2.9 million urine tests in the January to June 2009 period (Quest Diagnostics, Inc. 2011).[30] In addition, a substantial, and growing number of tests for drugs of abuse are being conducted on oral fluids, hair, and sweat, despite the prohibition on Federally-regulated employers to use these alternative specimens.[31][32]

---

[30] Overall, Quest Diagnostics report conducting over 7 million drug tests in calendar year 2009 for a variety of medical, sports, and workplace purposes ((Quest Diagnostics, Inc. 2011).

[31] The non-Federally-regulated workplace testing programs are increasingly using alternative specimens, particularly oral fluids. Quest Diagnostics reports conducting over 320,000 oral fluids tests and over 90,000 hair test in the January - June 2009 period (Quest Diagnostics, Inc. 2011).

[32] In the following discussion, it is important to note that the term "positive test" or "positive result" is often used by researchers to refer to laboratory results that exceed the specified cutoff level. In workplace drug testing these tests would be descr bed as "non-negative" until the results were reviewed by a qualified medical review officer (MRO).

Urine

Urine is collected from the urinary system. The urinary system is a set of organs whose primary function is to filter excess substances from the bloodstream and remove them from the body. The nephrons of the kidneys filter the blood[33] and produce urine, which is collected and stored in the bladder and excreted through the urethra. Through their ability to filter large amounts of blood (about 200 quarts per day), the kidneys play a key role in maintaining homeostasis, the acid/base balance, and the electrolyte balance of the blood. An individual typically produces between 1-2 liters of urine per 24 hours, although the volume varies considerably by individual, time of day, fluid consumption, and other variables. Most individuals can reliably provide a urine sample of 20-100 mL. The pH of normal urine is generally in the range of 4.6 to 8, with a typical average of about 6 (i.e., slightly acidic). Diet causes much of the variation in urine pH.[34]

Normal urine is about 95 percent water and 5 percent solutes, which include urea (a metabolite of dietary protein), creatinine (related to muscle mass), uric acid, ketone bodies, potassium, sodium, and chloride, along with other dissolved ions and compounds. Normal urine does not include significant amounts of protein or cellular components, which makes it a relatively clean matrix for analysis and reduces the complexity of specimen preparation for testing (Chyka 2009:48). However, urine can contain infectious agents and requires care in handling.

Consumption of some non-prohibited substances (such as poppy seeds) can produce the same substances (particularly metabolites) as drugs included in the test panel, and there is some potential that passive exposure[35] (e.g., second hand smoke) can be sufficient to cause a test result over the cutoff level (Röhrich et al. 2010). However, because the urinary system is less exposed to the external environment than oral fluids, hair, or skin/sweat, urine is less susceptible to inadvertent environmental contamination than these other matrices. However, it is not immune to the effects of intentional efforts to subvert the testing process through the consumption of excess fluids, other substances designed to alter the composition of the urine, or even the introduction of substitute urine into the bladder (Berge and Bush 2010).

The filtration and concentration process of urine formation yields relatively high concentrations of drug/metabolites in urine. This generally facilitates testing. However, the nature of the urinary system (containing primarily post-metabolism products – end-stage metabolites – and storage for an unknown amount of time in the bladder) reduces the utility of urine as an indicator of the amount of drug circulating in an individual's system at the time the specimen is collected.

Urine often contains primarily metabolites rather than the parent drug. Consequently, tests designed for urine often target drug metabolites rather than the parent drug itself. Urine has been and continues to be the primary specimen for workplace drug testing and, along with blood, is a principal specimen for clinical testing as well. Consequently, a large infrastructure for urine testing is now in place and the combined clinical and toxicological testing of urine has generated a very extensive base of information about the characteristics of urine, the pharmacokinetics of drugs and their metabolites in urine, and the performance of testing technologies as applied to urine. Given urine's status for many years as the only authorized

---

[33] This occurs through a complex chemical exchange as approximately a million tiny blood vessels intertwine with a million tiny urine-collecting tubes. Also, note that the term "blood" here also includes plasma, which in some studies is the specific fluid measured and tested, rather than blood in its entirety.

[34] Increasing the acidity of urine can increase the concentration of some basic drugs/metabolites.

[35] Passive drug exposure in workplace testing is considered exposure to a drug of abuse unintentionally, for example by inhalation of drug smoke, vapor, or dust; contamination of the skin or hair by contact with the drug; or by ingestion of food that contains the drug (e.g., poppy seeds) (Cone et al. 2007).

specimen for Federally-regulated workplace testing, immunoassays developed for workplace testing were designed primarily to test for drug metabolites rather than the parent drugs themselves. This limits the transferability of this technology to other specimens, in which the parent drugs were the more prevalent constituent (Crouch et al. 2004).

Because of the anatomy of the urinary system and social norms, urination is considered a personal, private function. Consequently, direct observation of urine collection is considered invasive, even when conducted by a collector who is the same sex as the donor. Subversion of urine testing through substitution, dilution, or adulteration is widely considered to pose a threat to the integrity of the testing process, despite the specific procedures implemented to control the collection process and test specimens for validity (Bush 2008a and 2008b; Crumpton and Sutheimer 2007; Dasgupta 2005 and 2008).

Oral fluids

Oral fluids are collected from the mouth. Oral fluids are a combination of saliva, the major constituent produced by the salivary glands, gingival crevicular fluid, a minor constituent produced by the epithelial cells along the gum line (the gingival crevices), and other substances present in the mouth. The salivary glands produce between 750 and 1500 mL of fluid per day. Saliva production is stimulated by both the sympathetic and parasympathetic nervous systems. The production rate is not constant, varying by physiological state (awake, asleep) and activity (eating, chewing). During waking hours, oral fluids are typically generated at a rate of between 0.5 mL/minute and 3 mL/minute, leading to a turnover of the oral fluids in the mouth approximately every 10 minutes (Aps and Martens 2005; Crouch et al. 2004). This is an advantage for testing, because it means a donor can be observed by the collector while a "fresh" oral fluids specimen is produced, thus reducing the potential for subversion.

However, salivary flow can be affected by disease and is reduced by some drugs of abuse (e.g., cocaine) (Crouch et al. 2004), making collection of an oral fluids specimen more difficult. The fact that oral fluids are, at least to some extent, collected as they are produced, can lengthen the time required to collect an oral fluids specimen. In addition, stimulation of saliva production (e.g., by chewing) changes the composition and pH of the resulting oral fluids and affects drug concentrations in the specimen (Bosker and Huestis 2009; Jenkins 2007; Crouch 2005). This effect was not immediately recognized, and some of the variability in early studies of oral fluids drug concentrations has been attributed to differences in the procedures and materials used in the collection process. (Aps and Martens 2005; Crouch 2005; Drummer 2005).

Saliva is an ultrafiltrate of plasma. Oral fluids are primarily water (~98 percent), with small amounts of plasma electrolytes, mucus, antibacterial compounds, and enzymes. The pH of saliva ranges from 6.2 to 7.4. Drugs enter saliva primarily through passive diffusion from the bloodstream, although ultrafiltration and active transport also play a role (Crouch et al. 2004; Drummer 2008; Jenkins 2003). The pH of oral fluids is normally somewhat lower (more acidic) than plasma. This typically leads to an oral fluids/plasma ratio for drugs greater than 1 (i.e., a higher concentration of the drug in oral fluids than in plasma) (Cone 1993). Variation in the pH of the oral fluids leads to variation in the concentration of many drugs in oral fluids, and alters the oral fluids/plasma ratio.

Because drugs enter oral fluids primarily by passive filtration from the blood stream, oral fluids tend to contain primarily the parent drug rather than its metabolites, with drug concentrations in the oral fluids reflecting the unbound fraction of drug circulating in the blood.[36] Consequently, the concentrations of drugs in oral fluids are often relatively similar to those in plasma, but much lower than the concentrations of drugs/metabolites in urine (due to the concentrating processes of the kidneys). Drugs of abuse typically appear in oral fluids very shortly after ingestion, though the timing is affected by mode of administration. Ingestion of the drug through oral, intranasal (insufflation), or smoking may result in "shallow depots" of the drug in the buccal cavity that increase the concentration of the drug in oral fluids, sometimes quite dramatically. This effect tends to be greatest immediately following ingestion and declines substantially over a several hour period, depending upon the drug (Cone 1993).[37]

The time profile of drugs of abuse in oral fluids tends to be similar to that in plasma. Oral fluids therefore are considered better indicators of the current levels of the psychoactive drug in the system than any of the other matrices except plasma. This has created great interest in using oral fluids in testing conditions where knowledge about current levels of psychoactive drug is important (e.g., drugged driving, post-accident testing or for-cause) (Cone 1993; Drummer 2006). A consequence of this close relationship with the parent drug, however, is that the window of detection for drugs in oral fluids is generally shorter than in urine, sometimes quite substantially.[38]

A substantial number of studies have been conducted to better understand the pharmacokinetics of each of the drugs of abuse in oral fluids and to develop and validate technologies capable of collecting, separating, and measuring drug concentrations in oral fluids specimens. This work falls into the following categories:

- Refining and validating the technologies' abilities to identify and quantify drugs in oral fluids specimen accurately and reliably, given the small volume of the specimens and the low concentrations of the drugs/metabolites.
- Delineating the pharmacokinetics of drugs of abuse in oral fluids, including controlled administration studies of the different drugs to establish the concentration of the drug/metabolites in oral fluids over time. This information is needed to establish cutoff levels. For some time it was an open question whether marijuana could actually pass into oral fluids or whether it was present there only through contamination of the buccal cavity as a residue from administration (shallow depots).
- Studies testing the performance of the collection and immunoassay devices and procedures. Because of the high interest in using oral fluids as a specimen for roadside testing for drugged drivers, a number of studies were conducted to test the procedures and evaluate and compare the performance of different collection and screening devices. These studies included a very large study in Europe – The Roadside Testing Assessment (ROSITA). Many of these studies not only compared the performance of different devices, but also compared the results of the screening tests with those of the confirmatory tests.

---

[36] Delineating the relationship between the concentration of a drug in oral fluids and in plasma has been the focus of numerous studies. If the plasma/oral fluids ratio of a drug is consistent over the time, within and between individuals, oral fluids levels can be used for therapeutic drug monitoring/management and as a variable in assessing the relationship between drug levels and impairment. This work is still underway (see Bosker and Huestis 2009; Choo and Huestis 2004; Gjerde and Verstraete 2010; Laloup et al. 2006; Ramaekers et al. 2006; Samyn et al. 2002; Verstraete 2005b).

[37] This "shallow depots" effect has been observed particularly with marijuana, which otherwise is found in oral fluids only at very low concentrations.

[38] It is important to note that the half-life and window of detection vary widely from one drug to another and they are affected by many factors, including dose, mode of administration, age, time of day, etc.

Persistent issues regarding oral fluids testing include: a) the relatively short window of detection for many drugs of abuse; b) the impact of collection procedures and collection devices on test results as a consequence of stimulation of saliva and recovery of the drugs from the collection device; and c) the potential for very high concentrations of some drugs in tests conducted close to the time of drug administration because of drugs contained in the "shallow depots." Subsequent studies appear to have put to rest initial concerns over the ability of oral fluids to provide an appropriate specimen for marijuana testing (Kauert et al. 2007; Milman et al. 2010). Other studies have delineated conditions for assuring stability of the drugs in oral fluids specimens (Moore 2009; Verstraete 2005a). Validation of point-of-collection devices for oral fluids, which have shown considerable variability in performance, is an on-going effort, as discussed in Section 2.4.

2.3.3 Sweat

Sweat as a specimen is collected from the skin, with the consequence that the matrix called sweat is actually a combination of sweat, secretions from sebaceous glands, and substances that have been transported to the skin surface by transdermal liquid transport (also called insensible perspiration) (Jenkins, 2003; Kintz et al. 2007).[39] Sweat is produced by sweat glands that are distributed unevenly throughout the body's dermis layer of skin. Many sweat glands are in close proximity to hair follicles. The sweat glands are controlled by the autonomic nervous system and circulating hormones. Transdermal liquid transport (insensible perspiration) also occurs throughout the body.

The pH of sweat ranges from 4 to 6 and is influenced by the amount of lactic acid being excreted (Kintz et al. 2007). Exercise is reported to increase the pH of sweat (Kadehjian 2005). As with oral fluids, changes in the pH may affect the transfer of drugs from the bloodstream and the resulting concentration of drugs/metabolites, although no studies were identified examining this effect.

Sebaceous glands are located in the skin and are distributed unevenly throughout the body, absent from the palms, soles and tops of feet, and lower lip and most dense on the face, scalp, upper neck, and chest. Sebaceous glands are often connected to hair follicles and hair shafts. They produce sebum, a waxy/oily substance that is deposited on the hairs and brought to the skin surface along the hair shaft. Sebum, which is made of lipids, wax, and the debris of the fat-producing cells, helps lubricate the skin and hair.[40] On the surfaces of the skin and hair, sebum mixes with sweat and insensible perspiration (Fortner 2008; Sachs 2000).

Neither the mechanisms by which drugs are distributed into sweat, sebum, and insensible perspiration nor the pharmacokinetics of drugs in sweat are well understood (Jenkins, 2003; Kadehjian 2005; Sachs 2000).[41] However, it is generally thought that drugs diffuse from the bloodstream into sweat, sebum, and insensible perspiration at rates and in concentrations affected by the pharmacokinetic factors discussed in sections 2.3.1 and 2.3.3 (Chawarski et al 2007; de Martinis 2007; Marchei et al. 2010). As with oral fluids, the parent drug is the primary constituent in sweat specimens, with metabolites appearing in much lower concentrations, if at

---

[39] This is similar to the earlier practice of calling the specimens composed of oral fluids, "saliva," which, though the major constituent of oral fluids, is not the sole one.

[40] As discussed below, sebum and sweat provide additional pathways for drugs to enter and/or adhere to hair.

[41] Sweat appears to be the least studied specimen in the pharmaceutical/medical, analytical toxicology, and forensic science areas, where most of the work on pharmacokinetics and physiological mechanisms seems to occur.

all (Sachs 2000). Greater specificity about the pharmacokinetics of drugs is needed to establish protocols for sweat collection and interpretation of test results.[42]

Most pharmacokinetics information about drugs in sweat is derived from studies examining the various sweat sampling technologies (particularly the sweat patch and the sweat wipe) with subjects in treatment/recovery programs or criminal justice monitored drug abstinence programs. The interest in these programs is to identify any drug use that occurs during the extended observation period (usually a week to ten days). Consequently, most studies have not collected detailed information about the time-concentration pattern of the drugs in sweat specimens.[43] Therefore, the windows of detection for the different drugs of abuse in sweat specimens are not well understood. From the limited data available, it appears that drugs of abuse may typically not appear in sweat specimens until several hours after the drug is administered and that detectable levels of the drug/metabolites may persist for days or weeks.[44] As with all test specimens, the windows of detection vary by drug, dose, mode of application, use patterns, and other factors. When reviewing discussions of detection windows in studies on drugs of abuse in sweat, it is important to clarify whether the description is about the pharmacokinetics of the drug (i.e., when and at what levels it is present in sweat) or the attributes of the testing strategy (i.e., continual monitoring that allows detection of any drug use that occurs during the period the patch is being worn).

To protect the subject's skin, sweat patches allow the water from the sweat to evaporate, leaving behind the drugs and their metabolites. Instead of measuring the drug/metabolite concentration in the sweat sampled, the measurements are made on the basis of the amount of drugs on the sweat patch. Some studies have attempted to develop a method for estimating the volume of sweat (e.g., on the basis of the salts contained in the patch (see Appenzeller et al. 2007)), but most studies seem to be using a cutoff based on amount per patch, with cutoff levels based on data obtained from known users or volunteers (Levisky et al. 2000). One advantage of this strategy is that tests using the sweat patch are not affected by the variability that fluid consumption can cause in the concentration of drugs in urine (DuPont and Selavka 2008).

Other concerns about the sweat patch are the potential for passive contamination, intrusiveness, and the potential for subversion. Because sweat specimens are collected from locations on the body that are open to the external environment, there is concern that this exposure creates the potential for contamination through many pathways. Kidwell and Smith (2001) and Long and Kidwell (2002) have demonstrated that drug residue on the skin at the time the patch is applied can contaminate the patch, even if a washing protocol has been followed prior to application. One response to this is for the collector to retain the wash water/material to be tested if the sweat patch test is non-negative. Others have argued that any effect from environmental contamination would be small and that the sweat patch is effectively

---

[42] For example, because sweat glands and sebaceous glands are unevenly distributed throughout the body, studies report variable results when the sweat patch is applied to different locations on the body. As one example, Uemura et al. (2004) report that cocaine levels varied by the location of the sweat patch, with the highest amounts recovered from patches placed on the back rather than the shoulders. However, the literature search did not find any studies addressing the impact of patch (or wipe) location.

[43] Cone et al. (1994), Liberty and Johnson (2004), and Marchai et al. (2010) are among the exceptions.

[44] The Marchei et al. (2010) study (sample of 1) was of methylphenidate, a drug used to treat attention-deficient hyperactivity disorder. This study found that although the drug began to appear in plasma and oral fluids shortly after administration, it did not appear in the sweat specimen (tested with a sweat patch) until about 5 hours after administration. Cone et al. (1994) found that the first traces of cocaine appeared in the sweat specimen 1 to 2-hours after administration. Liberty and Johnson (2004) found that sweat patches did not show positive results for cocaine for more than 2 hours (and less than 1 day). None of the studies reviewed explained why the time profile in sweat was so different from that of oral fluids – why there was such a long delay before drugs begin to appear in sweat when they appear so quickly in oral fluids.

tamper evident. Most of the sweat patch tests have been done on individuals participating in either a treatment/recovery program or subject to criminal justice testing requirements. The continuous monitoring provided by the patch also raises concerns about intrusiveness and privacy.

2.3.4 Hair

Hair specimens are collected by cutting a group of hair strands (typically 150-200 strands or 50 mg, about the diameter of a pencil) as close to the skin/scalp as possible, using sharp scissors or a razor that has been cleaned with alcohol (Jenkins 2003). In general, hair samples are typically collected from the back of the head (the vertex posterior region) where hair growth is most consistent and the loss of hair will be least noticeable (Cone et al. 2007; Kintz 2008). Scalp hair's typical growth rate (elongation of the hair shaft) is 0.3 to 0.5 mm per day, or between 9 and 15 mm per month. Therefore, approximately 5 to 7 days after it has been formed, the hair shaft emerges through the surface of the scalp (Coulter et al. 2010). As the hair continues to grow, this portion of the hair shaft moves outward from the scalp. When waiting to obtain the portion of a hair shaft formed at the time of a suspected drug-facilitated crime (to serve as evidence), Kintz (2007) recommends waiting 4-5 weeks before collecting the hair sample. In this way, hair can provide a retrospective record of drug use.

Growing interest in using hair as a specimen in forensic investigations, treatment/recovery monitoring, criminal justice (including custody cases), and non-Federally-regulated workplace testing programs has prompted considerable research to improve the scientific basis for tests of drugs of abuse in hair and the interpretation of results.[45] The international organization, the Society of Hair Testing, was established to provide expert judgments and guidance about hair testing.

Much more research has been conducted on hair than on sweat. In part, this is because hair is a complex mini-organ, with a complicated physiology and structure (Krause and Foitzik 2006) and an alternating cycle of growth and quiescence.[46] The hair shaft is an epidermal outgrowth of the hair follicle. The hair bulb, located 3-4 mm below the surface of the skin, is in close contact with the capillaries and serves as the factory for the hair shaft (Baliková 2005; Krause and Foitzik 2006). During the growth phase, the cells in the hair bulb are the fastest dividing cells in the body (Boumba et al. 2006:144). As they divide, the newly formed cells move up the follicle into the keratinogenous zone where they synthesize pigment (melanin) and begin to form protein fibers (keratin) that arrange themselves into three layers, the medulla, the cortex, and the cuticle. The hair follicle is physiologically very active. In addition to producing the pigmented hair shafts (composed of keratins and melanin), it also synthesizes or metabolizes a large variety of hormones, neurotransmitters, and other biological substances (Krause and Foitzik 2006).[47]

The process by which drugs enter hair is complex and is still not well understood. One avenue of drug entry is passive diffusion from the bloodstream, similar to the process described for oral fluids and sweat. It is thought that drugs and metabolites that diffuse into the hair follicle are incorporated into the hair shaft where they are "trapped," and protected by the surrounding layers of keratin (Schaffer and Hill 2005). Demonstration that some drugs (e.g., cocaine) with

---

[45] Schaffer and Hill (2005) found more than 750 reports on drugs in hair published between 1984 and 2002.

[46] Hair goes through a three-phase cycle: rapid growth (anagen), cell-death driven regression (catagen), and relative quiescence (telogen) (Krause and Foitzik 2006).

[47] It is not clear how this level of physiological activity affects the composition and concentration of drugs in hair, if it affects it at all.

an affinity to bind to melanin are found in higher concentrations in hair with high melanin content, indicates that at least some drugs and metabolites enter the hair follicle early in the growth process (Musshoff and Madea 2007a,b; Schaffer and Hill 2005). A number of studies were conducted to determine how precisely the time of drug administration is reflected in the location of the drug/metabolites in the hair shaft and to establish dose-related concentration relationships for the various drugs of abuse. They found that drugs and metabolites may be transferred into or onto the hair shaft by sweat, sebum, and the skin, and also by contamination from other external sources (e.g., second hand smoke or airborne particles). As described above, hair follicles and hair shafts are exposed to secretions from the sweat and sebaceous glands both before and after the hair shaft emerges from the scalp (Auwärter et al. 2010; Baliková 2005; Jenkins 2003; Kintz 2008; Nakahara 1999). Baliková (2005) identifies melanin content of the hair, pH of the hair matrix,[48] and lipophilicity and basicity of the drug/metabolite as the key factors influencing drug incorporation into hair.

The relatively long time interval, which is generally unobserved, between drug administration and specimen collection that can occur with hair specimens creates the potential for uncontrolled exposure to external environmental conditions and contaminations, and for efforts at subversion. This has raised concerns about the integrity of the specimen and the ability to interpret test results. A growing number of studies are being completed to address these issues and to characterize the retention and stability of drugs/metabolites in hair and their vulnerability to the effects of exposure to sun, repeated washing, bleaching, and other strategies to remove drugs from hair (Baliková 2005; DuPont and Baumgartner 1995). Such studies typically are needed for each drug of interest.

Because hair is a solid, high-protein substance, composed of about 65-95 percent protein (keratin, melanin), 15-35 percent water, and 1-9 percent lipids, it requires more preparation for analysis than other specimens. Preparation steps include careful washing to remove contaminants from the hair surface and extraction (which may include alkaline digestion, acidic extraction, or enzymatic digestion) (Boumba et al. 2006; de la Torre et al. 2010). A challenge has been to identify preparation procedures that provide reliable results while reducing the time required for the process (Coulter et al. 2010). Kintz and Agius (2009) published draft European Workplace Drug Testing Services Guidelines to help standardize the procedures for testing hair.

Controversy continues about how to interpret results, particularly concerning external contamination, cosmetic treatments, washing, and melanin differences (Kintz and Mangin 1995; Musshoff and Madea 2007a,b). However, there is general agreement that screening and confirmatory analytical technologies are capable of reliably determining the nature and concentration of drugs of abuse in hair specimens (de la Torre et al. 2010; Kintz 2007). The long window of detection provided by testing hair raises intrusiveness concerns similar to those regarding sweat testing and the sweat patch as a collection device.

2.3.5 Considerations Regarding Alternative Specimens for Drugs of Abuse: Pros, Cons, and Issues

Table 2.3 summarizes the attributes of alternative specimens that affect their utility/desirability for workplace D&A testing and the technologies needed to use them. As discussed in Section 2.4, there is little question that analytical tools are now available to conduct reliable tests for

---

[48] The hair matrix is more acidic than plasma, creating a pH gradient that favors the transfer of basic drugs/metabolites (Baliková 2005).

drugs of abuse in all the specimens and the growing body of studies is making progress in building the empirical basis for interpreting test results with reliability.

Table 2.3.   Specimen Attributes that Affect Utility/Desirability for Workplace Testing

Attribute	Why It Matters	Comments about Specimens (broad summary)[49]
Accessibility (ease and invasiveness of collection)	Collection of an accessible specimen is typically considered less invasive Collection of an accessible specimen typically requires less specialized skills in the collector However, accessibility may increase exposure to contamination (passive exposure) from the environment	Blood/plasma – Low (invasive) Urine – Moderate (social norm) Oral fluids – Low (invasive) Sweat – Low (invasive) Hair – Moderate (cosmetic considerations)
Volume of specimen available for collection and production/ turnover rate	Although current technologies can deal with very small sample volumes, high production volume facilitates split samples and typically reduces collection time A rapid production or turnover rate, creates the potential for collection protocols that collect the specimen from a "refreshed" supply or collect a second specimen from a refreshed supply (to thwart subversion)	Blood/plasma – High, fast Urine – High, moderately fast Oral fluids – Moderate, fast Sweat – Moderate, moderately fast Hair – Moderate, slow
Ease of observation of collection process	Full and continuous observation reduces the potential for substitution or adulteration	Blood/plasma – High Urine – Low Oral fluids -- High Sweat – Low Hair – High (Any prior treatment of the hair and environmental exposure is not observed)
Intra- and interpersonal variability in amount, production rate, or growth rates of the specimen	The less variability, the more straightforward the protocols for collection, setting of testing parameters, and interpretation of results	Blood/plasma – Low Urine – Moderate Oral fluids – Moderate Sweat – High (but may not affect testing results) Hair – Moderate
Risk of infection,	Increases the complexity of the collection and handling protocols	Blood/plasma – Yes Urine – Moderate

---

[49] Because the factors that determine these attributes can be numerous and complex, these comments should be recognized as broad summaries that may not hold in all cases (e.g., different drugs can have widely different windows of detection in the same matrix).

Attribute	Why It Matters	Comments about Specimens (broad summary)[49]
complications, or hazards from collection process or specimen itself	and increases need for training	Oral fluids – Moderate Sweat – No Hair – No (except lice)
Composition of the fluid or tissue, particularly the amount of protein present	High levels of protein add complexity to sample preparation	Blood/plasma – High protein Urine – Low protein Oral fluids – Low protein Sweat – Low protein Hair – High protein
Timing and concentration of the drug and its metabolites in the fluid or tissue, variability in that timing and concentration; Window of detection	To serve as a specimen, the drug of interest, or an identifiable metabolite, must be distributed into the specimen at a concentration that can be measured How soon the drug and its metabolites can be detected in the specimen and for how long after drug administration establishes the window of detection and influences the applicability of the specimen for different testing conditions (e.g., for-cause, post-accident, follow-up) Specimen that contain the parent drug rather than its metabolites typically allow more definitive initial test results Variability complicates interpretation of results	Blood/plasma – Fast, low concentration (conc.) Urine – Moderate, high conc. Oral fluids – Fast, low conc. Sweat – Moderate, low conc. Hair – Slow, low conc. Blood/plasma – Short/moderate Urine – Moderate Oral fluids – Short Sweat – Moderate/long Hair – Long
Ratio of parent drug concentration in the specimen to its concentration in plasma, over time	The concentration of the parent drug (and psychoactive metabolites) in plasma is useful in determining the therapeutic dose and establishing dose-response effects (including impairment); most studies on the impairing effects of drugs of abuse use plasma concentration as the measure of drug level. A specimen with a stable ratio of analyte concentration to plasma	Blood/plasma – Base of comparison Urine – Mostly metabolites Oral fluids – Usually >1 Sweat – Unknown Hair – Unknown

Attribute	Why It Matters	Comments about Specimens (broad summary)[49]
	concentration allows application of that information to measures based on the specimen. For programs in which the purpose of the test is solely to determine evidence of use (i.e, applying "per se" policies) rather than consideration of impairment, this information is not pertinent	
The pH of the fluid/tissue and its variability;	The pH of the specimen can affect drug concentration and metabolite composition. Collection protocols may need to be designed to prevent modification of pH	Blood/plasma – 7.4, low variability Urine – Variable but inconsequential Oral fluids – Weakly acidic, some variability Sweat – Unknown Hair – Weakly acidic, variability unknown
Susceptibility of specimen to contamination through passive external exposure (e.g., drugs present in the environment or exposure external to the individual)	Creates the potential for false positive results	Blood/plasma – Low Urine – Low Oral fluids – Low/moderate Sweat – High Hair – High
Susceptibility to false positive tests for the drug or its metabolites from consumption of non-prohibited substances (e.g., poppy seeds)	Complicates interpretation of results and may affect decisions about cutoff levels	Blood/plasma – Moderate Urine – Moderate Oral fluids – Moderate Sweat – Moderate Hair – Moderate
Stability of the drug and its metabolites in the matrix over time	If a drug/metabolite is not stable in the matrix, the concentration at the time of the test (or retest) may differ from the concentration at the time the specimen is collected	Blood/plasma – OK Urine – OK Oral fluids – OK Sweat – Unknown Hair – Very good

Attribute	Why It Matters	Comments about Specimens (broad summary)[49]
Stability comments assume the recommended protocol is followed	(or first tested). Considerable research is needed to establish and validate the protocols that maintain stability throughout the collecting-testing-storage process; special requirements can add cost and complexity.	
Need for pretreatment to clean specimen and/or to release drug from specimen or prepare specimen for testing technology	Adds complexity to the testing process May add time and cost	Blood/plasma – Yes, requires centrifuge Urine – No Oral fluids – No Sweat – Yes, remove from pad Hair – Yes, remove from hair
Availability of validated, cost-effective technologies	Prerequisite for consideration as a required or permitted technology	Blood/plasma – High Urine – High Oral fluids – Med High Sweat – Med Hair – Unknown

Sources: Derived in part from Kerrigan and Goldberger (2008); Aps and Martens (2005).

An important consideration in decisions about requiring or allowing more than one specimen to be used in workplace testing programs is the necessity of and ability to establish comparability across test results. Table 2.4, derived from Cone et al. (2007), provides an illustration of the kinds of disparate results that might result from testing one individual with two different specimens. Among other issues, these disparate results create a challenge for establishing cutoff levels that are comparable across specimens. Lack of equivalence in the consequences of similar drug-taking behavior because of difference in detection methodology could become the basis of challenges to the fairness of the program.

Table 2.4.   Disparities that Might Occur when Testing Different Specimens for Drugs of Abuse[abc]

"P" represents a positive result; "N" represents a Negative result

Test #	Specimen Blood	Urine	Oral fluids	Sweat	Hair	Possible Reason(s) for Disparities (Assuming the collection procedures were followed correctly for all matrices)
1	P	N				Urine collected too soon after drug use
2	P		N			Low dose of highly protein-bound drug (e.g., benzodiazepines) may not appear above cutoff in oral fluids
3	P			N		Low drug dose; time of collection missed overlap of detection windows (e.g., patch applied too late)
4	P				N	Low drug dose; low binding affinity to hair specimen; time of collection missed overlap of detection windows; hair treatment
5	N	P				Concentration effect of kidneys; time of collection missed overlap of detection windows (e.g., long interval after administration)
6		P	N			Concentration effect of kidneys; low dose; highly protein bound drug; time of collection missed overlap of detection windows (e.g., long interval after administration)
7		P		N		Concentration effect of kidneys; low dose; highly protein bound drug; time of collection missed overlap of detection windows (e.g., long interval after administration)
8		P			N	Concentration effect of kidneys; low dose; low binding affinity to hair; time of collection missed overlap of detection windows
9	N		P			Time of collection missed overlap of detection windows (e.g., too soon after administration); drug residue from mode of administration in oral fluids (i.e., "shallow depot" effect)
10		N	P			Time of collection missed overlap of detection windows (e.g., too soon after administration for metabolism and excretion); shallow depot effect
11			P	N		Time of collection missed overlap of detection windows (e.g., too soon after administration for metabolism and excretion); shallow depot effect
12			P		N	Low dose, low binding to hair, time of collection missed overlap of detection windows (e.g., too soon after administration and

| Test # | Specimen | | | | | Possible Reason(s) for Disparities (Assuming the collection procedures were followed correctly for all matrices) |
	Blood	Urine	Oral fluids	Sweat	Hair	
						excretion)
13	N			P		Time of collection missed overlap of detection windows (e.g., too long after administration)
14		N		P		Time of collection missed overlap of detection windows (e.g., too long after administration)
15			N	P		Time of collection missed overlap of detection windows (e.g., too long after administration)
16				P	N	Low dose, low binding to hair, time of collection missed overlap of detection windows (e.g., too soon after administration for metabolism and excretion)
17	N				P	Time of collection missed overlap of detection windows (e.g., too long after administration)
18		N			P	Time of collection missed overlap of detection windows (e.g., too long after administration)
19			N		P	Time of collection missed overlap of detection windows (e.g., too long after administration)
20				N	P	Time of collection missed overlap of detection windows (e.g., too long after administration)

[a] Source: Derived from Cone et al. 2007:829, with additional observations by the PNNL research team.
[b] The results are assumed to be for the same individual; same sample collection time.
[c] As mentioned, non-equivalency of cutoff levels could create a difference in test results between specimens.

2.4 Testing Technologies to Prepare, Separate, Detect, Identify, and Measure Substances of Interest in Drug Testing Programs

The drug testing process entails a series of steps, from the selection of a donor and collection of a specimen, through the implementation of one or more test methods or technologies to provide analytic results relative to the drugs in question, to the interpretation of those results to determine the actual test outcome. Although innovation occurs at each step of the process, the following four steps have been subject to the greatest efforts at innovation: specimen collection; validity testing; initial (or "screening") testing;[50] and confirmatory testing. This section reviews the methods, technologies, and devices used in these four program components, discusses issues associated with their use or performance, and describes some of the new and innovative methods, technologies, and devices being developed and tested for application. A separate subsection discusses testing for alcohol, which relies on a different set of technologies and

---

[50] Throughout this chapter, we follow the NRC convention of referring to the first testing step as the "initial test." However, many other organizations use the term "screening test" for this step. To clarify this equivalence, we will periodically refer in the text to the "initial (screening) test."

protocols, and is managed separately from testing for other drugs of abuse in most workplace testing programs.

2.4.1 Introduction to Drug Testing Measurement Technologies

The underlying technologies for testing the validity of a specimen and performing the initial and confirmatory tests for drugs of abuse (other than alcohol) remain fundamentally the same as those described in a previous update on technical issues for fitness for duty (Barnes et al., 1988). Urine continues to be the primary specimen used, with testing methods and technologies remaining generally the same as those in use in the 1980s. Tests to determine the validity of urine specimens involve a series of analytic chemistry methods to determine the chemical and physical properties of the specimen. In addition, the technologies to generate these measurements are well established. Consequently, increasing ease of use and reducing cost are the primary strategies for innovation in validity testing.

Testing for drugs of abuse in urine specimens continues to follow the established sequence of immunoassay testing for initial testing, with a combination of chromatography (for separation of compounds) and mass spectrometry (for detection of specific compounds) for confirmatory testing. Although future technological breakthroughs might succeed in modifying what is essentially a one-to-one relationship between certain testing technologies and certain prescribed process steps (with immunoassay used for initial testing, and chromatography/mass spectrometry used for confirmatory testing), the information reviewed for this report does not indicate that this change will occur in the near future.

Although the basic technologies used in the testing process have not changed, there have been substantial improvements in the scope, sensitivity, speed, and cost of these testing methods in the past 30 years. These improvements have increased the types of drugs that can be identified and quantified in both initial and confirmatory tests. In addition, a number of collection devices have been developed that include either instrumented or non-instrumented processes to perform initial testing at the point of collection. Consequently, although innovations have not changed the fundamental type of technology used in validity, initial, or confirmatory testing, they have changed the look, feel, and performance of the methods, devices, and equipment used to perform these tests.

Drug tests using specimens other than urine also rely on immunoassay and chromatography/mass spectrometry techniques. For example, these same technologies are routinely used for blood testing in medical settings. In addition, improvements in the materials, methods, and equipment systems over the past dozen or so years have made these techniques sensitive and reliable enough to give valid results for the lower concentrations of drugs present in hair, oral fluids, and sweat specimens. Along with the compilation of pharmacokinetic information on the time course of drugs in these alternative specimens, these advances in testing technology are demonstrating the potential for workplace testing. Technology improvements, along with an extensive array of validation studies and FDA clearances, have enabled manufacturers to successfully market collection devices and testing services for specimens other than urine, and a variety of point-of-collection validity and initial testing devices to those implementing non-Federally-regulated workplace testing programs.

2.4.2 Considerations and Criteria for Testing Methods and Technologies

Drug testing measurement technologies must not only have the sensitivity and specificity to provide accurate and reliable results for the specimen and drug being tested, their performance must also be validated for each collection device and preparation protocol. The volume of studies comparing alternative devices and protocols illustrates the effort expended in the process of establishing "best practices" and standard procedures for innovative technologies.

Table 2.5 summarizes attributes of performance identified frequently in the literature as important for methods used to detect and measure drugs in body fluids. With minor variations, these are the attributes used to describe and evaluate the methods, devices, and equipment for drug testing, for example, by Jones et al. (2003) in their extensive review of the state of knowledge in drug testing. The characteristics in Table 2.5 provide a practical set of performance criteria that include both technical and usability characteristics. These terms are defined in the glossary.[51]

---

[51] Note that there are other related lists of technical attributes for analytical methods. For example, Smith et al. (2007) list: Accuracy, precision, interference, robustness, signal-to-noise ratio, LOD (Level of Detection), LOQ (Level of Quantitation), ULOQ (Upper Limit of Quantitation).

Table 2.5.   Attributes That Affect Acceptability of Workplace Drug Testing Methods, Devices, and Technologies[a]

Characteristic	Definition
Sensitivity	The ability of a method to detect the presence of drugs or classes of drugs (lower limit of detection).
Specificity/Selectivity	The ability of a method to differentiate and quantify the analyte in the presence of other compounds in the sample.
Linearity	The ability of an analytical procedure (within a given range) to produce results that are directly proportional to the concentration (amount) of the analyte in the sample.
Speed	The time from start to end of the analytical process using a method.
Simplicity	Usually related to the speed of a method; often reflecting the requirement for little training for technicians and few steps in preparing the specimen and instruments for testing; often associated with procedures that can be highly automated and that have limited potential for carryover (i.e., contamination of a sample by contents of the previous one).
Reliability	The dependability of a method; its ability to reproduce accurate and precise results day-to-day.
Accuracy	The degree to which a method produces results consistent with actual values (i.e., the closeness of the mean test results to the true value).
Precision	The consistency with which a method reproduces results when measuring the same sample (intra-run, inter-run, total).
Economy/Cost	Economic considerations such as: time of analysis, number of samples processed in a single run, degree of training required of personnel, price of obtaining (and maintaining) instrumentation, price of chemicals and other reagents used in analytical procedure, and overhead of analytical laboratory or other facility.
Safety	The degree to which personnel using a procedure are exposed to risk of injury or long-term toxicity associated with chemicals required by a method.
Ruggedness	The ability of the device/equipment to work correctly within its established performance parameters over a large number of operations and/or long period of time under field conditions.
Stability	The extent to which an analyte in a biological fluid or tissue remains unchanged during collection, transportation, analysis, and storage. Analyte stability in a biological fluid is a function of the storage conditions, the chemical properties of the analyte, the matrix, and the container system. Conditions used in stability experiments should reflect those likely to be encountered during actual sample handling and analysis.

[a] Source: After Joscelyn et al. (1980); and Isenschmid and Goldberger (2007)

Isenschmid and Goldberger (2007:785) provide a clear discussion of the meaning of each of the analytical attributes and why they are important for validating methods, and device and instrument performance. Specificity, for example, describes the ability of an analytical method to distinguish the target compound (e.g., drug being tested) from other compounds, including those whose structure is similar to that of the target compound. In testing for drugs of abuse, test specificity is influenced by the extent to which the specimen contains the parent drug or its metabolites and also by the number and frequency of use in the tested population of other materials (foods or drugs) that yield the same or similar metabolites. As mentioned previously, blood, oral fluids, hair, and sweat typically contain primarily the parent drug, while urine often contains primarily drug metabolites. The presence of the parent drugs tends to increase the specificity of testing; by contrast, if only the drug metabolites are being measured, this decreases specificity because multiple drugs can result in the same metabolite.

In addition to the set of characteristics and criteria used by researchers and technology developers to evaluate technologies and methods, government agencies also have delineated criteria for adequate performance. In particular, the HHS Guidelines define a framework of extensive quality assurance and quality control of the testing process and testing technologies. Prior to authorizing an analytical technique or technology for use in drug testing, SAMHSA requires a battery of analytical performance specifications and characteristics to be established, validated, and verified. These requirements include: a) demonstrating that the technique or technology shows both sensitivity and specificity for HHS target drugs or drug classes, including acceptable performance around the HHS testing cutoff levels (i.e., samples 25 percent above and below the cutoff can be discriminated and are correctly reported); b) determining that controls are available to ensure correct in-laboratory operation of the technique or technology; and c) requiring, as applicable, clearance by the FDA for diagnostic medical devices (HHS SAMHSA 1998).

The FDA clearance process requires that a series of principles and procedures to validate the performance of methods, devices, or products are met prior to approval for commercial sale. The validation principles that the FDA requires for bioanalytical methods are shown in Table 2.6.

Table 2.6.   FDA Principles for Validating and Establishing Bioanalytical Methods[a]

Validation Principles
The fundamental parameters to ensure the acceptability of the performance of a bioanalytical method validation are accuracy, precision, selectivity, sensitivity, reproducibility, and stability.
A specific, detailed description of the bioanalytical method should be written. This can be in the form of a protocol, study plan, report, and/or Standard Operating Procedure (SOP).
Each step in the method should be investigated to determine the extent to which environmental, matrix, material, or procedural variables can affect the estimation of analyte in the matrix from the time of collection of the material up to and including the time of analysis.
It may be important to consider the variability of the matrix because of the physiological nature of the sample. In the case of LC-MS/MS-based procedures, appropriate steps should be taken to ensure the lack of matrix effects throughout the application of the method, especially if the nature of the matrix changes from the matrix used during method validation.
A bioanalytical method should be validated for the intended use or application. All experiments used to make claims or draw conclusions about the validity of the method should be presented in a report (method validation report).
Whenever possible, the same biological matrix as the matrix in the intended samples should be used for validation purposes. (For tissues of limited availability, such as bone marrow, physiologically appropriate proxy matrices can be substituted.)
The stability of the analyte (drug and/or metabolite) in the matrix during the collection process and the sample storage period should be assessed, preferably prior to sample analysis.
For compounds with potentially labile metabolites, the stability of analyte in matrix from dosed subjects (or species) should be confirmed.
The accuracy, precision, reproducibility, response function, and selectivity of the method for endogenous substances, metabolites, and known degradation products should be established for the biological matrix. For selectivity, there should be evidence that the substance being quantified is the intended analyte.
The concentration range over which the analyte will be determined should be defined in the bioanalytical method, based on evaluation of actual standard samples over the range, including their statistical variation. This defines the standard curve.
A sufficient number of standards should be used to adequately define the relationship between concentration and response. The relationship between response and concentration should be demonstrated to be continuous and reproducible. The number of standards used should be a function of the dynamic range and nature of the concentration-response relationship. In many cases, six to eight concentrations (excluding blank values) can define the standard curve. More standard concentrations may be recommended for nonlinear than for linear relationships.
The ability to dilute samples originally above the upper limit of the standard curve should be demonstrated by accuracy and precision parameters in the validation.
In consideration of high throughput analyses, including but not limited to multiplexing, multicolumn, and parallel systems, sufficient quality control (QC) samples should be used to ensure control of the assay. The number of QC samples to ensure proper control of the assay should be determined based on the run size. The placement of QC samples should be judiciously considered in the run.
For a bioanalytical method to be considered valid, specific acceptance criteria should be set in advance and achieved for accuracy and precision for the validation of QC samples over the range of the standards.

[a] Source: HHS FDA 2001:8-9

Taken together, the characteristics and criteria detailed here provide a quick snapshot of what is required to determine that a newly emerging technology or method will perform successfully. In the following discussions of new or emerging technologies for specimen collection, validity testing, and initial and confirmatory testing, these characteristics and criteria are highlighted as they pertain to performance issues or advantages of the technologies being reviewed.

2.4.3 Specimen Collection

Specimen collection includes the procedures and the devices, equipment, and technologies used to remove or collect the specimen from the donor's body, measure its quantity and physical characteristics, stabilize it (if necessary), and contain it for initial testing or transfer to a packaging/transport device. The devices, equipment, and technologies used for specimen collection range from standard laboratory equipment, such as scissors/razors and thermometers, to highly specialized devices. The collection process must ensure that an adequate quantity and quality of specimen is collected to support the testing process and that the specimen, once collected, is protected from deterioration, modification, or subversion.

Specimen collection methods and technologies are important because they affect the attributes of the specimen that are the basis for subsequent testing. Standardization of collection methods helps avoid bias by ensuring consistency on factors that affect matrix or drug characteristics, such as location on the body from which the hair specimen is collected or the composition of the oral fluid collection device. In many cases, factors that introduce variability were only discovered during the testing and validation process. Specimen collection methods and technologies vary by the specimen being tested, as detailed in Table 2.7.

Table 2.7.   Basic Collection Process By Specimen[a]

Specimen	Removal Process	Additional Notes
Blood	Collected by trained phlebotomist following American Medical Association (AMA)-phlebotomy procedures (alcohol/syringe); involves puncturing the skin with a syringe and removing the blood sample	Protocols must be followed in collection, storage, and handling of blood samples to prevent contamination of collector by blood-borne pathogens Requires use of sterile equipment and sterilization of the injection site to reduce the risk of infection
Breath	Provided by donor by blowing into a tube attached to the measuring device	No retention of the breath specimen; collection is done by measurement device
Urine	Provided by donor by urinating into collection vial	Requires examination to detect substitution or adulteration
Oral fluids	Provided by donor by spitting into collection vial; absorbent pad or swab placed in mouth until saturated	Recent studies have found that the collection method (stimulated versus non-stimulated) and materials used in the collection process (swab and container) can affect results Requires mixing with buffer, refrigerated storage, and measurement as soon as possible to avoid bacterial growth and breakdown of drug constituents
Hair	Collector cuts approximately 50 mg hair strands (typically from the back of the head) with a razor/scissors as close to the scalp as possible	May be stored at ambient temperature in envelope, plastic container, etc.
Sweat	Collected via: Swab of donor's body area with towel, filter paper, or other material Sweat patch that is worn for a specified period of time (typically between 3 and 10 days)	

[a] Source: Drawn in part from Caplan and Huestis (2007)

Considerations and Challenges

Part 26 requires collection procedures and equipment to be as minimally intrusive and protective of the donor's privacy as possible, consistent with securing a specimen of adequate size or volume to permit testing. In addition, the collection process/device must also support efforts to ensure that donors do not substitute or adulterate specimens. In the case of urine collection, this includes allowing the collector to check the temperature of the specimen and do a visual inspection for unusual color or other abnormal properties[52]. Collection procedures for alternative specimens have similar considerations.

Recent and Emerging Developments

Overall, there have not been significant changes in the collection technology for urine, blood, or hair specimens.[53] However, two developments that deserve review are the sweat patch used for the collection of sweat, and oral fluids collection devices.

Sweat Patch for Collection of Sweat

DuPont and Selavka (2008) describe the development and functioning of the sweat patch, which is the only method currently in widespread use for this specific specimen. The patch is a waterproof, adhesive pad that is similar to a nicotine patch. The sweat patch is designed to show noticeable puckering if removed, which provides "reasonable integrity" to the collection process. The patch functions by allowing water to evaporate through an outer membrane while the drugs/metabolites accumulate on the absorbent pad. The patch collects evidence of drug use prior to the application of the patch (for the window of detection of the drug in sweat) and contemporaneously during the period the patch is worn by the donor. Because skin is continuously renewed, patches can be worn for periods of a few hours to a few weeks before they loosen and fall off. The patches do not record the volume of sweat from which the drugs retained on the patch are derived; consequently, they do not allow accurate determination of the concentration of drug in the sweat itself. However, sweat patches, like hair, allow rough quantitation of the level of drug use, and are able to distinguish heavy from light drug users. Both sweat and hair are more capable than other specimens of reliably producing positive test results for opiates after heroin ingestion.[54]

Although sweat patches have not been widely marketed for detection of chronic, excessive alcohol use, DuPont and Selavka (2008:659) state that the use of alcohol detection patches for this purpose may grow. The alcohol metabolite ethyglucuronide (EtG) has become a useful target metabolite in sweat because it indicates alcohol use in the prior 3-5 days.[55] There is some dispute about the specificity of this test (Thierauf et al., 2010), although it is generally considered reliable as an indicator of chronic alcohol consumption.

---

[52] This requirement is included in both SAMHSA guidelines and NRC regulatory requirements. Although guidelines for initial validity testing by collectors for other specimens have not been established (either by SAMHSA or NRC), it is probable that such procedures would need to be developed. Any such guidelines would be a consideration for collection devices and technologies.

[53] Technologies to collect and test blood for medical monitoring purposes (e.g., diabetes) have undergone significantly miniaturization and advancement, requiring very small quantities of blood and providing test results very quickly and automatically. To date, these technologies have not been widely applied to the testing of drugs of abuse.

[54] Detection of heroin is challenging in oral fluids because of its extremely rapid metabolism; the window of detection of heroin itself is very short. In urine, it is challenging to distinguish heroin from other drugs that yield the same metabolites.

[55] The metabolite is also found in urine.

The FDA approved several sweat collection patches in 1995 for use in drug abuse screening programs to test for amphetamines, cocaine, marijuana, phencyclidine, heroin, methamphetamines, their metabolites, and other opiates. Products include the Sudormed Sweat Specimen Container, made by Sudormed, of Santa Anna, California and the PharmaChek™ Sweat Patch from Pharmachem Laboratories Inc. of Menlo Park, California[56] (Patentstorm US 2010).

Oral Fluids Collection Devices

Assessments of oral fluids as a viable specimen identified the importance of the collection device and collection procedures in ensuring the reliability and validity of test results (Bosker and Huestis 2009). Oral fluids collection devices primarily consist of a sorbent material or pad that is placed in the mouth to become saturated. Once the pad or material is fully saturated (some devices provide a visual identifier when this occurs), the material is removed and placed in a sealed container with a buffer solution, which is sent to a laboratory for analysis. Some examples of commercially available devices are Intercept® (OraSure Technologies), Quantisal™ (Imunalysis Corporation), Salivette® (Sarstedt AG), Saliva-Sampler™ (StatSure), Omni-Sal® (Cozart Biosciences Ltd.), Oralstat ®(American Bio Medica), Salicule™ (Acro Biotech), and Bio-One Saliva Collection System® (Greiner).

These devices have been subjected to numerous evaluations, which have generally found variations in performance across the available devices. Interpretation of results concerning device performance is complicated by the rapid change in the devices – many studies note that modifications to the devices have already been made before the study was complete. Two devices used frequently in scientific studies and thus subject to extensive review are Intercept® (OraSure Technologies) and Quantisal™ (Immunalysis). In a study by Quintela et al. (2006), Quantisal™ received high marks because it was able to meet the cutoffs in the HHS Guidelines, including those for tetrahydrocannabinol (THC). In a separate study, Intercept® performed well, although procedural modifications were necessary for adequate detection of THC (Kauert et al. 2006). By contrast, in an evaluation of nine different oral fluids collection devices, Langel et al. (2008) found that StatSure's Saliva-Sampler™ was the only device with recoveries (accurate detection across multiple samples) of more than 80 percent for all of the eight analytes being tested (amphetamine, MDMA, THC, cocaine, morphine, codeine, diazepam, and alprazolam). The recoveries for Intercept®, Quantisal™, Greiner's Bio-One Saliva Collection System®, and Acro Biotech's Salicule™ were above 80 percent for all drugs except THC. Meanwhile, other devices showed lower levels of recovery. For example, Salivette® showed only 16 percent recovery for diazepam and 27 percent recovery for MDMA.

Overall, the literature indicates that the performance of oral fluids collection devices continues to improve over time. However, two remaining issues involve (1) the stability of drugs in the specimens after days of storage, and (2) consistent performance, particularly relative to the HHS Guidelines cutoff levels. As indicated above, some devices have been shown to perform more consistently than others.

---

[56] Note that these sweat patch must be paired with appropriate testing devices to complete the drug screening process. Both of these patches rely on subsequent immunoassay (IA) tests; details on IA test devices are presented in Section 2.4.5. The Sudormed Sweat Specimen Container was co-developed with a specific IA test, the EIA Microplane Assay by SolarCare Technologies Corporation, of Bethlehem, PA (Patentstorm US 2010).

2.4.4 Validity Testing

The purpose of validity testing (also referred to as "specimen validity testing" or SVT in the literature) is to identify individuals who are attempting to subvert the drug testing process, thereby violating drug-testing policies. Onsite and laboratory validity testing is conducted to:

- obtain information about the physical characteristics of the specimen at the time of collection (e.g., for urine: temperature, appearance) to verify that they are consistent with the expected physical parameters of the matrix provided by the individual being tested (i.e., not substituted);
- obtain information about the chemistry of the specimen (e.g., for urine; specific gravity), also to verify that the specimen is consistent with expected parameters;
- identify efforts to dilute the concentration of drugs/metabolites in the specimen (e.g., diluting a urine specimen); and/or
- identify adulterants that have been added to the specimen (i.e., chemicals or other materials added to mask the presence of drugs).

Validity testing technologies vary by the physical and chemical characteristics of the particular specimen and the types of materials or processes used in subversion attempts. They must not adversely affect the ability to test for the presence of drugs/metabolites in the specimen. To date, validity-testing procedures have focused primarily on urine specimens, given the prevalence of urine testing and the plethora of strategies and technologies used to subvert urine testing. Because most urine specimens are not collected under direct observation, it is particularly important to verify that the specimen is indeed urine, and that it has come from the donor's body. The 2008 HHS Guidelines specify validity-testing procedures for urine specimens. For liquid specimens, validity-testing technologies may include instruments to measure the physical characteristics of the specimen (e.g., temperature, specific gravity, volume, etc.) and some of the chemical characteristics that may be altered in substituted or adulterated specimens (e.g., pH; the concentration of creatinine, a naturally-occurring by-product of muscle metabolism in urine; and the concentration of protein in oral fluids). The utility of these tests depends upon the availability of reliable parameters for these characteristics in an appropriate "normal" population. Because the technologies to generate these measurements are common and well established, the primary opportunities for innovation involve identifying additional characteristics that provide indications of validity or subversion, increasing ease of use, and reducing cost (while meeting precision and documentation requirements). Combining multiple tests into a single process or device, miniaturizing the test to reduce the amount of specimen and reagents needed, and automating the testing and reporting processes are prominent ways of creating "better, faster, cheaper" technologies for this component.

Considerations and Challenges

The NRC has defined two categories of initial validity tests: non-instrumented validity screening tests and instrumented initial validity tests. The non-instrumented tests are cheaper and easier to implement than the instrumented tests. The NRC provides detailed performance requirements and quality assurance/quality control (QA/QC) requirements for validity screening tests in Part 26. At the time of this review, no screening devices met this standard.

The continuous challenge for validity testing techniques and devices is to remain current and responsive to changes in subversion techniques. As Wu (2002) states: "The Internet provides drug abusers with an education in common drug testing practices and how to exploit the limitations of screening assays, as well as a shopping center for sample adulterants. Laboratories constantly need assays to detect the presence of ever more deceptive adulterants." Similarly, SAMHSA (2005) reports that manufacturers of subversion products continuously develop and modify their products, such that each successive version is more effective in masking drug presence without being detectable. In addition, manufacturers change their product formulas regularly to stay ahead of the drug testing labs. As a result, Caplan (2008:128) points out: "the task of comprehensively searching for normal elements that comprise…[the specimen] and ascertaining that no foreign materials have been added can be more complex and costly than the drug testing itself."

Thus, validity testing techniques and devices are in constant competition with a well-funded and motivated subversion industry.

Recent and Emerging Developments

One area of significant change in the area of validity testing is the development of combined point-of-collection testing (POCT) devices that perform validity testing in conjunction with drug testing. Section 2.4.6, Point-of-collection testing (POCT), addresses this topic more fully, following an introduction to immunoassay techniques in Section 2.4.5. Otherwise, the devices used for onsite validity testing of urine demonstrate ongoing improvements in their precision, ease of use, and recordkeeping capabilities. For example, there are urine specimen collection cups now available that include a temperature recording strip.

In the laboratory setting, although these tests are based on long-standing laboratory procedures, researchers continue to develop new devices for performing tests in a more efficient manner. At the present time, most validity testing devices are designed for urine specimens. However, with the increasing use of oral fluids for non-Federally-regulated workplace testing and in roadside testing for individuals driving under the influence of drugs (DUID), researchers are also investigating the attributes of oral fluids that could serve as validity checks or screens (Crouch 2005; de la Torre et al. 2004).

Devices used for urine validity testing typically employ test strips that are placed in the specimen by the collector. These devices can test for pH, specific gravity, and a variety of compounds including creatinine, nitrate, glutaraldehyde, and bleach, although most devices do not test for all listed compounds. Some examples of commercially available devices are AdultaCheck 4® (Sciteck, Inc.), AdultaCheck 6® (Sciteck, Inc.), AdultaCheck 10® (Sciteck, Inc.), and Intect® 7 (Branan Medical).

In an evaluation of these devices by Peace and Tarnai (2000), Intect® 7 performed well. This device was compared to Adultacheck 4® and Multiple Adulterant Strip Kemistry (MASK) test strips.[57] A separate evaluation concluded that "Intect® 7 was superior…for detecting correctly the presence of adulterants in urine" (Dasgupta et al., 2004). AdultaCheck 6® also performed well, but was limited by testing for creatinine, pH, aldehyde, nitrite, chromate, and oxidant but not bleach (Dasgupta et al., 2004). Assessments of most of the other devices such as AdultaCheck® 4, which tests for creatinine, pH, aldehyde, and nitrite, have found them to be limited by failing to provide comprehensive coverage of all potential adulterating materials.

---

[57] The MASK test strip contains multiple 1-cm x 1-cm absorbent pads infused with an adulterant chromaphor substrate that produces a color reaction in the presence of a particular adulterant or urine analyte (Burrows et al., 2005).

Consequently, attention to evidence that candidate technologies meet validity test requirements remains important.

2.4.5 Initial (Screening) Testing for Drugs of Abuse

The purpose of initial testing for drugs and alcohol is to identify specimens that do not contain drugs of abuse or their metabolites and, increasingly, to identify and characterize any drugs of abuse present in the specimen. Combined with validity testing, this allows drug testing programs to identify individuals for whom there is no indication of subversion or use of drugs of abuse. By doing this, initial testing allows the testing program to reduce the number of specimens that require more elaborate and expensive confirmatory testing.

As mentioned previously, immunoassay techniques are the standard approach to initial testing. Because of their high sensitivity, coupled with low cost, rapid operation, and the ability to be automated for high throughput, immunoassays serve as an effective screening technique (Tsai and Lin, 2005). Current generation immunoassay technologies are rapid, with automated analyzers connected to a robotic track for optimum delivery of samples. To give a sense of the scale at which these technology systems operate, Huestis and Smith (2006) report that a workplace drug testing laboratory might perform immunoassay screening tests on up to 30,000 specimens per day.

General Description of Immunoassay[58]

Immunoassays are part of a broader category of molecular-recognition tests (or "assays"), and are based on the human system's immune response: the generation of a protein-based antibody that has high affinity to and binds with a foreign compound (called an "antigen") that has entered the system. Once an antibody and antigen come into contact, they become stably bound into what is called the antibody-antigen complex. Various techniques (described below) can then be used to measure the level of this bound complex, which indicates the amount of the foreign antigen in the system (Smith 2003:117).

Immunoassay techniques play a central role in medical diagnostic testing. For example, immunoassays are used to determine the presence of infectious diseases by detecting antibodies in a patient's serum sample. If the antibodies are present, it indicates that the infectious disease is present in the body as well. In the case of drug testing, although drugs are small-sized compounds that do not normally trigger an antibody response, this limitation can be overcome if they are bound to a larger protein molecule. This larger molecule is then what triggers the antibody response.

Before describing drug testing immunoassays in more detail, it is useful to understand a few of the key elements of this technology (see Glossary for further definitions):
- "Antigen" refers to the target drug or drug metabolite that is being detected.
- "Antibody" refers to the immune-response protein molecules that will bind with the specific antigen.
- "Analyte" is the substance measured by the immunoassay test[59].

---

[58] This general description of immunoassay is based on Tsai and Lin (2005).

[59] Technically, the actual substance measured can be either the ant body or the target drug antigen; however, in the literature, the target drug antigen is often referred to as the "analyte," because ultimately it is the substance of interest.

- "Label" is a substance that chemically attaches to either the antigen or antibody and conveys a measurable property to it, such as fluorescence or radioactivity.
- "Reagent" is the generic term for the various commercially-produced compounds used as inputs in the immunoassay process, including the relevant antibodies as well as target drug antigens[60].

Detecting the presence of a given drug or drug metabolite in a sample[61] is achieved by adding specifically targeted antibodies to the sample and then measuring changes (if any) that occur. Labels are often used to facilitate measurement of these changes.

Although the basic process is straightforward, there are many variations in how a given immunoassay procedure exploits the antigen-antibody interaction to detect and measure the presence of a drug in a sample. There are two basic types of immunoassays: competitive and non-competitive. In a competitive immunoassay, a labeled antigen and an antibody (an antigen reagent) are added to the sample (which may or may not contain the free drug antigen being tested). The free drug antigen and the labeled antigen compete for binding to the antibody. Once the reaction is complete, any unbound labeled antigen is removed/neutralized and the amount of labeled antigen-antibody complex is measured. If the sample contains none of the drug (no free drug antigen is present), the labeled antigen will monopolize the antibody binding process and the resulting labeled antibody-antigen complex will show a strong measurement signal. By contrast, if a significant amount of the drug (free drug antigen) is present, it will form unlabeled antibody-antigen complex, using up some or all of the antibody and reducing the amount of labeled antibody-antigen complex that can form. After the unbound labeled antigen is removed/neutralized, the measurement signal will be weak. In these systems, the measured response is inversely proportional to the amount of free drug antigen present in the sample.

In non-competitive immunoassays, the free drug antigen reacts with an excess of labeled antibody. The level of free drug antigen present determines how much of the labeled antibody becomes bound. Once the unbound labeled antibody is removed/neutralized, the strength of the measurement signal indicates how much free drug antigen was present. Often non-competitive immunoassays are designed as a "sandwich assay," wherein the free drug antigen becomes sandwiched between an antibody attached to a stationary plate and an antibody/label conjugate that is added to the assay. (This technique is further described in Table 2.8.)

Immunoassays are either homogeneous or heterogeneous. Homogeneous immunoassays use one type of medium (e.g., the assay occurs in the liquid phase only); heterogeneous immunoassays use more than one type of medium (e.g., the assay occurs with antibodies attached to a microplate which is then bathed in liquid containing the sample and other reagents).

Table 2.8 details the most common immunoassay techniques used in drug testing, which include (1) radioimmunoassay (RIA); (2) enzyme-based assays (EIA), of which there are several key types; (3) fluorescent-based immunoassays (FPIA); (4) particle immunoassays (known as KIMS); and (5) chemiluminescent immunoassay (CLIA). In addition to these basic formulations, Tsai and Lin (2005) describe various technologies that combine these approaches; for example,

---

[60] Antigen reagents are distinct from the "free antigens" (i.e., drugs that would have been consumed by the individual) that are being tested for in the individual's specimen.

[61] The term "sample" is used in the analytical chemistry literature as the generic form of material being analyzed. This is in contrast to the term "specimen," which the NRC uses to identify a specific unit of material provided by a donor. In the following technical discussion, both terms will be used. Sample will be used when the focus is on specific technology functioning; specimen, when the focus is on the material provided by the donor.

an immunoassay might involve enzyme and chemiluminescent reagents working together. In addition, immunoassays can be coupled with flow-injection or chromatograph techniques to create flow immunosensors.

Table 2.8.   Common Immunoassay Techniques[a]

Technology	Characteristics
Radioimmunoassay (RIA)	First type of immunoassay test developed for drug testing Uses radioactive isotopes (specifically Iodine-125) as a label General Process. Antibodies and radioactively-labeled antigen are added to a liquid sample, then the resulting antigen-antibody complex is precipitated into solid material. The liquid is poured off and the resulting precipitated solid material is subject to a Geiger counter; the amount of radioactivity measured is inversely proportional to the amount of free drug analyte in the sample. Performance. – Developed for all common drugs of abuse, and still considered the most sensitive and specific of all immunoassay methods – Radioactive reagents have limited shelf life as a result of the short half-life of isotopes used – Inherent complications of handling and disposing of radioactive materials; as a result, RIA has become less common than other methods for testing drugs in urine – Less sensitive to matrix effects than many enzyme-based immunoassays; thus may be more useful for alternative matrices such as hair
Enzyme immunoassay (EIA)	Uses enzyme labels in place of radioactive labels General Process. Wide range of approaches that rely on enzymes to effectuate a change in color, the emission of light, or other changes to determine the concentration of the target drug analyte Performance. Rapid, uses stable reagents and is readily adapted to automated instrumentation; but may lack specificity Key subcategories used in drug testing (described below) – Enzyme-multiplied immunoassay technique (EMIT) – Enzyme-linked Immunosorbent Assay (ELISA) – Cloned Enzyme Donor Immunoassay (CEDIA)
Enzyme-multiplied immunoassay technique (EMIT)	Introduced in the early 1970s by Syva Company under trade name EMIT[TM] Homogenous, liquid-phase assay that can be used in both competitive and non-competitive procedures General Process. Competitive EMIT is based on competition between the free drug and an enzyme-labeled reagent drug antigen for binding with the antibody. When the enzyme-labeled antigen binds to the antibody, the enzyme is rendered inactive. A colorless reagent is then added, and any unbound enzyme-labeled antigen converts the reagent to a form that emits a certain wavelength of color. With more free antigen, the solution will have decreased color emission. Performance. – Rapid assay that utilizes stable reagents and is easy to automate – Medium sensitivity relative to other assays
Enzyme-linked Immunosorbent Assay (ELISA)	First research published in 1971; currently widely used in pharmaceutical industry for drug discovery, clinical trials Heterogeneous assay with antibodies attached to a microplate; can

Technology	Characteristics
	be used in both non-competitive "sandwich" and competitive procedures Some variations use microparticles rather than a plate as the solid substrate General Process. – Sandwich technique. A liquid sample is poured over a microplate to which antibody molecules have been attached. Any free antigen in the sample will attach to the antibody on the plate. A second antibody with an attached enzyme is then added to the surface. The antigen that has bound to the first layer of antibody will now bind a second time to the newly added antibody/enzyme – causing the antigen to be "sandwiched" between antibodies. The plate is then washed of all excess reagents and a dye-containing substrate is added. The amount of free antigen in the sample will be directly proportional to the color generated by the attached enzyme. – Competitive technique. Also starts with antibody attached to a microplate. But rather than using additional enzyme-attached antibody, instead it adds enzyme-attached antigen reagent to compete with any free antigen in the sample. The amount of free antigen in the sample will be inversely proportional to the color generated by the enzyme, because the enzyme-attached antigen will be washed away if there is a significant amount of free drug antigen to bind to the antibody. Performance. – Rapid assay that utilizes stable reagents and is easy to automate – Less subject to matrix effects than homogenous enzyme assays – Used more commonly for oral fluids and blood testing than for urine testing
Cloned Enzyme Donor Immunoassay (CEDIA)	Research published in 1986; commercialized mid-1990s by Microgenics Corporation Homogenous, liquid phase competitive binding assay, based on splitting an enzyme into two component fragments, with one attached as a "label" to the reagent antigen General Process. The antibody and the reagent with attached enzyme fragment are added to the sample. If there is some free drug antigen in the sample, it will bind to the antibody and allow more of the reagent/enzyme to stay in the solution. Then the second enzyme fragment is added, along with a dye-containing substrate. Whatever enzyme fragments can reconnect will become active and cause the dye to be released from the substrate. The level of color is directly proportional to the amount of free drug antigen in the sample. Performance. – Rapid assay that utilizes stable reagents and is easy to automate
Fluorescence Polarization Immunoassay (FPIA)	Initially developed for therapeutic drug monitoring by Abbott Laboratories Homogeneous competitive binding immunoassay that uses a fluorescent tracer "label" on the antigen General Process. The tracer, when excited by polarized light, emits fluorescence with a degree of polarization inversely related to the rate of rotation. Because larger molecules rotate more slowly than smaller ones, any tracer-labeled antigen that binds to antibodies (creating a much larger molecule) will show an

Technology	Characteristics
	increase in the degree of polarization. The degree of fluorescent polarization is inversely related to the amount of free drug antigen in the sample. Performance. - Useful in the analysis of matrices other than urine - Very stable and resistant to a number of adulterants - Does not require the unbound fluorescent tagged antigen to be washed away, because the molecules are all rotating in random uncoordinated directions which does not affect the reading of the intense polarized light signal
Particle Immunoassay (KIMS)	Known as the "kinetic interaction of microparticles in solution" or KIMS, and marketed since the early 1990s by Roche Diagnostic Systems as the Abuscreen ONLINE[R] General Process. The assay is based on the competition between free antigen and an antigen/microparticle reagent. When the antigen/microparticle binds with antibodies, in the process light is scattered, which causes a reduction of light transmission. The absorbance change is inversely related to the amount of free drug antigen in the sample. Performance. - Rapid assay that utilizes stable reagents and is easy to automate - Resistant to a number of adulterants
Chemiluminescent Immunoassay (CLIA)	First described in 1976; now widespread in commercial applications because of its sensitivity Uses chemiluminescent (i.e., light-producing) compounds as labels, usually in noncompetitive sandwich configuration General Process. Similar to ELISA, using the chemiluminescent compound instead of the enzyme. The solid may be in the form of a microplate or microparticles. The amount of signal is directly proportional to the amount of free drug antigen in the sampled. Performance. - The high sensitivity of the chemiluminescent response means that smaller samples or dilute samples can be addressed; analytical sensitivity is equal to RIA - Easy to measure

[ᵃ]Source: Drawn from Isenschmid and Goldberger (2007); Wu (2006); and Flannagan et al, 2008.

Considerations and Challenges

Some considerations regarding the use of immunoassay testing in workplace drug testing programs are discussed below.

An integral part of workplace drug testing has been the establishment of cutoff levels for drug tests to ensure standardization and comparability. These cutoffs have been administratively determined to define the line between positive and negative test results. In immunoassay technology, the increasing sensitivity of emerging testing technologies and devices is driving increasingly lower limits of detection. In fact, Luzzi et al. (2004) describe commercial immunoassay systems able to detect the presence of drugs at concentrations below HHS Guidelines cutoff levels with acceptable and reliable accuracy. This may raise the question of whether established administrative cutoff levels should be revised to reflect the growing ability of the technology to measure drugs or metabolites at lower and lower concentrations.

Although the preceding discussion has focused on the specific functioning of immunoassay technologies as they detect drugs in a specimen, an integral part of the immunoassay test is preparation of the specimen. This is a significant step, because the preparation process can make a marked difference in the test results. HHS, along with FDA and other professional bodies, provide criteria for adequate specimen preparation for approved immunoassay techniques and devices. However, consideration of new matrices and/or other changes to the testing process (such as point-of-collection technologies), brings to the forefront the need to plan ahead in the development and validation of preparation protocols that ensure comparable results for test involving different specimens, tests, and devices.

To be effective for drug testing, testing methods must be capable of detecting drug concentrations at the levels typically found in human specimens. For example, drug concentrations in hair are typically in the picogram per milligram range. In general, this low concentration has not been an issue for immunoassay testing, given the sensitivity of the immune response to the presence of very small amounts of antigen.

The competition among developers of immunoassay techniques has been to reduce interference and increase specificity, thereby improving the ability to distinguish among drugs with overlapping or similar metabolites. Early immunoassay techniques were often limited to the detection of a class of drug (e.g., barbituates, opiates) rather than a specific drug of metabolite. This is generally no longer the case.

The effort to detect a drug or metabolite with immunoassay can be subject to cross-reactivity or interference from other chemical compounds in the specimen – either other drugs or other molecules endogenous to the specimen. Cross-reactivity refers to the potential of the assay to yield "false positive" results because of the structural similarity between the target drug and some other compound in the specimen being tested. Interference is the more general term for the potential of the assay to yield inaccurate results, due to the effect of other compounds on any part of the assay process. This includes, of course, interference caused intentionally by a specimen donor who is attempting to subvert the testing process.

Immunoassay tests show a wide range of performance relative to cross-reactivity or interference in the workplace drug testing setting. According to Wu (2002):

> The cocaine metabolite and THC immunoassays are model tests, producing no false positives or false negatives, but the failure rates for benzodiazepines and tricyclic antidepressants are so high that many physicians in emergency room settings recommend not ordering these urine tests because they produce more confusion than clinical value.

Finally, Wu (2002) also describes the challenge of developing methods to detect new drugs of abuse. Among other issues is the time lag in developing reagents and standardizing methods to address them:

> Changing patterns of drug abuse pose another challenge. New drugs like oxycodone, oxymorphone, Ecstasy (3,4-methylenedioxymethamphetamine), and gamma hydroxybutyrate become popular partly because they are not included in current drug testing programs. Regulatory agencies, drug testing laboratories, and manufacturers of diagnostic reagents have to keep up. Mandatory drug

testing guidelines are evolving, but slowly. Ecstasy has been proposed for addition to the menu of analytes, but not yet the semisynthetic opiates.

Recent and Emerging Developments

Over the last decade, manufacturers have developed a wide range of new immunoassays that are more specific, more sensitive, and that target a broader array of drugs. Indeed, the high-volume nature of workplace drug testing has been an important driver of the evolution of immunoassay techniques (Jones et al. 2003). Pharmaceutical and therapeutic applications have also been a major driver of technology development. The overall trend is toward increasing accuracy, sensitivity, specificity, and speed. Indeed, Chyka (2009) reports that the newer high-speed, fully automated immunoassay systems can test and report results within 1-2 hours of specimen receipt.

Recent advances in immunoassay testing include improvements upon existing technologies as well as developments of new approaches. However, by far the most extensive area of innovation is in miniaturization, with the development of microassays or "lab-on-a-chip" technologies that support point-of-collection procedures. The following provides a survey of some of these emerging technologies.

New Labeling Approaches

Kuma et al. (2010) describe research examining the use of magnetic markers as an alternative to the standard enzymes. The new detection method relies on competitive interaction between a magnetized reagent antigen and any free antigen (e.g., a drug in the specimen), as they compete to bind with antibodies. One advantage of this method is that it eliminates the need to rinse out unbound reagents because the magnetic signals from the unbound markers are nearly zero because of Brownian motion[62]. Eliminating the rinsing step reduces the time and cost of the process and researchers believe it will also increase accuracy.
Another promising approach is the use of DNA as a marker for antibodies in non-competitive immunoassays, taking advantage of the fact that an amplification of the signal can be accomplished through polymerase chain reaction (PCR) (Wu 2006). This amplification means that the technology will have increased sensitivity to lower concentrations of target analyte, thus allowing the evaluation of smaller specimen samples, other matrices, and additional drugs/metabolites.

New Measurement Technologies

Shankaran et al. (2007) report on the development of surface plasmon resonance (SPR) based immunoassays. These immunoassays use a surface-sensitive optical technique as the signal – specifically, the refractive index[63] changes associated with the antibody-antigen binding interaction. This approach obviates both the need for labeling of reagents as well as the need to rinse out unbound reagents prior to measurement. It thus allows real-time measurement of the binding interactions between an antibody (that is attached to a transducer surface) and the analyte in solution.

---

[62] A random movement of microscopic particles suspended in liquids or gases resulting from the impact of molecules of the surrounding medium.

[63] Refractive index is the measure of the bending, or refraction, of a beam of light as it enters a denser medium.

Reduction in Scale (Miniaturization)

Wu (2006) believes multiplex analysis (e.g., technologies that allow measurement of many analytes simultaneously) is a promising area of advancement for immunoassay, particularly coupled to increasing miniaturization. Indeed, many of the immunoassay devices currently on the market analyze for multiple drugs.

Recent progress in miniaturization has led to devices with micrometer-sized features that can allow for multiple assays and complete all steps in the assays on a single device – the "lab-on-a-chip" (Fortina and Kricka, 2010). Key advantages are the much lower sample volume required and the reduced need for reagents (saving costs). For example, Tachi et al. (2009) have taken the cloned enzyme technology (CEDIA) onto a microchip. They conclude that their device can support point-of-collection testing (discussed in the next section). Similarly, several researchers are exploring miniaturized SPR, including Kim et al. (2007), who present a miniaturized SPR immunosensor equipped with a multi-microchannel sensor that allows detection of low-molecular-weight analytes. This expands the types of drugs or drug metabolites that can be detected.

The next step in the process of miniaturization is the move from micrometer scale to the smaller nanometer scale. Fortina and Kricka (2010) report that a key area of ongoing research is the use of nanoscale materials (e.g., nanoparticles) supporting multiplexed immunoassay, including: gold nanoparticles, quantum dots, and magnetic nanoparticles.

Quantum dots, or Qdots, are semiconductor nanocrystals with diameters in the range of 2-10 nanometers (nm)[64] that can serve as a luminescent marker in an immunoassay. Compared to conventional marker dyes, Qdots have various advantages: (1) broad excitation and narrow emission; (2) color-tunable; (3) high fluorescence and photo-stability; and (4) excellent biological compatibility after encapsulation (Shen et al., 2007).

Another example of nanotechnology is Philips' magnetic biosensor platform, Magnotech™, which uses magnetic nanoparticles and is designed for point-of-collection testing (Philips.com 2010). The Magnotech™ can measure picomolar[65] concentrations of specific proteins in blood or saliva in minutes. It includes a disposable biosensor cartridge that inserts into a hand-held analyzer.

Finally, researchers at the University of Calgary recently announced the development of a new way to dispense and manipulate picoliter drops of fluids on a microchip (physorg.com, 2011). The method involves creating a structure called a micro-emulsion, which is a droplet of fluid captured inside a layer of another substance. Samples are dispensed electronically and tested by sensors on microchips. The chips then transfer data wirelessly to a computer.

Computational Efforts

Taking an alternative approach to address the issue of immunoassay specificity, some researchers are exploring the use of computational tools to predict cross-reactive molecules that would otherwise interfere with the assay by binding to the antibodies in a manner similar to the target molecules. The tools compare the structural similarity of target compounds to

---

[64] A nanometer is 1 billionth of a meter (1/1,000,000,000 meters).
[65] A picomolar is 1 trillionth molar (1/1,000,000,000,000 molar).

compounds listed in the FDA drug databases via whole-molecule "similarity analysis." The analysis generates "similarity coefficients" – i.e., the degree to which an unrelated molecule/compound is similar to the target compound. Two specific computational tools are (1) DiscoveryStudio 2.0, which allows similarity searching using mapping description language (MDL) public keys and functional class fingerprint description; and (2) Molecular Operating Environment, which calculates "pharmacophore fingerprints" (Krasowski et al., 2009). As such tools and techniques continue to develop they may serve to make the initial testing process and various immunoassay kits more accurate by highlighting potential false positives based on this similarity analysis.

Ongoing Evaluation

Finally, though it is not a breakthrough in approach, it is important to note the continuous development and evaluation of immunoassay testing products (sometimes called testing kits) for new categories of drugs. Researchers continue to evaluate and compare existing test products. For example, DeRienz et al. (2008) conducted a thorough evaluation of four commercial immunoassay kits for benzodiazepine detection in urine in which they examined linearity, precision, accuracy, carryover, reagent specificity, and confirmation rates. As part of the evaluation, the researchers screened more than 10,000 randomly collected urine samples. They found one assay – the Microgenics cloned enzyme donor immunoassay (CEDIA) high-sensitivity assay – that demonstrated exceptional response to the standards analyzed. Schwettmann et al. (2006) performed a comparison of two commercially available drug-screening assays: the Microgenics CEDIA and the Roche kinetic interaction of microparticles in solution (KIMS). Overall results of this latter study indicate good agreement between the two assays when testing for eight drugs in urine samples obtained from known drug-abuse patients.

As interest in the use of alternative specimens has grown, the number of studies evaluating and comparing the performance of test kits across specimens and across drugs has also increased, as illustrated by the number of publications cited in the publications documenting these assessments. Studies of these kinds are a key part of the maturation and validation process of new technologies.

2.4.6 Point-of-Collection Testing – An Emerging Category of Initial (Screening) Testing

As indicated in the previous section, one area of drug testing where there has been substantial development of new technologies and devices involves collecting and testing specimens at the point of collection – that is, fully avoiding requirements to send a specimen to a laboratory for initial testing. There is an active and competitive market for such devices, which are targeted to non-Federally-regulated workplace testing, law enforcement, and judicial, clinical, and home use. These technologies include improved collection devices that test for certain types of subversion (e.g., temperature; validity testing devices; and initial testing devices to identify non-negative specimens). Some combine both validity testing and initial screening in the same device. It should be noted that the underlying technology for these devices is the same as the technology used in laboratories; that is, in most cases the devices use well-established immunoassay technologies for drug detection.

Point-of-collection Testing (POCT)[66] devices can be instrumented or non-instrumented. An instrumented device reads the results for the person performing the test. A non-instrumented device requires that the person performing the test (while collecting the specimen in this case) interpret the results – for example, by comparing colors on the testing device result indicator with a chart. Some POCT devices also include the ability to generate a record of the results.

There are screening tests designed to detect drugs of abuse in urine, oral fluids, sweat, or other matrices[67]. Both instrumented and non-instrumented POCT devices typically use competitive binding immunoassays, the same approach used for initial testing at certified laboratories (HHS SAMHSA 2008). POCT devices vary in the number of drugs tested simultaneously. Some test for up to 12 drugs on a single strip set (Wong and Tse 2005). Some of these devices also include tests to determine the validity of the specimen with the identification of drugs of abuse. POCT validity tests use colorimetric assays, the same basic method used for validity testing in HHS-certified laboratories.

Dasgupta (2010) provides an overview of POCT devices and their role in workplace drug testing. Most POCT devices being marketed for use in on-site testing for drugs of abuse are based on lateral-flow immunochromatographic assay technology using colored microparticles rather than enzymes so that results can be read directly without requiring additional reagents. Colloidal gold or colored latex are the two main types of microparticles used (Tsai and Lin 2005).[68]

POCT devices are pre-calibrated during manufacturing, which means there is no need for onsite calibration, as well as no option for changing pre-determined cutoff levels (Tsai and Lin 2005).

Considerations and Challenges

Initial validity and screening tests are used to separate specimens that require testing with the more precise and expensive gas chromatography/mass spectrometry (GC/MS) or liquid chromatography/mass spectrometry (LC/MS) tests from the much larger number that require no further testing because they have been shown negative for both subversion and drugs by the initial tests. This sequential approach saves money by dramatically reducing the number of specimens sent for confirmation testing. The potential for greater convenience and lower cost are the major drivers of these technologies.

POCT devices are marketed primarily as a way to quickly and easily run initial validity and drug tests to identify true negatives, which then do not require further testing. However, some devices are being marketed as sufficiently sensitive and precise to be used as the sole, definitive test.[69] Because the POCTs are self-contained and designed to be administered by individuals with limited special training, their use can eliminate the need for a laboratory and a cadre of highly trained staff. However, POCT devices have not been approved by the HHS for use in Federally-regulated testing programs.

---

[66] POTC is also referred to in the literature as "point of care testing", reflecting a therapeutic application. Another term used by researchers from the medical diagnostic arena is "near patient technologies" or NPT.

[67] Again, the term "screening" is used in the generic sense, distinct from the specific term used in 10 CFR Part 26 for validity screening tests.

[68] One variant of the multiple-drug POCT device using colloidal gold is the "ascending multi-immunoassay technique, descr bed by Dasgupta (2010:96) and Carlberg (2005). It requires an approximately 10 minute incubation and shows the presence of drugs by the presence or absence of lines on a test strip.

[69] The conclusion of experts (e.g., Wong and Tse 2005) is that the fundamental attr butes of immunoassay tests make them inappropriate for use without a confirmatory test (using chromatographic separation and mass spectrometric measurement).

Some employers find it more convenient and less costly to conduct their own validity and initial drug tests using POCT devices. Some employers find the ability to complete the initial validity and drug tests on-site, or in the field, and to have test results quickly very desirable. Other employers do not want the responsibility and accountability of recordkeeping and QA/QC documentation required for effective implementation of a POCT-based program and find it more convenient and cost effective to send their specimens to an off-site laboratory for initial validity and drug testing.

These devices also have disadvantages, a principal one being limited shelf life. Most devices have a shelf-life of 12 to 15 months, which requires careful inventory management and procedures to ensure that devices are not used beyond their expiration date. Another disadvantage is that although the devices are designed to provide easy interpretation of results, the manual tests can sometimes be difficult to read. Colors can vary by drug and intensity. This creates an opportunity for misinterpretation and inconsistency. Wu (2002) points to the need for controls and proficiency testing of individuals administering POTC devices – including those designed for easy interpretation.

Another important issue is the protection of individual privacy. Because these tests provide results shortly after the specimen is collected, specific procedures are needed ensure that the donor is anonymous to those reading the results. Wu (2002) observes that inappropriately administered onsite drug tests could subject the collection site to legal challenge of test results. Although no instances of this type of failure were cited, the risks associated with wrongful discharge in connection with tests based on POCTs was identified as one of the factors to be considered in weighing the advantages and disadvantages of the different technologies and administrative options.

Wu (2002) presents the following table to compare automated high-throughput laboratory assays with POCT testing devices. It is important to note that this table dates from 2002 and the performance of POCT devices has improved significantly since that time. It is also important to note that reliance on any single test method (e.g., immunoassay) eliminates the protection obtained from the long-standing requirement to validate the results of the first test with a test employing a substantially different analytical method and MRO review to interpret test results.

Table 2.9.   A Comparison of Automated and POCT Immunoassay Screening Tests[a]

Attribute	Automated Test	POCT Test
Cost of each	Low (<$1 each)	High ($5-25)
Instrumentation	Expensive	None or marginal
Sample Delivery	Required	Not required
Sensitivity	High	Moderate
Specificity	Antibody dependent	Antibody dependent
Adulteration Testing	Available	Separate dipsticks available

[a] Source: Wu (2002)

Recent and Emerging Developments

Walsh et al. (2007) report that over 50 non-instrumented POCT devices are commercially available for urine, comprising three different types: (1) combination collection/test cups; (2) card- or cassette-types with pipette; and (3) dipstick-type tests. In addition, more than 12 non-instrumented POCT devices for oral fluids are commercially available. The test component in all these devices is an absorbent strip impregnated with an antibody-dye complex that is specific for the test analyte(s). Single and multiple analyte versions are being marketed. Test results are interpreted by visual reading (usually a line appears if the test analyte is not present, although sometimes the test works in the opposite manner with the absence of a line indicating presence of the test analyte).

Walsh et al. (2007) also report there are about the same number (e.g., over 50) of instrumented POCT immunoassay devices commercially available. Most of these devices are for testing urine; some newer devices are designed to test oral fluids. In general, the POCT devices used to test urine have performed well in detecting drugs and have proven effective in tests for adulterants and creatinine, though comparative evaluations have shown considerable variability across devices (Walsh et al. 2007).[70]

POCT devices for oral fluids, in which the concentration of drugs and metabolites is much lower than in urine, have shown greater variability in performance. A series of evaluations of oral fluids devices (Walsh et al. 2007, Crouch et al. 2008, Verstraete and Puddu 2000; Blencowe et al. 2011; Pehrsson et al. 2011) indicate the variability in performance among different devices. Particularly in the earlier tests, most oral fluids devices perform well for some drugs but poorly for others. More recent studies have identified some devices that demonstrate consistently good performance. Oral fluids devices are increasingly accurate in detecting methamphetamine, amphetamines, and opiates, but have shown less reliability in detecting marijuana, which is typically present in very low concentrations in oral fluids. Also, detection of cocaine varies significantly across devices, while the detection of marijuana is generally poor for many devices. Furthermore, oral fluids devices are still limited by the short window of detection (a few hours after administration of the drug). Many of the researchers working in this field warn that the technology for these devices is changing relatively rapidly, competition among suppliers is high, and new generation devices are introduced frequently. Consequently, they recommend checking the most recent evaluation information to determine which devices, if any, meet the user's performance requirements.

In 2000, NHTSA completed a project in which police officers in Houston, Texas and Long Island, New York evaluated five POCT urine test kits (Triage®, TesTcup5®, AccuSign®, Rapid Drug Screen®, and TesTstik®) with driving-under-the-influence (DUI) suspects. The officers participating in this project were certified "Drug Recognition Experts" (DREs) who had been trained in the NHTSA-approved "Drug Recognition and Classification Program." Overall results indicated a 36 percent positive rate among the population evaluated for illegal drugs (mostly cannabis, cocaine, and MDMA). GC/MS confirmation of all on-site test positives (and some negatives) indicated that the kits performed well, and the DRE officers participating in the study "favored the use of on-site devices in the enforcement of impaired driving laws" (Hersch et al., 2000). Indeed, road-side efforts to detect impaired drivers are a significant driver in the development of POCT devices for testing oral fluids.

One further emerging development is the deployment of the Philips Magnotech™ nanotechnology device, which is a device targeted for roadside drug testing (Phillips.com 2010). The Philips device is designed to collect oral fluids via expectoration into a small receptacle that is then inserted into the measurement chamber containing magnetic nanoparticles coated with ligands that test for cocaine, heroin, cannabis, amphetamines and methamphetamine. The

---

[70] Studies evaluating the performance of these devices typically retest the specimen using standard confirmatory test procedures (GC/MS or LC/MS) and compare the results of the two tests.

device has been marketed since 2008, and provides color-coded test results in about 90 seconds. Studies validating the performance of this device in field conditions are still to be conducted.

2.4.7 Confirmatory Testing for Drugs of Abuse

Confirmatory testing technologies are based on different testing principles than the screening technologies. Confirmatory tests employ technologies to separate the components in the specimen and technologies that identify and quantify the specific drugs of abuse and their metabolites contained in the specimen. Because confirmatory testing is used to validate initial test results and provide a final test outcome to be reported, the emphasis in confirmatory testing is on methods that are highly sensitive and highly specific.

Until the 2008 update, the HHS Guidelines required the use of GC/MS for confirmatory testing. This reinforced GC/MS as the "gold standard" of practice in confirmatory workplace drug testing. GC/MS is a combination of two analytical techniques. Chromatographic procedures are used to separate the different components in a specimen. Mass spectrometry is used to identify, very specifically, each of the components of the specimen.

The current HHS Guidelines (2008) continue to require a confirmatory procedure to identify the presence of a specific drug or metabolite. The confirmatory procedure must be independent of the initial test and use a different technique and chemical principle from that of the initial test. The purpose is to enhance assurance of the reliability and accuracy of test results. The HHS Guidelines have approved several new mass spectrometry methods for use in confirmatory testing.

The following is a review of each of the main components of confirmatory testing (specimen preparation, separation via chromatography and measurement via spectrometry). This review is followed by a discussion of their combination into coupled systems that highlights recent advancements and areas of research. Tsai and Lin (2005) provide a succinct and clear overview of these technologies.

Specimen Preparation (Hydrolysis, Extraction, Derivatization)

The differing chemical properties of the alternative specimens and the drugs and metabolites that may be in the specimens sent for testing pose distinct challenges to the establishment of procedures that provide consistent and comparable results. A critical step in the confirmatory testing process is preparing the specimen so that any relevant components (e.g., any target drugs and metabolites present) are in a form that is suitable for analysis with a particular confirmatory testing technology. As indicated in the bibliography included in Section 2.5, developing, testing, and standardizing specimen preparation procedures, and validating their impact on test results, are essential components of the technology innovation process that require concerted research. Standardization of specimen preparation procedures is a prerequisite for adoption of new matrices, new drugs, or new separation and measurement technologies into testing programs. Tsai and Lin (2005), Segura et al. (1998), Isenschmid and Goldberger (2007), and Ojanperä and Rasanen (2008) all provide reviews of preparatory methods and procedures.

The primary purpose of specimen preparation is to separate and pre-concentrate the target compounds and to purify the extract as much as possible. Appropriate specimen preparation increases the speed, sensitivity, and selectivity of the testing process, by inducing specific chemical changes to the target compounds (Janicka et al., 2010). In general, the preparation requirements for parent drugs, which are typically lipophilic, differ from those for drug metabolites, which frequently are hydrophilic, polar molecules that dissolve in water (Segura et al., 1998).

There can be many steps in the specimen preparation process, including the conversion of the specimen into a solution for solid specimens such as hair. However, the main components are hydrolysis, extraction, and derivatization. Hydrolysis is a chemical process use to break bonds of a target compound via acid/base or enzymatic catalysis (Tsai and Lin 2005). The process also involves water molecules being split in two, hence the name of the process.

Extraction involves isolating the target compounds from the background solution. The most common extraction techniques are liquid-liquid solvent extraction (LLE) and solid-phase extraction (SPE) (Tsai and Lin 2005). LLE uses a solvent that is immiscible with the specimen solution to extract compounds via differences in solubility. SPE uses absorbent solid cartridges that are tuned to the target compounds. The specimen solution is loaded onto the cartridges, and the target compounds adhere while the rest of the solution is washed off (or vice versa depending on the particular cartridges) (Janicka 2010). As reported by Tsai and Lin (2005), there is a wide array of both solvents and solid-phase materials commercially available and researchers are constantly working to refine the materials and processes to increase effectiveness and speed, and to reduce cost.

Derivatization is a general set of chemical processes that change the character and often increase the size of the target compounds to make them more amenable for analysis by the confirmatory testing technology. The specific type of derivatization used must be tailored both to the target compound and to the separation and measurement technology. For gas chromatography, derivatization is used to make compounds more volatile, less polar, and more stable under high temperatures – all of which are chemical attributes required for effective chromatography (Isenschmid and Goldberger 2007). Derivatization is also used to resolve interference (i.e., compounds so similar they interfere with one another in a process) by creating derivatized forms of the target compounds that are more distinct relative to other compounds. In addition, derivatization can increase the stability of the target compounds during storage and isolation (Isenschmid and Goldberger 2007). For liquid chromatography, derivatization can be used to render substances fluorescent, increasing the sensitivity of detection (Chromatography-online.org 2011). Finally, in some cases derivatization increases the specificity of ions detectable in mass spectroscopy systems. For example, amphetamines require derivatization to decrease the potential for interference with a range of other compounds that generate similar ion fragments (Isenschmid and Goldberger 2007). In summary, derivatization is a tuning process to optimize detection and minimize interference. As with other steps in the preparation process, efforts in biological and medical research, and technology development are continuously exploring ways to improve the derivatization process to increase specificity and reduce complexity. As with other aspects of workplace testing, the requirement for extensive validation and standardization tends to make workplace testing the recipient rather than the developer of such advances.

Although hydrolysis, extraction and derivatization are all used to make target compounds more readily separated and measured, these processes can have side effects. For example, they can create other compounds that interfere with measurement or that affect the stability of the intended derivative compounds (Segura et al. 1998). One technique to manage this problem is the use of internal standards. Internal standards are reference compounds that have a similar structure and possess similar behavior in the testing equipment as the target compound. Because they are added at a known concentration, they allow for calibration of results. Internal standards are also used to ensure that the instruments are operating correctly (Tsai and Lin 2005).

Chromatography[71]

Chromatography is a term for a family of techniques that separates compounds in the gaseous or liquid phase based on differences in their volatility and solubility. It involves a mobile phase mixture (either gas or liquid) moving over or through a stationary phase (e.g., a strip of paper, a column of beads). The different compounds in the mixture will have different reactions to the stationary phase, allowing them to be separated in space and/or time. Some novel types of chromatography involve liquid-liquid interaction.

The term chromatography, which means "color writing," was first used in the 19th century to describe a technique used to separate plant pigments (e.g., chlorophyll), producing a visible separation of colors (and the materials that produced them). Recognized as an effective way to separate materials, chromatography became an essential laboratory process, and, through continued use, evolved significantly. It now includes many variations. Chromatography can be preparative or analytical. Preparative chromatography is used to separate compounds for further analysis; analytical chromatography is used to both separate and detect or measure compounds in a mixture.

A range of separation mechanisms are used in chromatography. Gas chromatography relies on differences in vapor pressure and solubility of the analytes. Ion exchange chromatography uses a charged stationary phase to separate charged compounds. Size exclusion chromatography separates molecules according to their size or hydrodynamic diameter. A key term is "retention time" which is the time it takes a particular compound to elude (i.e., exit) from the system. Another key term is "eluent" which is the material that is exiting the system.

There are over 20 types of chromatography technologies (Tsai and Lin 2005). Gas chromatography (GC) has been used for many years in a wide range of laboratory settings, and is well established. Other forms of chromatography important for drug testing include: liquid chromatography (LC) and the variation of high and ultra-high performance liquid chromatography (HPLC; Ultra-HPLC), and thin layer chromatography (TLC). Key goals in the development of improved chromatographic methods are to reduce the amount of material needed, reduce the loss of test material, increase the precision of the separations, and reduce the time needed to complete the separation. Substantial progress is being made on all these dimensions.

Although chromatography is established as a standard technique in drug testing, separation techniques other than chromatography are gaining prominence in other detection and measurement arenas. In particular, capillary electrophoresis (CE) has emerged in the past decade as a promising alternative separation technique (Tsai and Lin 2005). CE uses an electric field to induce differential movement of charged compounds through a capillary.

---

[71] This general description of chromatography is based on Tsai and Lin (2005).

Table 2.10 details the most common chromatography techniques along with capillary electrophoresis for comparison.

Table 2.10. Common Separation Techniques Relevant to Workplace Drug Testing, Including Chromatography and Capillary Techniques[a]

Technology	Attributes
Gas Chromatography (GC)	Also referred to as gas-liquid partition chromatography (considered the most accurate name) or columns (after the stationary support) Standard chromatographic technique for drug testing; a multitude of commercial devices are available Used for thermally stable, volatile compounds General Process. GC uses an inert or non-reactive carrier gas, such as helium, for the mobile phase, and a microscopic layer of liquid or polymer as the stationary phase on a support referred to as a column. Under high temperature, the gaseous compounds being analyzed interact with the walls of the column, and partition according to their particular vapor pressure and solubility in the stationary phase. Performance. – Requires the sample to be vaporized and subject to heat
Liquid Chromatography (LC)	General Process. LC uses a liquid for the mobile phase, and a column of small solid particles as the stationary phase. Compounds are separated based on their differential solubility. Performance. – Requires less intense specimen preparation that GC, including no need to derivatize the analytes to improve volatility – Can be used for thermally unstable compounds as well as polar compounds
High Performance Liquid Chromatography (HPLC)	Formerly referred to as high pressure liquid chromatography Modification of LC technique, but using smaller solid particles (on the order of 3-5 microns in diameter) and higher pressures (used to compensate for the smaller particles to maintain reasonable retention times) General Process. As with LC, HPLC uses a liquid for the mobile phase, and a column of small solid particles as the stationary phase. Compounds are separated based on their differential solubility. Performance. – HPLC achieves better and more efficient separation in a shorter time with less use of solvents
Ultrahigh Performance Liquid Chromatography (UPLC)	Similar to HPLC, but will particle sizes as small as 1.5 microns and operating pressures correspondingly increased to 5000 psi
Thin Layer Chromatography (TLC)	Planar device General Process. TLC uses a solvent for the mobile phase, and a thin layer of polar absorbent material coated on plate as the stationary phase. The sample being analyzed is applied near the lower edge of the plate and the plate is inserted into a solvent chamber. As the solvent is drawn up through the stationary phase via capillary action, the components in the sample partition based on how they interact with the stationary phase relative to the mobile phase. The polarities of both the component and the

Technology	Attributes
	solvent affect partition rates. Performance. – Relatively inexpensive for screening – Relatively higher and variable detection limits – Labor-intensive
Capillary Electrophoresis (CE)	General Process. The sample is added to a conductive liquid in a capillary tube, which is then subject to an electric field. Compounds are partitioned based on their size to charge ratio. Performance. – High separation efficiency – Minimal sample preparation and minimal use of reagents – Broad analytical spectrum – Can analyze basic, neutral, and acidic compounds in the same assay

[a] Source: Drawn from Smith et al. (2007), and Tsai and Lin (2005).

While their principles are similar, each chromatographic procedure has distinctive advantages and disadvantages in drug testing of urine, as described by Wu (2002) and shown in Table 2.11. TLC and LC are useful for nonforensic purposes because they are faster and cheaper than GC/MS. Many laboratories use these technologies to support clinical toxicology.

Table 2.11.   Comparison of TLC and HPLC With GC/MS As Assays for Testing Urine for Drugs of Abuse[a]

Attribute	TLC	HPLC	GC/MS
Sensitivity	Fair (1000 ng/ml)	Good (500 ng/ml)	Excellent (<10 ng/ml)
Specificity	Fair	Fair	Definitive
Labor required	Medium	Low	Very high
Assay turnaround time	Slow (3 hours)	Fast (20 minutes)	Very slow (>8 hours)
Menu of tests	Wide (>200)	Wide (>200)	Very wide

[a] Source: Wu (2002).

Mass Spectrometry[72]

Mass spectrometry (MS) is the established technology for confirming the presence and identity of drugs or their metabolites in a test specimen. Because of its ability to provide detailed structural information, it is widely used in forensic toxicology (Tsai and Lin 2005).

MS measures the mass-to-charge ratio of charged particles that have been injected into a chamber. The typical steps in mass spectrometry include:

1. vaporizing the sample, if it is not already in gaseous phase;
2. ionizing the components of the sample, using a variety of methods (e.g., by impacting them with an electron beam);
3. separating the ions using electromagnetic fields according to their mass-to-charge ratio in what is referred to as the "analyzer";
4. detecting the ions, usually by a quantitative method; and

---

[72] This general description of mass spectrometry is based on Tsai and Lin 2005.

5. processing the signal into a mass spectra.

Identification of target analytes occurs by comparison of the particular mass spectra with those in databases (similar to fingerprinting).

Ionization of the sample is a critical component of the process. The most common ionization technique is electron ionization, which includes bombarding the sample with an electron beam. Chemical ionization, called "soft" ionization, uses a charged reagent gas to ionize the compounds. Chemical ionization is more stable and less extensively fragmented than electronic ionization, making it more sensitive but less selective (because the fewer number of ions mean the analyte may be harder to identify) (Smith et al. 2007). A third technique, electrospray ionization, can be used for liquid samples (e.g., material exiting an LC or HPLC device) (Tsai and Lin 2005).

Mass spectrometers can be operated in full scan or selective ion monitoring mode. As the name implies, with selective ion monitoring only a selected number of ions are used to compare with databases. In full scan mode, the mass spectrometer analyzes the entire signature.

There are variations in the type of analyzer used in mass spectrometers. The most common versions are quadrupole mass spectrometers, which use four rods to create an oscillating electrical field that selectively affects the paths of ions. The quadropole MS (QMS) is the standard analyzer used in GC/MS systems. Other analyzers include ion traps, where the ions are first trapped and then selectively ejected from the analyzer, and fourier transform MS, which "measures the image current produced by ions cyclotroning in a magnetic field" (Tsai and Lin 2005).

Putting mass spectrometers in sequence (i.e., "tandem mass spectrometers) provides several advantages. Tandem mass spectrometers can be arranged in space (i.e., one unit physically following the previous) or in time (i.e, the same unit used, but a portion of the sample is passed through a second time). The benefit of tandem mass spectrometers is that it allows the quantitation of low levels of target compounds and otherwise improves the signal to noise ratio.

The ability of mass spectrometers to identify the substances present in a sample has made them a key technology in the development of instruments for use on manufacturing production lines, in security screening, and in forensics. This has provided impetus for the miniaturization of instruments and the incorporation of mass spectrometers in portable devices.

Combined Chromatographic/Spectrometric Systems

As mentioned previously, the "gold standard" for confirmatory testing is a coupled system using GC and MS. GC/MS instruments are typically an integrated platform with a wide array of assays and applications available.

Other coupled systems include liquid chromatography/mass spectrometry (LC/MS) as well as liquid chromatography with tandem mass spectrometry (LC-MS/MS). LC-MS/MS is particularly useful when analyzing a large number of compounds in small sample volumes with suitable sensitivity. It also is useful for more polar or heat sensitive compounds. LC-MS/MS has been a standard in the pharmaceutical industry for years (Ghosh 2010). Maurer (2007) believes LC/MS may become the new "gold standard" once some technical issues (such as irreproducibility of fragmentation) are fully overcome and an apparatus standard is established.

Some examples of available GC/MS and LC/MS instruments are the 1200 Series (Agilent Technologies), 6410 LC (Agilent Technologies), COBAS INTEGRA® 800 (Roche Diagnostics), 1200 LC/MS (Varian, Inc.), and Waters® Alliance 2695 (Waters Corporation).

In addition, capillary electrophoresis (CE) has been coupled to MS. According to Smith et al. (2007), although a number of methods for CE/MS have been published, more sensitive and robust procedures are required before this technology can be successfully deployed in the forensic toxicology setting.

Considerations and Challenges

The standard confirmatory test technology system (GC/MS) is rugged and well established. Indeed, Smith et al. (2007:247) report that some GC/MS instruments "are still operating in high-volume laboratories after two decades of use." While research continues on alternative separation and detection systems, it seems likely that GC/MS will continue to dominate the confirmatory testing field. Although the basic configuration of the two technologies has not changed, there has been a dramatic increase in their capabilities in the last 20 years, particularly for MS (Gallardo et al. 2009). Consequently, as laboratories update their equipment, these advances will be represented in the equipment used for a growing proportion of drug tests.

As discussed above, one of the key challenges of confirmatory testing is ensuring proper sample preparation. This is particularly the case with alternative specimens and new drugs (Isenschmid and Goldberger 2007). As part of the extraction and derivatization processes, there are a range of preparation steps that vary based on the specific drug of abuse being tested and specific requirements of the GC/MS system used to complete the test. Indeed, given the maturity of both chromatography and spectrometry technologies, the most critical area for successful technology performance is not with the instruments themselves, but with the development and implementation of appropriate sample preparation techniques.

Some other considerations include:

- Over the past few decades, there was been signification debate within the forensic toxicology community regarding whether selected ion monitoring provided sufficient accuracy, or if full scan MS should be used. More recently, a consensus has formed that selected ion monitoring is accurate and reliable (Isenschmid and Goldberger 2007).
- One limitation of GC/MS is that the practical limits of detection (LOD) for systems currently in use is parts-per-billion. With the increasing interest in alternative specimens, lower quantification limits are required, as a result of the lower concentrations of drugs or metabolites that are found in those matrices (Smith et al. 2007).
- By contrast, a limitation with LC/MS is that results continue to be non-reproducible between instruments, delaying the development of shared spectral libraries (Smith et al. 2007). CE/MS instrumentation protocols are even less established. However, as detailed below, this research area is continuing to receive significant attention, which suggests this limitation is likely to be overcome in the relative near term.

Recent and Emerging Developments

Several noteworthy developments in confirmatory testing include:

- As mentioned previously, CE is gaining renewed attention as an alternative separation technology. For example, Lurie et al. (2004) performed a study of CE performance for analysis of a series of drugs of abuse.
- Two-dimensional gas chromatography (GC×GC) with MS has been developed in the past several years as a means to enhance sensitivity of the combined system. An example is the Deans Switch[R], manufactured by Agilent Technologies (Smith et al. 2007). This method functions similarly to a tandem MS, wherein only a small segment of the GC eluent is transferred to a second column, which eliminates many interfering substances.
- The greatest amount of research has been focused on tandem MS systems (particularly LC/MS/MS or HPLC/MS/MS). Gallardo et al. (2009) review LC/MS and LC/MS/MS systems for application in workplace drug testing, and conclude that these technologies are "allowing analyte detection at concentrations that were unthinkable just a few years ago ...[with the consequence that] ... many laboratories are finally at the point where they are considering acquisition of these capabilities" (Gallardo et al. 2009:114). Similarly, Eichhorst et al. (2009:1531), reporting on their test of UPLC/MS/MS for high-throughput screening of 200 urine specimens a day for 40 drugs/metabolites, conclude that the new procedure is "a viable alternative" to previous immunoassay methods. Advantages they reported include: acceptable turnaround, simple sample preparation, and analytic reliability. As one further example, Jagerdeo et al. (2010) investigated the use of an automated LC/MS/MS system as a fast method for screening marijuana and its metabolites, and were able to achieve reasonable accuracy within a total analysis time of 10 minutes, including sample preparation, separation, and detection. These developments imply an emerging technical ability to support the analysis of alternative matrices and low concentrations of drug analytes (Tsai and Lin 2005).
- As a general trend, MS and MS/MS systems are becoming smaller, more rugged, and easier to use. In fact, although the size, weight, and power consumption of standard laboratory configurations prevents mobile applications, various efforts to develop smaller more portable instruments are being made. An example is the handheld tandem mass spectrometer, the Mini 11 by Aston Labs, a research group at Purdue University. As described by Gao et al. (2007:1): "This instrument employs a rectilinear ion trap mass analyzer and weighs 10 lbs, has a size of 10" L x 6" W x 5" H and a power consumption of 35W. A digital control board with wireless communication capability was developed to execute pre-programmed scan functions, collect spectra and transfer data to the remote computer."
- Following the terrorist attacks of September 11, 2011, focus on the development of technology capable of screening for and identifying potential weapons and hazardous materials outside of the laboratory has accelerated research and development of sensors and detectors. This has lead to the deployment of technologies in airport security systems and equipment used for the stand-off detection of improvised explosive devices. Although these technologies are not yet applicable to workplace testing for drugs and alcohol, they are driving advancements in the sensing and detection technologies that may provide a base for adaptation to testing programs. This includes concerted efforts to develop substantially smaller and more portable devices with high sensitivity.

2.4.8 Alcohol Testing as a Special Category of Testing

Part 26 provides for all alcohol testing to be done on site, using an evidential breath testing device (EBT) or an oral fluids testing device for the initial test, and an EBT as a confirmatory testing device. Alcohol breath tests do not require validity tests. The test for alcohol is complete with the collection and reading of the results by the technician. Breath or oral fluids specimens being used to test for alcohol do not need to be sent for confirmatory testing at a laboratory. If the initial test is positive (either using oral fluids or breath), a second, confirmatory test is done immediately with a NHTSA-approved EBT device that is capable of producing documentation of the test result. At that point, the result is reported to the fitness-for-duty manager who takes administrative action.

The underlying technology for alcohol testing has not fundamentally changed since the NRC first published Part 26 in 1989 (see Moore et al. 1989), although substantial improvements in the convenience and reliability of these devices have been made. These are discussed below.

Breath Testing Devices

In the late 1940s alcohol breath testing replaced blood and urine testing as the main method for both screening and evidentiary testing. In 1954, R.F. Borkenstein invented an instrument called the Breathalyzer™ (or alternately Breathalyser™) that used chemical oxidation and photometry to determine alcohol concentration in breath. The term is now used generically for breath alcohol measurement technologies, regardless of their design (Intoximeter, Inc. website).

Since the mid-1980s, the principal technology for breath alcohol testing has been infrared light (IR) measurement systems. Recent improvements include use of optical filters. In these devices, a narrow band of infrared light, with a wavelength selected to maximize absorption by alcohol, is passed through one side of the breath sample chamber. A detector on the opposite side measures the emerging light. The amount of alcohol in the sample can be calculated based on the amount of light absorbed (Cao and Duan 2006; Intoximeters, Inc).

Another approved technology for breath alcohol testing is the alcohol fuel cell. In the 1960s, researchers in Austria developed a fuel cell specific to alcohol. It forms the basis of all current fuel cell breath alcohol measurement devices. The alcohol fuel cell is comprised of a porous, chemically inert block coated on both sides with platinum. Platinum wire electrical connections are made to each side of the block. The block is impregnated with an acidic electrolyte solution. When a breath sample is introduced into the cell, any alcohol in the sample is oxidized by the electrolytic solution. This process releases H+ ions from the upper surface of the cell, which then migrate to the lower surface and combine with atmospheric oxygen to form water. If the two surfaces are connected electrically, a current flows through this external circuit to neutralize the charge. The amount of current is directly proportional to the amount of alcohol oxidized by the cell. Given appropriate signal processing, the alcohol level in the sample can be accurately determined. (Intoximeters, Inc. website.) An advantage of alcohol fuel cells is their accuracy at low blood alcohol concentrations. As reported by (Intoximeters, Inc):

> The National Highway Transportation Safety Administration (NHTSA) conducted tests on seven models of evidential breath testers that met NHTSA Model Specifications. The results showed that six of the seven instruments demonstrated accuracy within NHTSA Model Specifications for evidential breath

testers. It is noteworthy that the two instruments using fuel cells showed greater accuracy at low BACs than the instruments using infrared techniques.

Desktop analyzers can use IR technology, electrochemical fuel cell technology, or a combination of both. In the U.S., the most common breathalyzer devices are Alcosensor®, Datamaster™, Alcotest®, Intoxilyzer®, and Intoximeter® (Alcohol Test Info 2011).

Oral fluids Devices for Alcohol Testing

Regarding oral fluids testing devices, the techniques used to measure alcohol in the field are similar to those for other drugs. These devices are typically composed of an "absorbent strip impregnated with an antibody dye complex that is specific for the test analyte(s)" (Walsh, 2008). Results are produced by either an instrument that does the analysis and provides readout or "hand-held cartridges requiring visual identification" (Drummer 2006).

Considerations and Challenges

A significant advantage of the breathalyzer technology (as well as oral fluids testing) over blood and urine testing is that it allows real-time and non-invasive measurements and results. This technology has been well established and accepted as part of workplace (and roadside) testing efforts (Mashir et al. 2011).

However, there are still some issues with breathalyzer technology performance, including the fact that test conditions can affect test results. These conditions include: the person's temperature (elevated body temperature can increase results), the person's breathing rate just prior to the test (hyperventilation will depress results, while holding breath will increase results), other substances in the mouth (which may contain chemical groups that are confounded with alcohol or which might otherwise interfere with the measurement) or in the air (e.g., smoke). (Alcohol Test Info 2011). In addition, on a technical level, the IR detectors output is nonlinear with alcohol concentration, and requires correction (Intoximeters, Inc., website).

For historical reasons, the cutoff level for alcohol is usually stated in terms of blood alcohol concentration (BAC). Breath testing for alcohol therefore also involves applying a mathematical algorithm to convert the observed breath alcohol concentration into its equivalent in blood alcohol concentration. The validity of these algorithms, which are often considered proprietary information by the manufacturers, has been the subject of legal challenge, which prompted research to calibrate and document the relationship between breath and blood concentrations (Stowell et al. 2008).

Recent and Emerging Developments

In general, the present survey of the literature did not identify any significant emerging technology innovations for alcohol testing either in the area of breath alcohol testing, or in the testing of other specimens. Although there has been recent interest in testing for a metabolite of ethanol, Ethyl glucuronide (EtG), in hair, urine, and other specimens, this approach is less relevant for workplace testing than for treatment/recovery and criminal justice because it does not detect recent alcohol consumption (Erowid.org 2011). Studies indicate the detection window for EtG begins 80 hours or more after alcohol consumption and that EtG is more appropriate as an indicator of excessive, chronic alcohol use than episodic or recent consumption. There is still debate in the scientific community about the specificity of this test and the ability to reliably interpret test results (Concheiro et al. 2009; Høiseth et al. 2010; Morini et al. 2010).

2.5 Summary and Sources for Continuing Updates

The combined efforts of scientists, manufacturers, service providers, and regulators in the multiple sectors identified in Section 2.2 are driving advancements in the technologies and knowledge useful to workforce testing. The literature reviewed for this report highlights the range and complexity of the issues to be identified, researched, and validated in the course of introducing a new method or technology into a linked technology system like workplace drug testing. The extensive research on methods to collect, prepare, and test the various drugs of abuse in alternative specimens that has been conducted over the last several decades has made considerable progress in identifying best practices and establishing standards, procedures, and devices that provide consistent, interpretable results. This work has already transformed the non-Federally-regulated drug testing environment, where point-of-collection testing devices are widely marketed for home and workplace use. The demand for a single test that can be administered at the roadside to identify drivers who are under the influence of either drugs or alcohol is driving improvements in point-of-collection oral fluids testing devices and research on their effectiveness. Considerable challenges lie ahead before these new technologies can be fully institutionalized into Federally-regulated testing programs, but a strong base has been established and additional research is underway. The future for these technologies looks promising.

The capital intensive laboratory testing equipment used in commercial testing laboratories is durable and reliable, and consequently turns over slowly. Nevertheless, advances in chromatography and mass spectrometry, including equipment capable of reliably detecting the low concentrations of drug analytes in alternative matrices, are finding their way into these laboratories. Driven by research to serve markets that demand less validation, standardization, and documentation than workplace drug testing, advancements are being made in a variety of instruments to detect and identify substances at low concentrations and in diverse situations and materials. Although it is less clear how they will impact the workplace drug testing industry, advances in chromatography and mass spectrometry are leading to equipment with greater capability, smaller size, and more portability.

Table 2.12 identifies some of the organizations and sources of information that are likely to be reporting on or engaged in the projects, research, and meetings that will be driving these innovations during the upcoming years and that may be of continuing interest to NRC staff.

Table 2.12.   Organizations and Sources to Monitor for Continuing Updates

Topic	Sources
Organizations	• U.S. HHS SAMHSA and its Drug Testing Advisory Committee; SAMHSA News website • U.S. DOT and its agencies: U.S. DOT Office of Drug and Alcohol Policy and Compliance Website • U.S. Nuclear Regulatory Commission; 10 CFR Part 26 – Fitness for Duty Programs • National Institute on Drug Abuse (NIDA) • Office of National Drug Control Policy (ONDCP) • Society of Forensic Toxicologists (SOFT) • American Association of Medical Review Officers (AAMRO) • Quest Diagnostics, Inc. Annual Drug Testing Index and website • Society of Hair Testing (SOHT) • Drug and Alcohol Testing Industry Association (DATIA)
Journals Application of analytical techniques to drugs and/or matrices; characteristics of drugs and/or matrices	• *Forensic Science International* • *Journal of Analytical Toxicology* • *Therapeutic Drug Monitoring* • *International Journal of Toxicology* • *Clinica Chimica Acta* • *Analytical and Bioanalytical Chemistry* • *Clinical Chemistry*
Journals Development, features, and performance of methods and equipment	• *Journal of Chromatography A; B* • *Journal of Immunological Methods* • *Sensors and Actuators* • *Drug Testing Analysis* • *Review of Scientific Instruments* • *Clinical Chemistry Laboratory Methods* • *Biomedical Chromatography* • *Journal of Mass Spectrometry*
Conferences	• Forensic Toxicology Association Drug and Alcohol Program National Conference • FTA Drug and Alcohol Program National Conference • Annual Meeting of the American Association for Clinical Chemistry • Drug and Alcohol Testing Association Conference
Other Countries and Their Research Areas	• UK: Roadside testing using oral fluids • Europe: Alternative matrices

2.6   Bibliography (including References)

Ackermann, Bradley L., Michael J. Berna, James A. Eckstein, Lee W. Ott, and Ajai K. Chaudhary. 2008. Current Applications of Liquid Chromatography/Mass Spectrometry in Pharmaceutical Discovery After a Decade of Innovation. *Annual Review of Analytical Chemistry* 1:357-396.

Alcohol Test Info. 2011. Website. Alcohol Blood Tests vs. Breathalyzers. Accessed 01.2011 at http://www.alcohol-test-info.com/Alcohol_Blood_Tests_vs_Breathalyzers.html

Allen, K.R., R. Azad, H.P. Field, and D.K. Blake. 2005. Replacement of Immunoassay by LC Tandem Mass Spectrometry for the Routine Measurement of Drugs of Abuse in Oral fluids. *Annals of Clinical Biochemistry* 42(4):277-284.

Appenzeller, Brice M.R., Claude Schummer, Sophie Boura Rodrigues, and Robert Wennig. 2007. Determination of the Volume of Sweat Accumulated in a Sweat-Patch Using Sodium and Potassium as Internal Reference. *Journal of Chromatography B* 852:333-357.

Aps, Johan K.M., and Luc C. Martens. 2005. Review: The Physiology of Saliva and Transfer of Drugs into Saliva. *Forensic Science International* 150:119-131.

Armstrong, Scott C., Kelly L. Cozza. 2003. Pharmacokinetic Drug Interactions of Morphine, Codeine, and Their Derivatives: Theory and Clinical Reality, Part 1. *Psychosomatics* 44:167-171.

Ashton, C. Heather. 2001. Pharmacology and Effects of Cannabis: A Brief Review. *The British Journal of Psychiatry* 178:101-106.

Auwärtera, Volker, Ariane Wohlfartha, Jessica Trabera, Detlef Thiemeb, and Wolfgang Weinmann. 2010. Hair Analysis for Δ9-tetrahydrocannabinolic Acid A: New Insights into the Mechanism of Drug Incorporation of Cannabinoids into Hair. *Forensic Science International* 196(1-3):10-13.

Awad, Tamer, Tarek Belal, Jack DeRuiter, Kevin Kramer, and C. Randall Clark. 2008. Comparison of GC-MS and GC-IRD Methods for the Differentiation of Methamphetamine and Regioisomeric Substances. *Forensic Science International* 185:67-77.

Awosika-Olumo, A., Trangle, K., and Fallon Jr., L. 2009. Drug Testing: A Case Report. *Journal of Controversial Medical Claims* 16(3):13-15.

Badawi, N., K.W. Simonsen, A. Steentoft, I.M. Bernhoft, and K. Linnet. 2009. Simultaneous Screening and Quantification of 29 Drugs of Abuse in Oral fluids by Solid-Phase Extraction and Untraperformance LC-MS/MS. *Clinical Chemistry* 55(11):2004-2018.

Baliková, Marie. 2005. Hair Analysis for Drugs of Abuse, Plausibility of Interpretation. *Biomedical Papers of the Medical Facility of the University of Palacký Olomouc Czech Republic* 149(2):199-207.

Barnes, Allan J., Bruno S. De Martinis, David A. Gorelick, Robert S. Goodwin, Erin A. Kolbrich, and Marilyn A. Huestis. 2009. Disposition of MDMA and Metabolites in Human Sweat Following Controlled MDMA Administration. *Clinical Chemistry* 55(3):454-462.

Barnes, A.J., M.L. Smith, S.L. Kacinko, E.W. Schwilke, E.J. Cone, E.T. Moolchan, and M.A. Huestis. 2008. Excretion of Methamphetamine and Amphetamine in Human Sweat Following Controlled Oral Methamphetamine Administration. *Clinical Chemistry* 54(1):172-180.

Barnes, A.J., I. Kim, R. Schepers, et al. 2003. Sensitivity, Specificity, and Efficiency in Detecting Opiates in Oral fluids with the Cozart Opiate Miroplate EIA and GC-MS Following Controlled Codeine Administration. *Journal of Analytical Toxicology* 27:402-407.

Barnes, Valerie, India Fleming, Thomas Grant, Joseph Hauth, J. Hendrickson, B. Kono, C. Moore, J. Olson, L. Saari, J. Toquam, D. Wieringa, P. Yost, P. Hendrickson, D. Moon, and W. Scott. 1988. *Fitness for Duty in the Nuclear Power Industry: A Review of Technical Issues.* Seattle, WA: Battelle Human Affairs Research Centers and the Pacific Northwest Laboratory. NUREG/CR-5227.

Bates, M., J. Brick, and H. White. 1993. Correspondence between Saliva and Breath Estimates of Blood Alcohol Concentration: Advantages and Limitations of the Saliva Method. *Journal of Studies on Alcohol* 54(1):17-22.

Baumgartner, A.M., P.F. Jones, W.A. Baumgartner, and C.T Blank. 1979. Radioimmunnoassay of Hair for Determining Opiate Abuse Histories. *Journal of Nuclear Medicine* 20:748-752.

Baxter, Louis, and Alan Trachtenberg. 2003. Report of the CSAT National Advisory Council (NAC) Subcommittee on Oral fluids Testing. Washington, DC: U.S. Deparment of Health and Human Services.

Bell, Suzanne. 2009. Forensic Chemistry. *Annual Review of Analytical Chemistry* 2:297-319.

Bennett, G.A., E. Davies, and P. Thomas. 2003. Is Oral fluids Analysis as Accurate as Urinalysis in Detecting Drug Use in a Treatment Setting? *Drug Alcohol Dependency* 72(3):265-269.

Berge, Keith H., and Donna M. Bush. 2010. The Subversion of Urine Drug Testing. *Minnesota Medicine* 93(8):45-47.

Blencowe, Tom, Anna Pehrsson, Pirjo Lillsunde, Kari Vimpari, Sjoerd Houwing, Beitske Smink, René Mathijssen, Trudy Ven der Linden, Sara-Ann Legrand, Kristof Pil, and Alain Verstraete. 2011. An Analytical Evaluation of Eight On-Site Oral fluids Drug Screening Devices Using Laboratory Confirmation Results from Oral Fluids. *Forensic Science International* 208(1-3):173-179.

Bogusz, M.J. 2008. Opioids: Methods of Forensic Analysis. In *Forensic Science Handbook of Analytical Separations, Vol 6.* M.J. Bogusz, Editor. Amsterdam: Elsevier, B.V. Pp. 3-72.

Bogusz, M.J., and A. Carracedo. 2004. Forensic Analysis. In *Chromatography, 6th Edition.* Erich Heftmann, Editor. Amsterdam: Elsevier B.V. Pp. 1073-1134.

Bones, Jonathan, Kevin V. Thomas and Brett Paull. 2007. Using Environmental Analytical Data to Estimate Levels of Community Consumption of Illicit Drugs and Abused Pharmaceuticals. *Journal of Environmental Monitoring* 9:701-701.

Bonnano, L.M., and L.A. DeLouise. 2010. Tunable Detection Sensitivity of Opiates in Urine Via a Label-Free Porous Silicon Competitive Inhibition Immunosensor. *Analytical Chemistry* 82(2):714-722.

Bosker, Wendy M., and Marilyn A. Huestis. 2009. Oral Fluids Testing for Drugs of Abuse. *Clinical Chemistry* 55:*1910-1931*

Boumba, Vassilika A., Kallirroe S. Ziavrou, and Theodore Vougiouklakis. 2006. Hair as a Biological Indicator of Drug Use, Drug Abuse, or Chronic Exposure to Environmental Toxicants. *International Journal of Toxicology* 25(3):143-163.

Brahm, Nancy C., Lynn Yeager, Mark Fox, Kevin Farmer, and Tony Palmer. 2010. Commonly Prescribed Medications and Potential False-Positive Urine Drug Screens. *American Journal of Health Systems Pharmacists* 67:1344-1350.

Brewer, Colin. 1999. Novel Biological and Mechanical Analysis Techniques for Alcohol and Other Drugs of Abuse: Their Role in Addiction Management and Psychiatric Treatment. In *Drug Testing Technology.* Tom Mieczkowski, Editor. Washington, DC: CRC Press. Pp. 34-47.

Bryson, Peter D. 1996. *Comprehensive Review in Toxicology for Emergency Clinicians*, 3rd Edition. Boca Raton, FL: CRC.

Burrows, David L., Andrea Nicolaides, Peter J. Rice, Michelle Dufforc, David A. Johnson, and Kenneth E. Ferslew. 2005. Papaid: A Novel Urine Adulterant. *Journal of Analytical Toxicology,* Vol. 29(5): 296-300.

Bush, Donna M. 2008a. Overview of the Mandatory Guidelines for Federal Workplace Testing Programs. In *Workplace Drug Testing.* Steven B. Karch, Editor. Boca Raton, FL: CRC Press. Pp. 7-20.

Bush, Donna M. 2008b. U.S. Mandatory Guidelines for Federal Workplace Drug Testing Programs: Current Status and Future Considerations. *Forensic Science International* 174:111-119.

Bush, Donna M. 2007. Federal Regulation of Workplace Drug and Alcohol Testing: An Overview of the Mandatory Guidelines for Federal Workplace Drug Testing Programs. In *Drug Abuse Handbook, 2nd Edition.* Steven B. Karch, Editor. Boca Raton, FL: CRC Press. Pp. 736-747.

Cairns, Thomas, Virginia Hill, Michael Schaffer, and William Thistle. 2004. Levels of Cocaine and its Metabolites in Washed Hair of Demonstrated Cocaine Users and Workplace Subjects. *Forensic Science International* 145:175-181.

Campbell, Nancy D. 2005. Suspect Technologies: Scrutinizing the Intersection of Science, Technology, and Policy. *Science, Technology, and Human Values* 30(3):374-402.

Cao, Wenqing, and Yixiang Duan. 2006. Breath Analysis: Potential for Clinical Diagnosis and Exposure Assessment. *Clinical Chemistry* 52(5):800-811.

Caplan, Yale H. 2008. Specimen Validity Testing. In *Workplace Drug Testing.* Steven B. Karch, Editor. Boca Raton, FL: CRC Press. Pp. 127-142.

Caplan, Yale H., and Marilyn A. Huestis. 2007. Drugs in the Workplace. In *Drug Abuse Handbook*, 2nd Edition. Steven B. Karch, Editor. Boca Raton, FL: CRC Press. Pp. 727-895.

Carlberg, David. 2005. Lateral-Flow Assays. In *Drugs of Abuse: Body Fluid Testing*. Raphael C. Wong and Harley Y Tse, Editors. Totowa, NJ: Humana Press. Pp. 99-114.

Cary, Paul L. nd. Drug Testing Basics for Courts-Mandated Testing. University of Missouri. Accessed 12.2010 at http://www.co.el-paso.tx.us/243dc/documents/conference2010/PCary

Catlin, D., D. Cowan, M. Donike, D. Fraisse, H. Oftebro, and S. Rendic. 1992. Testing Urine for Drugs. *Clinica Chimica Acta 207:S13-S26*.

Chamberlain, R.T. 2007. Legal Review for Testing of Drugs in Hair. *Forensic Science Review* 19:85-94.

Charvat, A., E. Lugovoj, M. Faubel, and B. Abel. 2004. New Design for a Time-of-Flight Mass Spectrometer with a Liquid Beam Laser Desorption Ion Source for the Analysis of Biomolecules *Review of Scientific Instruments* 75(5):1209-1218.

Chawarski, Marek, David Fiellin, Patrick O'Connor, Mathew Bernard, Richard Schottenfeld. 2007. Utility of Sweat Patch Testing for Drug Use Monitoring in Outpatient Treatment for Opiate Dependence. *Journal of Substance Abuse Treatment* 33(4):411-415.

Choo, Robin E., and Marilyn A. Huestis. 2004. Oral fluids as a Diagnostic Tool. *Clinical Chemistry and LaboratoryMedicine* 42(11):1273-1287.

Chopra, Nitin, Vasilis G. Gavalas, Leonidas G. Bachas, and Bruce J. Hinds. 2007. Functional One-Dimensional Nanomaterials: *Applications in Nanoscale Biosensors*. Analytical Letters 40(11):2067-2096.

Chou, Su-Lien, and Yun-Seng Giang. 2008. Influences of Seven Taiwan-Produced Adulterants on the Fluorescence Polarization Immunoassay (FPIA) of Amphetamines in Urine. *Forensic Science Journal* 7(1):1-12.

Chou, Su-Lien, and Yun-Seng Giang. 2007. Elucidation of the US Urine Specimen Validity Testing (SVT) Policies and Performance Evaluation of Five Clinical Parameters for Pre-Screening Adulterants in Taiwan's Opiates Urinalysis. *Forensic Science Journal* 6(2):45-58.

Christrup, L.L. 1997. Morphine Metabolites. *Acta Anaesthesiology Scandinavia* 41(1):116-22.

Chromatography-online.org. 2011. Website. Derivatization. Accessed 01.2011 at http://www.chromatography-online.org/topics/derivatization.html.

Chyka, Peter A. 2009. Substance Abuse and Toxicological Tests. In *Basic Skills in Interpreting Laboratory Data, 4th Edition*. Mary Lee, Editor. Bethesda, MD: American Society of Health-System Pharmacists, Inc. Pp. 47-72.

Cirimele, Vincent, Marion Villain, Patrick Mura, Marc Bernard, and Pascal Kintz. 2006. Oral fluids Testing for Cannabis: On-site OraLine®IV s.a.t. device versus GC/MS. *Forensic Science International* 161:180-184.

Clarke, Joe, and John F. Wilson. 2005. Proficiency Testing (External Quality Assessment) of Drug Detection in Oral fluids. *Forensic Science International* 150:161-164.

Cody, John T. 2008. Amphetamines. In *Forensic Science Handbook of Analytical Separations, Vol 6*. M.J. Bogusz, Editor. Amsterdam: Elsevier B.V. Pp. 127-174.

Committee on Substance Abuse and Council on School Health. 2007. Policy Statement: Testing for Drugs of Abuse in Children and Adolescents: Addendum—Testing in Schools and at Home. *Pediatrics* 119(3):627-630.

Concheiro, Marta, Teresa R. Gray, Diaa M. Shakley, and Marilyn A. Huestis. 2010. High-Throughput Simultaneous Analysis of Buprenorphine, Methadone, Cocaine, Opiates, Nicotine, and Metabolites in Oral fluids by Liquid Chromatography Tandem Mass Spectrometry. *Analytical and Bioanalytical Chemistry 398:915-924.*

Concheiro, Marta, Angelines Cruz, Marison Mon, Ana de Castro, Oscar Quintela, Angeles Lorenzo, and Manuel López-Rivadulla. 2009 Ethylglucuronide Determination in Urine and Hair from Alcohol Withdrawal Patients. *Journal of Analytical Toxicology* 33(3):155-161.

Concheiro, Marta, Ana de Castro, Oscar Quintela, Angelines Cruz, and Manuel López-Rivadulla. 2007 Determination of Illicit Drugs and Their Metabolites in Human Urine by Liquid Chromatography Tandem Mass Spectrometry Including Relative Ion Intensity Criterion. *Journal of Analytical Toxicology* 31(9):573-580.

Cone, E.J. 2001. Legal, Workplace, and Treatment Drug Testing with Alternate Biological Matrices on a Global Scale. *Forensic Science International* 121:7-15.

Cone, E.J. 1996. Mechanisms of Drug Incorporation in Hair. *Therapeutic Drug Monitoring* 18(4):438-443.

Cone, E.J. 1993. Saliva Testing for Drugs of Abuse. *Annals of the New York Academy of Science* 694:91-127.

Cone, Edward J., Joe Clarke, and Lolita Tsanaclis. 2007. Prevalence and Disposition of Drugs of Abuse and Opioid Treatment Drugs in Oral fluids. *Journal of Analytical Toxicology* 31(8):424-433.

Cone, Edward J., and Marilyn A Huestis. 2007. Interpretation of Oral fluids Tests for Drugs of Abuse. *Annals of the New York Academy of Science* 1098: 51-103.

Cone, E.J., J. Oyler, and W.D. Darwin. 1997. Cocaine Disposition in Saliva Following Intravenous, Intranasal, and Smoked Administration. *Journal of Analytical Toxicology* 21:465-475.

Cone, E. J., L. Presley, M. Lehrer, W. Seiter, M. Smith, K. Kardos, D. Fritch, S.J. Salamone, and R. Niedbala. 2002. Oral fluids Testing for Drugs of Abuse: Positive Prevalence Rates by

Intercept Immunoassay Screening and GC-MS-MS Confirmation and Suggested Cutoff Concentrations. *Journal of Analytical Toxicology* 26(8):541-546.

Cone, Edward J., Angela Sampson-Cone, and Marilyn A. Huestis. 2007. Interpreting Alternative Matrix Test Results. In *Drug Abuse Handbook*, 2nd Edition. Steven B. Karch, Editor. Boca Raton, FL: CRC Press. Pp.814-842.

Cone, Edward J., Yale H. Caplan, Frank Moser, Tim Robert, Melinda K. Shelby, and David L. Black. 2009. Normalization of Urinary Drug Concentrations with Specific Gravity and Creatinine. *Journal of Analytical Toxicology* 33(1):1-7.

Cone, Edward J., Anne Zichteman, Rebecca Heltsley, David Black, Beverly Cawthon, Tim Robert, Frank Moser, Yale H. Caplan. 2010. Urine Testing for Norcodeine, Norhydrocodone, and Noroxycodone Facilitates Interpretation and Reduces False Negatives. *Forensic Science International* 198:58-61.

Cone, E.J., M.J. Hillsgrove, A.J. Jenkins, R.M. Keenan, and W.D. Darwin. 1994. Sweat Testing for Heroin, Cocaine, and Metabolites. *Journal of Analytical Toxicology* 18(6):298-305.

Cook, Janine D., Kathy A. Strauss, Yale H. Caplan, Charles P. LoDico, and Donna M. Bush. 2007. Urine pH: The Effects of Time and Temperature After Collection. *Journal of Analytical Toxicology* 31:486-496.

Cooper, Gail, Manfred Moeller, and Robert Kronstrand. 2008. Current Status of Accreditation for Drug Testing in Hair. *Forensic Science International* 176:9-12.

Corbett, Michael R. 2006. Pharmacokinetics. Chapter 3. In *Medical-Legal Aspects of Drugs,* 2nd Edition. Marcelline Burns and Donald J. Bartel, Editors. Tucson, AZ: Lawyers and Judges Publishing. Pp. 23-44.

Coulter, Cynthia, Elizabeth Miller, Katherine Crompton, and Christine Moore. 2008. Tetrahydrocannabinaol and Two of Its Metabolites in Whole Blood Using Liquid Chromatography-Tandem Mass Spectrometry. *Journal of Analytical Toxicology* 32:653-658.

Coulter, Cynthia, Margaux Taruc, James Tuyay, and Christine Moore. 2010. Antidepressant Drugs in Oral fluids Using Liquid Chromatography-Tandem Mass Spectrometry. *Journal of Analytical Toxicology* 34(2):64-72.

Coulter, Cynthia, James Tuyay, Margaux Taruc, and Christine Moore. 2010. Semi-Quantitative Analysis of Drugs of Abuse, Including Tetrahydrocannabinol in Hair Using Aqueous Extraction and Immunoassay. *Forensic Science International* 196(1-3):70-73.

Couper, Fiona J., and Barry K. Logan. 2004. Drug and Human Performance Fact Sheets. Seattle, WA: Washington State Patrol Forensic Laboratory Services Bureau. Prepared for the National Highway Traffic Safety Administration.

Crooks, C. Richard, and Sue Brown. 2010. Roche DAT Immunoassay: Sensitivity and Specificity Testing for Amphetamines, Cocaine, and Opiates in Oral fluids. *Journal of Analytical Toxicology* 34(2):103-109.

Crouch, Dennis J. 2005. Oral fluids Collection: The Neglected Variable in Oral fluids Testing. *Forensic Science International* 150:165-173.

Crouch, Dennis J., Jayme Day, Jakub Baudys, and Alim A. Fatah. 2004 Evaluation of Saliva/Oral fluids as an Alternate Drug Testing Specimen. NIJ Report 605-03. U.S. Department of Justice.

Crouch, Dennis J., Royer F. Cook, James V. Trudea, and David C. Dove. 2002. An Evaluation of Innovative Sweat-Based Drug Testing Techniques for Use in Criminal Justice Drug Testing. NISTIR 6825. National Institute of Standards and Technology.

Crouch, D.J., J.M. Walsh, L. Cangianelli, and O. Quintela. 2008. Laboratory Evaluation and Field Application of Roadside Oral fluids Collectors and Drug Testing Devices. *Therapeutic Drug Monitoring* 30(2):188-195.

Crumption, S.D., and C.A. Sutheimer. 2007. Specimen Adulteration and Substitution in Workplace Drug Testing. *Forensic Science Review* 19:1-27.

Current, William. 2010. 2010 Trends in Drug Testing: 11th Annual Industry Survey. WFC and Associates. Powerpoint presentation. Accessed 11.2010 at http://www.billcurrent.com/site/wp-content/uploads/2010/06/2010-Drug-Testing-Trends-V2.ppt

Dams, Riet, Robin E. Choo, Willy E. Lambert, Hendree Jones, and Marilyn A. Huestis. 2007. Oral fluids as an Alternative Matrix to Monitor Opiate and Cocaine Use in Substance-Abuse Treatment Patients. *Drug Alcohol Dependence* 87(2-3):258-267.

Dams, Riet, Marilyn A. Huestis, Willy E. Lambert, and Constance M. Murphy. 2003. Matrix Effect in Bio-Analysis of Illicit Drugs with LC-MS/MS: Influence of Ionization Type, Sample Preparation, and Biofluid. *Journal of the American Society of Mass Spectrometry* 14:1290-1294.

Dasgupta, Amitava. 2010. *A Health Educator's Guide to Understanding Drugs of Abuse Testing.* Sudbury, MA: Jones and Bartlett Publishers.

Dasgupta, Amitava, Editor., 2008. Handbook of Drug Monitoring Methods: Therapeutics and Drugs of Abuse. Totowa, NJ: Humana Press.

Dasgupta, Amitava and Susan D. Crumpton. 2007. Adulterants can Interfere with Urine Drugs-of-Abuse Tests. *Clinical and Forensic Toxicology News* (March):1-12.

Dasgupta, Amitava. 2005. Adulteration of Drugs-of-Abuse Specimens. In *Drugs of Abuse: Body Fluid Testing.* Raphael C. Wong and Harley Y Tse, Editors. Totowa, NJ: Humana Press. Pp.215-232.

Dasgupta, Amitava. 2003. Urinary Adulterants and Drugs-of-Abuse Testing. *MRO* February:28-31.

Dasgupta, Amitava, Omar Chughtai, Christina Hannah, Bonnette Davis, and Alice Wells. 2004. Comparison of Spot Tests with AdultaCheck 6 and Intect 7 Urine Test Strips for Detecting the Presence of Adulterants in Urine Specimens. *Clinica Chimica Acta* 348:19-25.

Davey, Jeremy, James Freeman, Anita Lavelle. 2009. Screening for Drugs in Oral fluids: Illicit Drug Use and Drug Driving in a Sample of Urban and Regional Queensland Motorists. *Transportation Research Part F* 12:311-316.

DATIA (Drug and Alcohol Testing Industry Association). Website homepage: http://www.datia.org/.

Davey, Jeremy, James Freeman, and Anita Lavelle. 2009. Screening for Drugs in Oral fluids: Illicit Drug Use and Drug Driving in a Sample of Urban and Regional Queensland Motorists *Transporation Research Part F* 12:311-316.

Day, David, David J. Kuntz, Michael Feldman, and Lance Presley. 2006. Detection of THCA in Oral fluids by GC-MS-MS. *Journal of Analytical Toxicology* 30(November/December):645-650.

De La Torre, Rafael, Magi Farré, Monica Navarro, Roberta Pacifici, Piergiorgio Zuccaro, and Simona Pichini. 2004. Clinical Pharmacokinetics of Amfetamine and Related Substances: Monitoring in Conventional and Non-Conventional Matrices. Clinical Pharmacokinetics 43(3): 157-185.

De la Torre, Rafael, E. Civita, F.Svaizerc, A. Lottic, M. Gottardic, and M. Miozzoc. 2010. High Throughput Analysis of Drugs of Abuse in Hair by Combining Purposely Designed Sample Extraction Compatible with Immunometric Methods Used for Drug Testing in Urine. *Forensic Science International* 196(1-3):18-21.

DeLuca, Alexander. 2002 (updated 2006). An Evaluation of Fitness-for-Duty Testing. 103rd Annual Convention of the American Psychological Association.

DeMartinis, Bruno, Allan Barnes, Karl Scheidweiler, and Marilyn Huestis. 2007. Development and Validation of a Disk Solid Phase Extraction and Gas Chromatography-Mass Spectrometry Method for MDMA, MDA, HMMA, HMA, MDEA, Methamphetamine and Amphetamine in Sweat. *Journal of Chromatography B*: 852(1-2): 450-458.

DeRienz, Rebecca T., Justin M. Holler, Megan E. Manos, John Jemionek, and Marilyn R. Past. 2008. Evaluation of Four Immunoassay Screening Kits for the Detection of Benzodiazepines in Urine. *Journal of Analytical Toxicology* 32(6):

Dickson, Stuart, Alexandra Park, Susan Nolan, Sarah Kenworthy, Cheryl Nicholson, Julie Midgley, Rowena Pinfold, and Scott Hampton. 2007. The Recovery of Illicit Drugs from Oral fluids Sampling Devices. *Forensic Science International* 165:78-84.

Ditton, Jason. 2002. Technical Review. Hair Testing: Just How Accurate Is It? *Surveillance and Society* 1(1):86-101.

Djurendic-Brenesel, Maja, Neda Mimica-Dukic, Vladimir Pilja, and Milos Tasic. 2010. Gender-Related Differences in the Pharmokinetics of Opiates. *Forensic Science International* 194:28-33.

Dougherty, Ronald J. 1987. Controversies Regarding Urine Testing. *Journal of Substance Abuse Treatment* 4:115-117.

Douglas, Cathy. 2010. Oral fluids Drug Testing. Human Resources, City of Yuma, Arizona. E-mail. http://lists.tempe.gov/admin/WA.EXE?A2=ind1009&L=jims&T=0&F=P&P=9786

Dresen, S., N. Ferreirós, H. Gnann, and R. Zimmermann. 2010. Detection and Identification of 700 Drugs by Multi-Target Screening with a 3200 Q TRAP® LC-MS-MS System and Library Searching. *Analytical and Bioanalytical Chemistry* 396:2425-2434.

Drewnick, Frank, Silke S. Hings, Peter DeCarlo, John T. Jayne, Marc Gonin, et al. 2005. A New Time-of-Flight Aerosol Mass Spectrometer (TOF-AMS)—Instrument Description and First Field Deployment. *Aerosol Science and Technology* 39:637-658.

Drug Policy Alliance. 2011. Drugs, Police, and the Law: Drug Testing Technologies: Sweat Patch. Accessed 12.2010 at http://www.drugpolicy.org/law/drugtesting/sweatpatch

The Drug Testing Workplace Act of 1988. Public Law 100-71. 101 STAT. 391. 11 JULY 1987.

Drummer, Olaf H. 2008. Introduction and Review of Collection Techniques and Applications of Drug Testing of Oral fluids. *Therapeutic Drug Monitoring* 30(2):203-206.

Drummer, Olaf H. 2006. Drug Testing in Oral fluids. *Clinical Biochemical Review* 27(3):147-159. [Includes review of suppliers' technologies]

Drummer, Olaf H. 2005. Review: Pharmacokinetics of Illicit Drugs in Oral fluids. *Forensic Science International* 150:133-142.

Duer, Wayne C., Paul J. Ogren, Alison Meetze, Chester J. Kitchen, Ryan Von Lindern, Dustin C. Yaworsky, Christopher Boden, and Jeffery A. Gayer. 2008. Comparison of Ordinary, Weighted, and Generalized Least-Squares Straight-Line Calibrations for LC-MS-MS, GC-MS, HPLC, GC, and Enzymatic Assay. *Journal of Analytical Toxicology* 32(5):329-338.

DuPont, Robert L. and Werner A. Baumgartner. 1995. Drug Testing by Urine and Hair Analysis: Complementary Features and Scientific Issues. *Forensic Science International* 70:63-76.

DuPont, Robert L. and Carl M. Selavka. 2008. Testing to Identify Recent Drug Use. Chapter 46 in *The American Psychiatric Publishing Textbook of Substance Abuse Treatment*, 4th Edition. Marc Galanter and Herbert D. Kleber, Editors. Arlington, VA: American Psychiatric Publishers. Pp. 655-664.

Durbin, Nancy and Thomas Grant. *1996. Fitness for Duty in the Nuclear Industry: Update of the Technical Issues 1996.* Seattle, WA: Battelle Seattle Research Center. NUREG/CR-6470.

Durbin, Nancy, C. Moore, T. Grant, I Fleming, P. Hunt, R. Martin, S. Murphy, J. Hauth, R. Wilson, A. Bittner, A. Bramwell, J. Macaulay, J. Olson, E. Terrill, and J. Toquim. 1991. *Fitness for Duty in the Nuclear Power Industry: A Review of the First Year of Program Performance and an Update of the Technical Issues.* Seattle, WA: Battelle Human Affairs Research Centers. NUREG/CR-5784.

Dyer, K.R., and C. Wilkinson. 2008. The Detection of Illicit Drugs in Oral fluids: Another Potential Strategy to Reduce Illicit Drug-Related Harm. *Drug Alcohol Review* 27(1):99-107.

Edgell, Kenneth C. 2008. The U.S. Department of Transportation's Workplace Testing Program. In *Workplace Drug Testing.* Steven B. Karch, Editor. Boca Raton, FL: CRC Press. Pp. 21-36.

Eichhorst, Jeff C., Michele L. Etter, Nadine Rousseaux, and Denis C. Lehotay. 2009. Drugs of Abuse Testing by Tandem Mass Spectrometry: A Rapid, Simple Method to Replace Immunoassays. *Clinical Biochemistry* 42:1531-1542.

Elliott, Simon, Helen Woolacott, and Robin Braithwaite. 2009. The Prevalence of Drugs and Alcohol Found in Road Traffic Fatalities: A Comparative Study of Victims. *Science and Justice* 49(1):19-23.

El Sohly, Mahmoud, Waseem Gui, and Maissa Salem. 2008. Cannabinoids Analysis: Analytical Methods for Different Biological Specimens. In *Handbook of Analytical Separations, 2nd Edition, Vol 6, Forensic Science.* M.J. Bogusz, Editor. Amsterdam: Elsevier, B.V. Pp. 203-242.

Engblom, Charlotta, Teemu Gunnar, Anna Rantanen, and Pirjo Lillsunde. 2007. Driving Under the Influence of Drugs – Amphetamine Concentrations in Oral fluids and Whole Blood Samples. *Journal of Analytical Toxicology* 31(5):276-280.

Enos, Gary A. 2008. Backing a Variety of Testing Vehicles. *Addiction Professional.* Accessed 07.2009 at http://findarticles.com/p/articles/mi_m0QTQ/is_7_6/ai_n32107823/?tag=content;col1.

ENT Specialists. Salt Lake City, UT. MyEntSpecialist.com website. Diagram of Salivary Glands. Accessed 09.2010 at http://www.myentspecialist.com/images/salivary_glands.gif. [Picture]

Erowid.org. 2011. Website. Alcohol Drug Testing. Accessed 01.2011 at http://www.erowid.org/chemicals/alcohol/alcohol_testing.shtml.

Executive Order 12564. 1986. Drug-Free Federal Workplace. *Federal Register* 51(180):32889-32893.

Fang, W., W. Xie, J. Hsieh, and B. Matuszewski. 2005. Development and Applications of HPLC Methods with Tandem Mass Spectrometric Detection for the Determination of Hydrochlorothiazide in Human Plasma and Urine Using 960Well Liquid-Liquid Extraction. *Journal of Liquid Chromatography and Related Technologies* 28:2681-2703.

Fanigliulo, Ameriga, Frederica Bortolotti, Jennifer Pascali, and Franco Tagliaro. 2008. Forensic Toxicological Screening with Capillary Electrophoresis and Related Techniques. In *Forensic Science Handbook of Analytical Separations, Vol 6.* M.J. Bogusz, Editor. Amsterdam: Elsevier, B.V. Pp. 513-534.

Felli, M., S. Martello, and M. Chiarotti. 2011. LC-MS-MS Method for Simultaneous Determination of THCCOOH and THCCOOH-Glucuronide in Urine: Application to Workplace Confirmation Tests. *Forensic Science International* 204:67-73.

Feng, J., L. Wang, I Dai, T. Harmon, and J.T. Bernert. 2007. Simultaneous Determination of Multiple Drugs of Abuse and Relevant Metabolites in Urine by LC-MS-MS. *Journal of Analytical Toxicology* 31(7):359-68.

Finkle, Bryan S., Robert V. Blanke, and J. Michael Walsh, Editors. 1990. *Technical, Scientific and Procedural Issues of Employee Drug Testing, Consensus Report.* U.S. Department of Health and Human Services, Public Health Service, Alcohol, Drug Abuse and Mental Health Administration.

Fishbain, D.A., R.B. Cutler, H.L. Rosomoff, and R.S. Rosomoff. 2003. Are Opioid-Dependent/Tolerant Patients Impaired in Driving-Related Skills? A Structured Evidence-Based Review. *Journal of Pain Symptom Management* 25(6):559-577.

Flanagan, Robert J., Andrew Taylor, Ian D. Watson, Robin Whelpton. 2008. *Fundamentals of Analytical Toxicology.* Wiley Online: http://onlinelibrary.wiley.com/resolve/openurl?genre=book&isbn=9780470516294

The Forensic Toxicology Council. 2010. Briefing: What is Forensic Toxicology? July. Accessed 11.2010 at http://www.abft.org/files/WHAT%20IS%20FORENSIC%20TOXICOLOGY.pdf.

Fortina, Paolo and Larry J. Kricka. 2010. Nanotechnology: Improving Clinical Testing? *Clinical Chemistry* 56:9 1384-1389.

Fortner, Neil A. 2008. The Detection of Drugs in Sweat. Chapter 6 in *Drug Testing in Alternate Biological Specimens.* Amanda J. Jenkins, Editor. Totowana, NJ: Humana Press. Pp.101-115.

Frajola, Walter. 2008. The DUI Professor. Accessed 09.2009 at http://www.rfrajola.com/duiprof/DUI_ProfJunkSciencePart3.htm.

French, Michael T., M. Christopher Roebuck, and Pierre Kébreau Alexandre. 2004. To Test or Not to Test: Do Workplace Drug Testing Programs Discourage Employee Drug Use? *Social Science Research* 33:45-63.

Fritch, Dean, Kristen Blum, Sheena Nonnemacher, Brenda J. Haggerty, Matthew P. Sullivan, and Edward J. Cone. 2009. Identification and Quantitation of Amphetamines, Cocaine, Opiates, and Phencyclidine in Oral fluids by Liquid Chromatography-Tandem Mass Spectrometry. *Journal of Analytical Toxicology* 33(9):569-577.

Frone, Michael R. 2006. Prevalence and Distribution of Alcohol Use and Impairment in the Workplace: A U.S. National Survey. *Journal of Studies on Alcohol* 67(1):147-156.

Frone, Michael R. 2008. Are Work Stressors Related to Employee Substance Use? The Importance of Temporal Context in Assessments of Alcohol and Illicit Drug Use. *Journal of Applied Psychiatry* 93(1):199-206.

Fu, Shanlin, and Lewis John. 2008. Novel Automated Extraction Method for Quantitative Analysis of Urinary 11-Nor-Δ^9-Tetrahydrocannabinol-9-Carboxylic Acid (THC-COOH). *Journal of Analytical Toxicology* 32(4):292-297.

Galanter, Marc, and Herbert D. Kleber. 2008. *The American Psychiatric Publishing Textbook of Substance Abuse Treatment*, 4th Edition. Arlington, VA: American Psychiatric Publishers.

Galinsky, R.E., and C.K. Svensson. 2006. Basic Pharmacokinetics and Pharmacodynamics. In *Remington: The Science and Practice of Pharmacy, 21st Edition*. Philadelphia, PA: Lippincott Williams and Wilkins.

Gallardo, E., M. Barroso, and J.A. Queiroz. 2009. LC-MS: A Powerful Tool in Workplace Drug Testing. *Drug Testing and Analysis* 1:109-115.

Gallardo, E., and J.A. Queiroz. 2008. The Role of Alternative Specimens in Toxicological Analysis. *Biomedical Chromatography* 22(208):795-821.

Galski, T., J.B. Williams, and H.T. Ehle. 2000. Effects of Opioids on Driving Ability. *Journal of Pain Symptom Management* 19(3):200-2008.

Gao, Liang, Andy Sugiarto, Jason Harper, Jason S. Duncan, Ray S. Milks, Bob Kline-Schoder, R. Graham Cooks, Zheng Ouyang. 2007. Mini 11 Handheld Mass Spectrometer with Glow Discharge Ion Source and Atmospheric Pressure Interface. *Journal of Mass Spectrometry* 42(5):675-680.

Garcia-Bournissen, F., M. Moller, M. Nesterenko, T. Karaskov, and G. Koren. 2009. Pharmacokinetics of Disappearance of Cocaine from Hair after Discontinuation of Drug Use. *Forensic Science International* 189(1-3):24-27.

George, Steve. 2004. Position of Immunological Techniques in Screening in Clinical Toxicology. *Clinical Chemistry & Laboratory Medicine* 42(11): 1288-1309.

George, Stephen, and Robin A. Braithwaite. 2002. Use of On-Site Testing for Drugs of Abuse. *Clinical Chemistry* 48(10):1639-1646.

Ghosh, Madhushree. 2010. Racing Ahead: The Future of Drugs-of-Abuse Testing and Drug Toxicology Products. *Clinical Lab Products Magazine Online* Accessed 11.2010 at http://www.clpmag/com/issues/articles/2010-10_04.asp.

Giaginis, Costas, and Anna Tsantili-Kakoulidou. 2008. Current State of the Art in HPLC Methodology for Lipophilicity Assessment of Basic Drugs: A Review. *Journal of Liquid Chromatography and Related Technologies* 31:79-96.

Gilman, Jessica. Electrospray Ionization Mass Spectrometry. Chemistry Class Presentation.

Gjerde, Hallvard, Per T. Normann, and Asbjørg S. Christophersen. 2010. The Prevalence of Alcohol and Drugs in Sampled Oral fluids is Related to Sample Volume. *Journal of Analytical Toxicology* 34(7):416-419.

Gjerde, Hallvard, Jon Mordal, Asbjørg S. Christophersen, Jørgen G. Bramness, and Jorg Morland. 2010. Comparison of Drug Concentrations in Blood and Oral fluids Collected with the Intercept® Sampling Device. *Journal of Analytical Toxicology* 34(4):204-209.

Gjerde, Hallvard, and Alain Verstraete. 2010. Can the Prevalence of High Blood Drug Concentrations in a Population Be Estimated By Analysing Oral fluids? A Study of Tetrahydrocannabinol and Amphetamine. *Forensic Science International* 195:153-159.

Goessaert, An-Sofie, Kristof Pil, Jolien Veramme, and Alain Verstraete. 2010. Analytical Evaluation of a Rapid On-Site Oral fluids Drug Test. *Analytical and Bioanalytical Chemistry* 396:2461-2468.

Goodwin, Robert S., William D. Darwin, C. Nora Chiang, Ming Shih, Shou-Hua Li, and Marilyn A. Huestis. 2008. Urinary Elimination of 11-Nor-9-Carboxy-Δ^9-tetrahydrocannabinol in Cannabis Users during Continuously Monitored Abstinence. *Journal of Analytical Toxicology* 32(9):562-569.

Gottlieb, Mark S. 2006. Drug Testing: An Industry Study. New York: MSG, Inc.

Gourlay, Douglas, Howard Heit, and Yale Caplan. 2006. *Urine Drug Testing in Clinical Practice: Dispelling the Myths and Designing Strategies,* 3rd Edition. California Academy of Family Physicians.

Grauwiler, Sandra B., Jürgen Drewe, and André Scholer. 2008. Sensitivity and Specificity of Urinary Cannabinoid Detection with Two Immunoassays after Controlled Oral Administration of Cannabinoids to Humans. *Therapeutic Drug Monitoring* 30(4):530-535.

Griffiths, Jennifer. 2008a. A Brief History of Mass Spectrometry. *Analytical Chemistry* 80 (21):5678-5683.

Griffiths, Jennifer. 2008b. A Mass Spectrometer in Every Hand. *Analytical Chemistry* 80 (21):7904.

Grotenhermen, Franjo. 2003. Pharmacokinetics and Pharmacodynamics of Cannabinoids. *Clinical Pharmacokinetics* 42(4):327-360.

Guerra, Maria Romero, Iva Chianella, Elena V. Piletska, Kal Karim, Anthony P.F. Turner, and Sergey A. Piletsky. 2009. Development of a Piezoelectric Sensor for the Detection of Methamphetamine. *Analyst* 134:1565-1570.

Hadfield, Lindsay. 2009. Testing Employees. *Drug Testing and Analysis* 1:116-117.

Hager, James W. 2004. Recent Trends in Mass Spectrometer Development. *Analytical and Bioanalytical Chemistry* 378:845-850.

Håkansson, Kristina, Roman A. Zubarev, Per Håkansson, Viktor Laiko, and Alexander F. Dodonov. 2000. Design and Performance of an Electrospray Ionization Time-of-Flight Mass Spectrometer. *Review of Scientific Instruments* 71(1):36-41.

Hall, Brad J., Mary Satterfield-Doerr, Aashish R. Parikh, and Jennifer S. Brodbelt. 1998. Determination of Cannabinoids in Water and Human Saliva by Solid-Phase Microextraction and Quadrupole Ion Trap Gas Chromatography/Mass Spectrometry. *Analytical Chemistry* 70:1788-1796.

Haller, D.L., M.C. Acosta, D. Lewis, D.R. Miles, T. Schiano, P.A. Shapiro, J. Gomez, S. Sabag-Cohen, and H. Newville. 2010. Hair Analysis versus Conventional Methods of Drug Testing in Substance Abusers Seeing Organ Transplant. *American Journal of Transplantation* 10:1305-1311.

Han, Eunyoung, Eleanor Miller, Juseon Lee, Yonghoon Park, Miae Lim, Heesun Chung, Fiona M. Wylie, and John S. Oliver. 2006. Validation of the Immunalysis® Microplate ELISA for the Detection of Methamphetamine in Hair. *Journal of Analytical Toxicology* 30(July/August):380-385.

Hartwell, Tyler, Paul D. Steele, and Nathaniel F. Rodman. 1998. Workplace Alcohol-Testing Programs: Prevalence and Trends. *Monthly Labor Review* June:27: 27-34.

Hashimoto, Kenji. 2007. New Research on Methamphetamine Abuse. In *New Research on Methamphetamine Abuse.* Gerald. H. Toolaney, Editor. NY: Nova Science Publishers. Pp. 1-51.

Hawks, Richard L., and C. Nora Chiang, Editors. 1986. Urine Testing of Drugs of Abuse. NIDA Research Monograph 73. Rockville, MD: National Institute on Drug Abuse.

Hegstad, S., H.Z. Khiabani, L. Kristoffersen, N. Kunøe, P.P Lobmaier, and A.S. Christophersen. 2008. Drug Screening of Hair by Liquid Chromatrography-Tandem Mass Spectrometry. *Journal of Analytical Toxicology* 32(5):364-372.

Helander, Anders, Charlotte Asker Hagelberg, Olof Beck, and Björn Petrini. 2009. Unreliable Alcohol Testing in a Shipping Safety Programme. *Forensic Science International* 189(1-3): e45-e47.

Henderson, G.L. 1993. Mechanisms of Drug Incorporation into Hair. *Forensic Science International* 63:19-29.

Henderson, G.L., M.R. Harkey, C. Zhou. 1998. Incorporation of Isotopically Labeled Cocaine into Human Hair: Race as a Factor. *Journal of Analytical Toxicology* 22:156-165.

Hersch, Rebekah K., Dennis J. Crouch, and Royer F. Cook. 2000. *Field Test of On-Site Drug Detection Devices.* Prepared for the U.S. Department of Transportation, National Highway Traffic Safety Administration. Washington, DC. Report No: DOT HS 809-192.

Hill, Virginia, Thomas Cairns, Michael Schaffer. 2008. Hair Analysis for Cocaine: Factors in Laboratory Contamination Studies and Their Relevance to Proficiency Sample Preparation and Hair Testing Practices. *Forensic Science International* 176:23-33.

Hill, Virginia, Thomas Cairns, Chen-Chih Cheng, and Michael Schaffer. 2005. Multiple Aspects of Hair Analysis for Opiates: Methodology, Clinical and Workplace Populations, Codeine, and Poppy Seed Ingestion. *Journal of Analytical Toxicology* 29:696-703.

Hoffmann, Thorsten. 2003. Real-Time Mass Spectrometry. *Analytical and Bioanalytical Chemistry* 375:36-37.

Høiseth, Gudrun, Borghild Yttredal, Ritva Karinen, Hallvard Gjerde, Jørg Mørland, and Asbjørg Christophersen. 2010. Ethyl Glucuronide Concentrations in Oral fluids, Blood, and Urine After Volunteers Drank 0.5 and 1.0 g/kg Doses of Ethanol. *Journal of Analytical Toxicology* 34:319-324.

Holland, Peter. 2003. Case-Study: Drug Testing in the Australian Mining Industry. *Surveillance and Society*1(2):204-209.

Huang, D.K., C. Liu, M.K. Huang, and C.S. Chien. 2009. Simultaneous Determination of Morphine, Codeine, 6-acetylmorphine, Cocaine and Benzoylecgonine in Hair by Liquid Chromatography/Electrospray Ionization Tandem Mass Spectrometry. *Rapid Communication Mass Spectrometry* 23(7):957-962.

Huang, Su-Hua. 2006. Gold Nanoparticle-Based Immunochromatographic Test for Identification of *Staphylococcus Aureus* from Clinical Specimens. *Clinica Chimica Acta* 373:139-143.

Huang M-H., R.H. Liu, Y-L. Chen, and S.L. Rhodes. 2006. Correlation of Drug-testing Results – Immunoassay versus Gas Chromatography-Mass Spectrometry. *Forensic Science Review* 18:9–41.

Huestis, M. 2009. A New Ultraperformance-Tandem Mass Spectrometry Oral fluids Assay for 29 Illicit Drugs and Medications. *Clinical Chemistry* 55(12):2079-2081.

Huestis, M. 2007. Human Cannabinoid Pharmacokinetics. *Chemical Biodiversity* 4(8):1770-1804.

Huestis, M. 1998. Judicial Acceptance of Hair Tests for Substances of Abuse in the United States: Scientific, Forensic, and Ethical Aspects. *Therapeutic Drug Monitoring* 18:456-459.

Huestis, M.A., K.B. Scheidweiler, T. Saito, N.Fortner, T. Abraham, R.A. Gustafson, M.L. Smith. 2008. Excretion of Delta 9-tetrahydrocannabinol in Sweat. *Forensic Science International* (174(2-3):173-177.

Huestis, Marilyn A., and Michael L. Smith. 2006. Modern Analytical Technologies for the Detection of Drug Abuse and Doping. *Analytical Chemistry* 3(1):49-57.

Intoximeters, Inc. Website. About Alcohol Testing. Accessed 1.2011 at http://www.intox.com/t-AboutAlcoholTesting.aspx.

Intoximeters, Inc. Website. Alcohol and the Human Body. Accessed 06.2009 at http://www.intox.com/physiology.asp.

Intoximeters, Inc. Website. Fuel Cell Technology Applied to Alcohol Breath Testing. Accessed 01.2011 at http://www.intox.com/t-FuelCellWhite Paper.aspx.

Isenschmid, Daniel S., and Bruce A. Goldberger. 2008. Analytical Considerations and Approaches for Drugs. In *Workplace Drug Testing*. Steven B. Karch, Editor. Boca Raton, FL: CRC Press. Pp. 56-77.

Isenschmid, Daniel S., and Bruce A. Goldberger. 2007. Analytical Considerations and Approaches for Drugs. In *Drug Abuse Handbook*, 2nd Edition. Steven B. Karch, Editor. Boca Raton, FL: CRC Press. Pp. 775-799.

Ives, Timothy J., Paul R. Chelminski, Catherine A. Hammett-Stabler, Robert M Malone, J Stephen Perhac, Nicholas M Potisek, Betsy Bryant Shilliday, Darren A DeWalt, and Michael P Pignone. 2006. Predictors of Opioid Misuse in Patients with Chronic Pain: A Prospective Cohort Study. *BMC Health Services Research* 6: 46-10.

Izake, Emad L. 2010. Forensic and Homeland Security Applications of Modern Portable Raman Spectroscopy. *Forensic Science International* 202:1-8.

Jaffee, William B., Elisa Trucco, Christian Teter, Sharon Levy, and Roger D. Weiss. 2008. Ensuring Validity in Urine Drug Testing. *Psychiatric Services* 59(2):140-142. Accessed 04.2009 at http://psychservices.psychiatryonline.org/cgi/reprint/59/2/140.

Jaffee, William B., Elisa Trucco, Sharon Levy, and Roger D. Weiss. 2007. Is This Urine Really Negative? A Systematic Review of Tampering Methods in Urine Drug Screening and Testing. *Journal of Substance Abuse Treatment* 33:33-42.

Jagerdeo, E., M.A. Montgomery,Roman Karas, and Martin Sibum. 2010. A Fast Method for Screening and/or Quantitation of Tetrahydrocannabinol and Metabolites in Urine by Automated SPE/LC/MS/MS. *Analytical and Bioanalytical Chemistry* 398:329-338.

Jagerdeo, E., M.A. Montgomery, M.A. Lebeau, and M. Siburn. 2008. An Automated SPE/LC/MS/MS Method for the Analysis of Cocaine and Metabolites in Whole Blood. *Journal of Chromatography B* 874(1-2):15-20.

Janicka, Monika, Agata Kot-Wasik, and Jacek Namieśnik. 2010. Analytical Procedures for Determination of Cocaine and Its Metabolites in Biological Samples. *Trends in Analytical Chemistry* 29(3):209-224.

Jatlow, Peter. 1988. Cocaine: Analysis, Pharmacokinetics, and Metabolic Disposition. *The Yale Journal of Biology and Medicine* 61:105-113.

Jemionek, John F., Curtis L. Copley, Michael L. Smith, and Marilyn R. Past. 2008. Concentration Distribution of the Marijuana Metabolite Δ^9-Tetrahydrocannabinol-9-Carboxylic Acid and the Cocaine Metabolite Benzoylecgonine in the Department of Defense Urine Drug-Testing Program. *Journal of Analytical Toxicology* 32(6):408-416.

Jenkins, Amanda J., Editor. 2008. *Drug Testing in Alternate Biological Specimens*. Totowa, NJ: Humana Press.

Jenkins, Amanda J. 2007a. Pharmacokinetics: Drug Absorption, Distribution, and Elimination. Chapter 3 in *Drugs of Abuse*. Steven B. Karch, Editor. Pp. 147-205. Boca Raton, FL: CRC Press.

Jenkins, Amanda J. 2007b. Pharmacokinetics of Specific Drugs. In *Pharmacokinetics and Pharmacodynamics of Abused Drugs*. Steven B. Karch, Editor. Pp. 25-64. Boca Raton, FL: CRC Press.

Jenkins, Amanda, J. 2003. Forensic Drug Testing. In *Principles of Forensic Toxicology, 2nd Edition*. Barry Levine, Editor. Washington, DC: AACC Press. Pp. 31-46.

Jenkins, Amanda J., and Bruce A. Goldberger, Editors. 2002. On-Site Drug Testing. Totowa, NJ: Humana Press.

Jenkins, A.J., J.M. Oyler, and E.J. Cone. 1995. Comparison of Heroin and Cocaine Concentrations in Saliva with Concentrations in Blood and Plasma. *Journal of Analytical Toxicology* 19:359-374.

Jones, Alan Wayne, Fredrik C. Kugelberg, Anita Holmgren, and Johan Ahlner. 2009. Five-Year Update on the Occurrence of Alcohol and Other Drugs in Blood Samples from Drivers Killed in Road-Traffic Crashes in Sweden. *Forensic Science International* 186:56-62.

Jones, Alan Wayne. 2007. Age- and Gender-Related Differences in Blood Amphetamine Concentrations in Apprehended Drivers: Lack of Association with Clinical Evidence of Impairment. *Addiction* 102:1085-1091.

Jones, Alan Wayne, Anita Holmgren, and Fredrik C. Kugelberg. 2007. Driving Under the Influence of Cannabis: A 10-Year Study of Age and Gender Differences in the Concentrations of Tetrahydrocannabinol in Blood. *Addiction* 103:452-461.

Jones, Alan Wayne, and Fredrik C. Kugelberg. 2010. Relationship Between Blood and Urine Alcohol Concentrations in Apprehended Drivers Who Claimed Consumption of Alcohol After Driving With and Without Supporting Evidence. *Forensic Science International* 194:97-102.

Jones, R.K., D. Shinar, and J.M. Walsh. 2003. State of Knowledge of Drug-Impaired Driving. DOT HS 809 642. Prepared by the Mid-America Research Institute for the National Highway Traffic Safety Administration, Office of Research and Technology, Washington, DC. Available www.nhtsa.dot.gov

Jones, Reese T. 1997. Pharmacokinetics of Cocaine: Considerations When Assessing Cocaine Use by Urinalysis. *Medication Development for the Treatment of Cocaine Dependence: Issues in Clinical Efficacy Trials*. Betty Tai, Nora Chiang, and Peter Bridge, Editors. NIDA Monograph 175. Pp. 221-234.

Joscelyn, K.B., A.C. Donelson, R.K. Jones, J.W. McNair, and P.A. Ruschmann. 1980. *Drugs and Highway Safety*. Final report. HS 805 461. Washington, DC: National Highway Traffic Safety Administration.

Joseph, Jr., Robert, Wei-Jen Tsai, Li-I Tsao, Tsung-Ping Su, and Edward J. Cone. 1997. *In Vitro* Characterization of Cocaine Binding Sites in Human Hair. *The Journal of Pharmacology and Experimental Therapeutics* 282:1228-1241.

Jufer, Rebecca, Sharon Walsh, Edward Cone, and Angela Sampson-Cone. 2006. Effect of Repeated Cocaine Administration on Detection Times in Oral fluids and Urine. *Journal of Analytical Toxicology* 30:458-462.

Jufer, R.A., A. Wstadik, S.L. Walsh, B.S. Levine, and E.J. Cone. 2000. Elimination of Cocaine and Metabolites in Plasma, Saliva, and Urine Following Repeated Oral Administration to Human Volunteers. *Journal of Analytical Toxicology* 24:467-477.

Jurado, Carmen, and Hans Sachs. 2003. Proficiency Test for the Analysis of Hair for Drugs of Abuse. *Forensic Science International* 133(1):175-178.

Kacinko, S.L., A.J. Barnes, E.J. Cone, E.T. Moolchan, and M.A. Huestis. 2005. Disposition of Cocaine and its Metabolites in Human Sweat After Controlled Cocaine Administration. *Clinical Chemistry* 51(11):2085-2094.

Kadehjian, Leo J. 2005. Specimens for Drugs-of-Abuse Testing. In *Drugs of Abuse*. Raphael C. Wong and Harley Y. Tse, Editors. Totowa, NJ: Humana Press. Pp. 11-28.

Kadehjian, Leo. J. 2004. Legal Issues in Oral fluids Testing. *Forensic Science International* 150:151-160.

Kala, S.V., S.E. Harris, T.D. Freijo, and S. Gerlich. 2008. Validation of Analysis of Amphetamines, Opiates, Phencyclidine, Cocaine, and Benzoylecgonine in Oral fluids by Liquid Chromatography-Tandem Mass Spectrometry. *Journal of Analytical Toxicology* 32(8):605-611.

Karch, Steven B. Editor. 2008. *Workplace Drug Testing*. Boca Raton, FL: CRC Press.

Karch, Steven B., Editor. 2007a. *Drug Abuse Handbook*, 2nd Edition. Boca Raton, FL: CRC Press.

Karch, Steven B. Editor. 2007b. *Pharmacokinetics and Pharmacodynamics of Abused Drugs*. Boca Raton, FL: CRC Press. Boca Raton, FL: CRC Press.

Karschner, Erin, Allan Barnes, Ross Lowe, Karl Scheidweiler, and Marilyn Huestis. 2010. Validation of a Two-Dimensionsl Gas Chromatography Mass Spectrometry Method for the Simultaneous Quantification of Cannabidiol, Δ^9-Tetrahydrocannabinol (THC), 11-Hydroxy-THC, and 11-Nor-9-Carboxy-THC in Plasma. *Analytical and Bioanalytical Chemistry 397:603-611.*

Kato, K., M. Hillsgrove, L. Weinhold, D.A. Gorelick, W.D. Darwin, and E.J. Cone. 1993. Cocaine and Metabolite Excretion in Saliva Under Stimulated and Nonstimulated Conditions. *Journal of Analytical Toxicology* 17:338-341.

Kauert, Gerold, Stefanie Iwersen-Bergmann, and Stefan W. Toennes. 2006. Assay of Delta-9-Tetrahydrocannabinol (THC) in Oral fluids – Evaluation of the OraSure Oral Specimen Collection Device. *Journal of Analytical Toxicology* 30:274-277.

Kauert, Gerold, Johannes Ramaekers, Erhard Schneider, Manfred Moeller, and Stefan Toennes. 2007. Pharmacokinetic Properties of Δ^9-Tetrahydrocannabinol in Serum and Oral fluids. *Journal of Analytical Toxicology* 31:288-293.

Kazmierczak, Steve C., and Hassan M.E. Azzazy. 2008. Alcohol Testing. In *Handbook of Drug Monitoring Methods*. Amitava Dasgupta, Editor. Totowa, NJ: Humana Press. Pp. 283-295.

Kean, Leslie, and Dennis Bernstein. 1999. Hair Tests Raise Doubts. *Baltimore Sun* May 30.

Keil, Adam, Nari Talaty, Christian Janfelt, Robert J. Noll, Liang Gao, Zheng Ouyang, and R. Graham Cooks. 2007. Ambient Mass Spectrometry with a Handheld Mass Spectrometer at High Pressure. *Analytical Chemistry* 79:7734-7739.

Kelly, Raymond C., Tom Mieczkowski, Stacy A. Sweeney, and James A. Bourland. 2000. Hair Analysis for Drugs of Abuse: Hair Color and Race Differentials or Systematic Differences in Drug Preferences? *Forensic Science International* 107:63-86.

Kempf, Jürgen, Thomas Wuske, Rolf Schubert, and Wolfgang Weinmann. 2009. Pre-Analytical Stability of Selected Benzodiazepines on a Polymeric Oral fluids Sampling Device. *Forensic Science International* 186(1-3):81-85.

Kerrigan, Sarah, and Bruce A. Goldberger. 2008. Specimens of Maternal Origin: Amniotic Fluid and Breast Milk. Chapter 1 in *Drug Testing in Alternate Biological Specimens*. Amanda J. Jenkins, Editor. Humana Press. Pp. 1-18.

Kidwell, David A., Emmelene H. Lee, and Saundra F. DeLauder. 2000. Evidence for Bias in Hair Testing and Procedures to Correct Bias. *Forensic Science International* 107:39-61.

Kidwell, D.A., and F.P. Smith. 2001. Susceptibility of PharmChek Drugs of Abuse Patch to Environmental Contamination. *Forensic Science International* 116(2-3):89-106.

Kidwell, D.A., J.C. Holland, and S. Athanaselis. 1998. Testing for Drugs of Abuse in Saliva and Sweat. *Journal of Chromatography B* 713:111-135.

Kim, Insook, Allan Barnes, Raf Schepers, Eric Moolchan, Lisa Wilson, Gail Cooper, Claire Reid, Chris Hand, and Marilyn Huestis. 2003. Sensitivity and Specificity of the Cozard Microplate EIA Cocaine Oral fluids at Proposed Screening and Confirmation Cutoffs. *Clinical Chemistry* 49(9):1498-1503.

Kim, Insook, Allan Barnes, A.J. Oylers, Raf Schepers, R.E. Joseph, Jr., E.J. Cone, D. Lafko, Eric Moolchan, and Marilyn Huestis. 2002. Plasma and Oral fluids Pharmacokinetics and Pharmacodynamics After Oral Codeine Administration. *Clinical Chemistry* 48:1486-1496.

Kim, Jin Young, Soon Ho Shin, and Moon Kyo In. 2010. Determination of Amphetamine-Type Stimulants, Ketamine, and Metabolites in Fingernails by Gas Chromatography-Mass Spectrometry. *Forensic Science International* 194:108-114.

Kim, Sook Jin, K. Vengatajalabathy Gobi, Hiroyuki Iwasaka, Hiroyuki Tanaka, and Norio Miura. 2007. Novel Miniature SPR Immunosensor Equipped with All-in-One Multi-Microchannel Sensor Chip for Detecting Low-Molecular-Weight Analytes. *Biosensors and Bioelectronics* 23:701-707.

Kintz, Pascal. 2010. Consensus of the Society of Hair Testing on Hair Testing for Chronic Excessive Alcohol Consumption 2009. Guest Editor. *Forensic Science International* 196:2.

Kintz, Pascal. 2008. Drug Testing in Hair. In *Drug Testing in Alternate Biological Specimens*. Amanda J. Jenkins, Editor. Humana Press. Pp. 67-81.

Kintz, Pascal. 2007. Bioanalytical Procedures for Detection of Chemical Agents in Hair in the Case of Drug-Facilitated Crimes. *Analytical and Bioanalytical Chemistry* 388:1467-1474.

Kintz, Pascal. 2006. Analytical and Practical Aspects of Drug Testing in Hair. Boca Raton, FL: CRC Press.

Kintz, Pascal, and Ronald Agius. 2009. Draft EWDTS (European Workplace Drug Testing System) Guidelines. Latest Update 20.01.2009. Accessed 10.2010 at www.ewdts.org/guidelines/hair.pdf.

Kinz, Pascal, Bertrand Brunet, Jean-Francois Muller, Wilfried Serra, Marion Villain, Vincent Cirimele, and Patrick Mura. 2009. Evaluation of the Cozart DDSV Test for Cannabis in Oral fluids. *Therapeutic Drug Monitoring* 31(1):131-134.

Kintz, Pascal, V. Cirimele, and B. Ludes. 2000. Detection of Cannabis in Oral fluids (Saliva) and Forehead Wipes (Sweat) from Impaired Drivers. *International Journal of Drug Testing* 24(7):557-561.

Kintz, Pascal, and Patrice Mangin. 1995. What Constitutes a Positive Result in Hair Analysis: Proposal for the Establishment of Cut-Off Values. *Forensic Science International* 70:3-11.

Kintz, Pascal, and N. Samyn. 2002. Use of Alternative Specimens: Drugs of Abuse in Saliva and Doping Agents in Hair. *Therapeutic Drug Monitoring* 23:239-246.

Kinz, Pascal, Marion Villain, and Vincent Cirimele. 2007. Analytical Approaches for Drugs in Biological Matrices Other than Urine. In *Drug Abuse Handbook,* 2nd Edition. Steven B. Karch, Editor. Roca Baton, FL: CRC Press. Pp. 800-813.

Klein, Julia, Tatyana Karaskov, and Gideon Koren. 2000. Clinical Applications of Hair Testing for Drugs of Abuse – the Canadian Experience. *Forensic Science International* 107(1-3):281-288.

Klugman, Anthony, and John Gruzelier. 2003. Chronic Cognitive Impairment in Users of 'Ecstacy' and Cannabis. *World Psychiatry* 2(3):184-190.

Knapp, Clifford M., Domenic A. Ciraulo, and Henry R. Kranzler. 2008. Neurobiology of Alcohol. Chapter 8 in *The American Psychiatric Publishing Textbook of Substance Abuse Treatment*, 4th Edition. Marc Galanter and Herbert D. Kleber, Editors. Arlington, VA: American Psychiatric Publishers. Pp. 111-128

Kopacek, Karen B. 2007. Absorption. *The Merck Manuals*. Updated November 2007. Accessed 07.2009 at http://www.merck.com/mmpe/sec20/ch303/ch303b.html.

Kraemer, Thomas, and Hans H. Maurer. 2008. Sedatives and Hypnotics. In *Forensic Science Handbook of Analytical Separations, Vol 6*. M.J. Bogusz, Editor. Amsterdam: Elsevier, B.V. Pp. 243-286.

Kraemer, Thomas, and Liane D. Paul. 2007. Bioanalytical Procedures for Determinatin of Drugs of Abuse in Blood. *Analytical and Bioanalytical Chemistry* 388:415-435.

Krasowski, Matthew, Mohamed Siam, Manisha Iyer, Anthony Pizon, Spiros Glannoutsos, and Sean Ekins. 2009. Chemoinformatic Methods for Predicting Interference in Drug of Abuse/Toxicology Immunoassays. *Clinical Chemistry* 55(6):1203-1213.

Krause, Karoline, and Kerstin Foitzik. 2006. Biology of the Hair Follicle: The Basics. *Seminars in Cutaneous Medicine and Surgery* 25:2-10.

Kronstrand, Robert, Ingrid Nyström, Malin Forsmana, and Kerstin Käll. 2010. Hair Analysis for Drugs in Driver's License Regranting: A Swedish Pilot Study. *Forensic Science International* 196(1-3):55-58.

Kronstrand, Robert, Ingrid Nyström, Joakim Strandberg, and Henrik Druid. 2004. Screening for Drugs of Abuse in Hair with Ion Spray LC-MS-MS. *Forensic Science International* 145:183-190.

Kronstrand, Robert, Sophie Förstberg-Peterson, Bertil Kågedal, Johan Ahlner, and Göran Larson. 1999. Codeine Concentration in Hair After Oral Administration is Dependent on Melanin Content. *Clinical Chemistry* 45(9):1485-1494.

Kricka, Larry J., and Stephen R. Master. 2005. Validation and Quality Control of Protein Microarray-Based Analytical Methods. *Microarrays in Clinical Diagnostics* 114:233-255.

Kruglick, Kim. Nd. A Beginner's Primer on the Investigation of Forensic Evidence. *Scientific Testimony: An Online Journal*. Accessed 12.2010 at http://www.scientific.org/tutorials/articles/kruglick/kruglick.html.

Kuma, Hiroyuki, Hiroko Oyamada, Akira Tsukamoto, Takako Mizoguchi, Akihiki Kandori, Yoshinori Sugiura, Kohji Yoshinaga, Keiji Enpulu, and Naotaka Hamasaki. 2010. Liquid Phase Immunoassays Utilizing Magnetic Markers and SQUID Magnetometer. *Clinical Chemistry and Laboratory Medicine* 48(9):1263-1269.

Labat, Laurence, Bernard Fontaine, Chantal Delzenne, Anne Doublet, Marie Marek, Dominique Tellier, Murielle Tonneau, Michel Lhermitte, and PaulFrimat. 2008. Prevalence of Psychoactive Substances in Truck Drivers in the Nord-Pas-de-Calais Region (France). *Forensic Science International* 174:90-94.

Lacey, John H., Tara Kelly-Baker, Debra Furr-Holden, Katharine Brainard, and Christine Moore. 2007. Pilot Test of New Roadside Survey Methodology for Impaired Driving. Prepared by the Pacific Institute for Research and Evaluation. Prepared for the U.S. Department of Transportation, National Highway Traffic Safety Administration. Washington, DC. Report No: DOT HS 810 704.

LaLoup, Marleen, Gert De Boeck, and Nele Samyn. 2008. Unconventional Samples and Alternative Matrices. In *Handbook of Analytical Separations, 2nd Edition, Vol 6, Forensic Science.* M.J. Bogusz, Editor. Pp. 653-697.

LaLoup, Marleen, Gaëlle Tilman, Miviane Maes, Gert De Boeck, Pierre Wallemacq, Jan Ramaekers, and Nele Samyn. 2005. Validation of an ELISA-Based Screening Assay for the Detection of Amphetamine, MDMA, and MDA in Blood and Oral fluids. *Forensic Science International* 153:29-37.

Langel, Kaarina, Charlotta Engblom, Anna Pehrsson, Teemu Gunnar, Kari Ariniemi, and Pirjo Lillsunde. 2008. Drug Testing in Oral fluids – Evaluation of Sample Collection Devices. *Journal of Analytical Toxicology* 32(July/August):393-401.

Langman, Loralie J. 2007. The Use of Oral fluids for Therapeutic Drug Management. *Clinical and Forensic Toxicology* 1098:145-166.

Langman, Loralie J., and Bhushan M. Kapur. 2006. Toxicology: Then and Now. *Clinical Biochemistry* 39:498-510.

Lappe, Murray. 2005. Instrumented Urine Point-of-Collection Testing Using the eScreen [copyright] System. In *Drugs of Abuse: Body Fluid Testing.* Raphael C. Wong and Harley Y Tse, Editors. Totowa, NJ: Humana Press. Pp. 201-214.

Larson, Scott J., Justin M. Holler, Joseph Magluilo, Jr., Christopher S. Dunkley, and Aaron Jacobs. 2008. Technical Note: Papain Adulteration in 11-Nor-Δ^9-Tetrahydrocannabinol-9-carboxylic Acid-Positive Urine Samples. *Journal of Analytical Toxicology* 32(6):438-443).

LeBeau, M.A., and M.A. Montgomery. 2009. Considerations on the Utility of Hair Analysis for Cocaine. *Journal of Analytical Toxicology* 33:343-344.

Lee, Sooyeun, Rosa Cordero, and Sue Paterson. 2009. Distribution of 6-Monoacetylmorphine and Morphine in Head and Pubic Hair from Heroin-Related Deaths. *Forensic Science International* 183:74-77.

Lee, Sooyeun, Eunyoung Han, Yonghoon Park, Hwakyung Choi, and Heesun Chung. 2009. Distribution of Methamphetamine and Amphetamine in Drug Abusers' Head Hair. *Forensic Science International* 190:16-18.

Lee, Sooyeun, Yonghoon Park, Wonkyung Yang, Eunyoung Han, Sanggil Choe, Miae Lim, and Heesun Chung. 2009. Estimation of the Measurement Uncertainty of Methamphetamine and Amphetamine in Hair Analysis. *Forensic Science International* 185:59-66.

Leino, A., J. Saarimes, M. Grönholm, and P. Lillsunde. 2001. Comparison of Eight Commercial On-Site Screening Devices for Drugs-of-Abuse Testing. Scandinavian *Journal of Clinical Laboratory Investigation* 61:325-332.

Levisky, Joseph A., David L. Bowerman, Werner W. Jenkins, and Steven B. Karch. 2000. Drug Deposition in Adipose Tissue and Skin: Evidence for an Alternative Source of Positive Sweat Patch Tests. *Forensic Science International* 110:35-46.

Levisky, Joseph A., David L. Bowerman, Werner W. Jenkins, Deborah G. Johnson, John S. Levisky, and Steven B. Karch. 2001. Comparison of Urine to Sweat Patch Test Results in Court Ordered Testing. *Forensic Science International* 122:65-68.

Li, Guohua, Joanne Brady, Charles DiMaggio, Susan Barker, and George Rebok. 2010. Validity of Suspected Alcohol and Drug Violations in Aviation Employees. *Addiction* 105:1771-1775.

Liberty, Hilary James, and Bruce D. Johnson. 2004. Detecting Cocaine Use Through Sweat Testing: Multilevel Modeling of Sweat Patch Length-of-Wear Data. *Journal of Analytical Toxicology* 28(8):667-673.

Lillsunde, Pirjo. 2008. Analytical Techniques for Drug Detection in Oral fluids. *Therapeutic Drug Monitoring* 30(2):181-187.

Lillsunde, Pirjo, Katariina Haavanlammi, Ritva Partinen, Kristiina Mukala, and Matti Lamberg. 2008. Finnish Guidelines for Workplace Drug Testing. *Forensic Science International* 174:99-102.

Lillsunde, Pirjo, Kristiina Mukala, Ritva Partinen, and Matti Lamberg. 2008. Role of Occupational Health Services in Workplace Drug Testing. *Forensic Science International* 174:103-106.

Liska, Ken. 2004. *Drugs and the Human Body: With Implications for Society*, 7th Edition. Upper Saddle River, NJ: Prentice Hall.

Liu, Jonathan, John Decatur, Gloria Proni, and Elise Champeil. 2010. Identification and Quantification of 3,4-Methylenedioxy-N-Methylamphetamine (MDMA, Ecstasy) in Human Urine by ^{1}H NMR Spectroscopy: Application to Five Cases of Intoxication.

Lloyd, David K. 2008. Capillary Electrophoresis Analysis of Biofluids with a Focus on Less Commonly Analyzed Matrices. *Journal of Chromatography B* 866:154-166.

Long, Melissa, and David A. Kidwell. 2002. Improving the Pharmcheck™ Sweat Patch: Reducing False Positives from Environmental Contamination and Increasing Drug Detection. Submitted to the U.S. Department of Justice.

Lowe, Rachel, Georgia Guild, Peter Harpas, Paul Kirkbride, Peter Haffmann, Micolas Voelcker, and Hilton Kobus. 2009. Rapid Drug Detection in Oral Samples by Porous Silicon Assisted Laser Desorption/Ionization Mass Spectrometry. *Rapid Communications in Mass Spectrometry* 23:3543-3548.

Lu, Yao, Ryan O'Donnell, and Peter Harrington. 2009. Detection of Cocaine and its Metabolites in Urine Using Solid Phase Extraction-Ion Mobility Spectrometry with Alternating Least Squares. *Forensic Science International* 189:54-59.

Lurie, Ira S., Patrick A. Hays, Kimberly Parker. 2004. Capillary Electrophoresis Analysis of a Wide Variety of Seized Drugs Using the Same Capillary with Dynamic Coatings. Wiley Online: http://onlinelibrary.wiley.com/doi/10.1002/elps.200405894/pdf.

Luzzi, Veronica, A.N. Saunders, John Koenic, John Turk, Stanley Lo, Uttam Garg, and Dennis Dietzen. 2004. Analytic Performance of Immunoassays for Drugs of Abuse Below Established Cutoff Values. *Clinical Chemistry* 50(4):717-722.

Majors, Ronald E. 2008. The Role of Polymers in Solid-Phase Extraction and Sample Preparation. *LC-GC North America* 26(11):1074-1090.

Makiko, Hayashida, Masahiko Takino, Masaru Terada, Emiko Kurisaki, Keiko Kudo, and Youkichi Ohno. 2009. Time-of-Flight Mass Spectrometry (TOF-MS) Exact Mass Database for Benzodiazepine Screening. *Legal Medicine* 11(Supplement 1):S423-S425.

Manns, Andreas, B. Lange, I. Kaneblei, A. Slomian, S. Steinmeyer, R. Polzius, J. Mahn, A. Reiter. 2007. Analytical Evaluation of a New Oral fluids Sample Drugs of Abuse Diagnostic System. Accessed 03.2010 at www.icadts2007.org/print/164analyticalevaluation.pdf.

Marais, Adriaan, and Johannes Laurens. 2009. Rapid GC-MS Confirmation of Amphetamines in Urine by Extractive Acylation. *Forensic Science International* 183:78-86.

Marchei, E., M. Farrè, M. Pellegrini, Ó. García-Algar, O. Vall, R. Pacifici and S. Pichini. 2010. Pharmacokinetics of Methylphenidate in Oral fluids and Sweat of a Pediatric Subject. *Forensic Science International* 196(1-3):59-63.

Mari, Francesco, Lucia Politi, Annibale Biggeri, Gabriele Accetta, Claudia Triganano, Marianna Di Padua, and Elisabetta Bertol. 2009. Cocaine and Heroin in Waste Water Plants: A 1-Year Study in the City of Florence, Italy. *Forensic Science International* 189:88-92.

Mashir, Alquam, Kelly Paschke, Daniel Laskowski, and Raed Dweik. 2011. Medical Applications of Exhaled Breath Analysis and Testing. *Pulminary, Critical Care, Sleep Update (PCCSU)* 25:1-4. Accessed 03.2011 at http://www.chestnet.org/accp/pccsu/medical-applications-exhaled-breath-analysis-and-testing?page=0,3.

Maurer, Hans H. 2010. Perspectives of Liquid Chromatography Coupled to Low- and High-Resolution Mass Spectrometry for Screening, Identification, and Quantification of Drugs in Clinical and Forensic Toxicology. *Therapeutic Drug Monitor* 32(3):324-327.

Maurer, Hans H. 2008. Forensic Screening with GC-MS. In *Forensic Science Handbook of Analytical Separations, Vol 6.* M.J. Bogusz, Editor. Amsterdam: Elsevier, B.V. Pp. 425-445.

Maurer, Hans H. 2007. Current Role of Liquid Chromatography-Mass Spectrometry in Clinical and Forensic Toxicology. *Analytical and Bioanalytical Chemistry* 388:1315-1325.

Maurer, Hans H. 2005. Advances in Analytical Toxicology: The Current Role of Liquid Chromatography – Mass Spectrometry in Drug Quantification in Blood and Oral fluids. *Analytical and Bioanalytical Chemistry* 381:110-118.

Maurer, Hans H. 2004. Position of Chromatographic Techniques in Screening for Detection of Drugs or Poisons in Clinical and Forensic Toxicology and/or Doping Control. *Clinical Chemistry & Laboratory Medicine* 42(11):1310-1324.

Maxwell, Jane Carlisle. 2005. Party Drugs: Properties, Prevalence, Patterns, and Problems. *Substance Use and Misuse* 40:1203-1240.

McLafferty, Fred W. 1981. Tandem Mass Spectrometry. *Science* 214: 280-287.

McCurdy, H.H., A.M. Morrison, and L.A. Holt. 2008. Liquid Chromatography-Tandem Mass Spectrometry Analysis of Opioids, Benzodiazepines, Cannabinoids, Amphetamines, and Cocaine in Biological and Other Specimens. *Forensic Science Review* 20:45-73.

McLafferty, Fred W., and František Tureček. 1993. *Interpretation of Mass Spectra.* Sausalito, CA: University Science Books.

McLellan, A. Thomas, David C. Lewis, Charles P. O'Brien, and Herbert D. Kleber. 2000. Drug Dependence, A Chronic Medical Illness. The Journal of the American Medical Association 284(13):1689-1695.

McQuay, H.J., and R.A. Moore. Nd. Opioid Problems, and Morphine Metabolism and Excretion. University of Oxford. Accessed 06.2009 at http://www.medicine.ox.ac.uk/bandolier/booth/painpag/wisdom/c14.html.

Melanson, Stacy E., Leland Baskin, Barbarajean Magnani, Tai C. Kwong, Annabel Dizon, and Alan Wu. 2010. Interpretation and Utility of Drug of Abuse Immunoassays. *Archives of Pathology and Laboratory Medicine* 134:735-739.

Meng, Pinjia, and Yanyan Wang. 2010. Small Volume Liquid Extraction of Amphetamines in Saliva. *Forensic Science International* 197(1-3):80-84.

Meririnne, Esa, Sirpa Mykkänen, Pirjo Lillsunde, Kimmo Kuoppasalmi, Risto Lerssi, Ilmo Laaksonen, Kyösti Lehtomäki, and Markus Henriksson. 2007. Workplace Drug Testing in a Military Organization: Results and Experiences from the Testing Program in the Finnish Defence Forces. *Forensic Science International* 170:171-174.

Mieczkowski, Tom. 1997. Distinguishing Passive Contamination from Active Cocaine Consumption: Assessing the Occupational Exposure of Narcotics Officers to Cocaine. *Forensic Science International* 84:87-111.

Mieczkowski, Tom. 1992. New Approaches in Drug Testing: A Review of Hair Analysis. *Annals of the American Academy of Political and Social Science* 521(1):132-150.

Mieczkowski, Tom. 1996. The Use of Hair Analysis for the Detection of Drugs: An Overview. *Journal of Clinical Forensic Medicine* 3:59-71.

Mieczkowski, Tom, Editor. 1999. Drug Testing Technology: Assessment of Field Applications. CRC Press.

Mieczkowski, Tom, and Kim Lersh. 2002. Drug-testing Police Officers and Police Recruits. The Outcome of Urinalysis and Hair Analysis Compared. *Policing: An International Journal of Police Strategies and Management* 25(3):581-601.

Mieczkowski, Tom, and R. Newel. 2000. An Analysis of the Racial Bias Controversy in the Use of Hair Assays. In *Drug Testing Technologies: Field Applications and Assessments*. T. Mieczkowski, Editor. Boca Raton, FL: CRC Press. Pp. 313-348.

Mikkelsen, Stephen and K. Owen Ash. 1988. Adulterants Causing False Negatives in Illicit Drug Testing. *Clinical Chemistry* 34(11):2333-2336.

Miller, E.I., F.M. Wylie, and J.S. Oliver. 2008. Simultaneous Detection and Quantification of Amphetamines, Diazepam and its metabolites, Cocaine and its metabolites, and Opiates in Hair by LC-ESI-MS-MS Using a Single Extraction Method. *Journal of Analytical Toxicology* 32(7):457-469.

Miller, Ted R., Eduard Zaloshnja, Rebecca S. Spicer. 2007. Effectiveness and Benefit-Cost of Peer-Based Workplace Substance Abuse Prevention Coupled with Random Testing. *Accident Analysis and Prevention* 39:565-573.

Milman, Garry, Allan J. Barnes, David M. Schwope, Eugene W. Schwilke, William D. Darwin, Robert S. Goodwin, Deanna L. Kelly, David A. Gorelick, and Marilyn A. Huestis. 2010. Disposition of Cannabinoids in Oral fluids after Controlled Around-the-Clock Oral THC Administration. *Clinical Chemistry* 56:1261-1269.

Milman, Garry, Allan J. Barnes, R.H. Lowe, and M.A. Huestis. 2010. Simultaneous Quantification of Cannabinoids and Metabolites in Oral fluids by Two-Dimensional Gas Chromatography Mass Spectrometry. *Journal of Chromatography A* 1217(9):1513-1521.

Mitchell, John M., and Francis M. Esposito. 2007. Quality Practices in Workplace Testing. In *Drug Abuse Handbook,* 2nd Edition. Steven B. Karch, Editor. Pp. 864-878. Roca Baton, FL: CRC Press.

Miyaguchi, Hajime, Hiroko Takahashi, Toshinori Ohashi, Kazuma Mawatari, Yuko T. Iwata, Hiroyuki Inoue and Takehiko Kitamori. 2009. Rapid Analysis of Methaphetamine in Hair by Micropulverized Extraction and Microchip-based Competitive ELISA. *Forensic Science International* 184(1-3):1-5.

Moeller, Karen, Kellly Lee, and Julie Kissack. 2008. Urine Drug Screening: Practical Guide for Clinicians. *Mayo Clinic Proceedings* 83(1):66-76.

Moeller, Manfred R. 1996. Hair Analysis as Evidence in Forensic Cases. *Therapeutic Drug Monitoring* 18(4):444-449.

Moeller, Manfred R., and Thomas Kraemer. 2002. Drugs of Abuse Monitoring in Blood for Control of Driving Under the Influence of Drugs. *Therapeutic Drug Monitoring* 24(2):210-221.

Moller, M., K. Aleska, P. Walasek, T. Karaskov, and G. Koren. 2010. Solid-Phase Microextraction for the Detection of Codeine, Morphine, and 6-Monoacetylmorphine in Human Hair by Gas Chromatography-Mass Spectrometry. *Forensic Science International* 196:64-69.

Moore, C., V. Barnes, J. Hauth, R. Wilson, J. Fawcett-Long, J. Toquam, K. Baker, D. Wieringa, J. Olson, J. Christensen. 1989. *Fitness for Duty in the Nuclear Power Industry: A Review of Technical Issues. Supplement 1.* Seattle, WA: Battelle Human Affairs Research Centers and the Pacific Northwest Laboratory. NUREG/CR-5227 Supplement 1.

Moore, Christine. 2010. Oral fluids and Hair in Workplace Drug Testing Programs: New Technology for Immunoassays. *Drug Testing and Analysis.*

Moore, Christine. 2009. Drugs of Abuse in Oral fluids. In *Handbook of Workplace Drug Testing, 2nd Edition.* Jeri D. Ropero-Miller and Bruce A. Goldberger, Editors. Washington, DC: American Association for Clinical Chemistry.

Moore, Christine, Cynthia Coulter, and Katherine Crompton. 2007. Achieving Proposed Federal Concentrations Using Reduced Specimen Volume for the Extraction of Amphetamines from Oral fluids. *Journal of Analytical Toxicology* 31:442-446.

Moore, Christine, Cynthia Coulter, Katherine Crompton, and Michael Zumwalt. 2007. Technical Note: Determination of Benzodiazepines in Oral fluids Using LC-MS-MS. *Journal of Analytical Toxicology* 31(9):596-600.

Moore, C., M. Vincent, S. Rana, C. Coulter, A. Agrawal, J. Soares. 2006. Stability of Delta-9-Tetrahdrocannabinol (THC) in Oral fluids using the Quantisal™ Collection Device. *Forensic Science International* 164:126-130.

Moore, Christine, Cynthia Coulter, Sumandeep Rana, Michael Vincent, and James Soares. 2006. Analytical Procedure for the Determination of the Marijuana Metabolite 11-Nor-Δ^9-Tetrahydrocannabinol-9-Carboxylic Acid in Oral fluids Specimens. *Journal of Analytical Toxicology* 30:409-412.

Morini, Luca, Alessandra Zucchella, Aldo Polettini, Lucia Politi, and Angelo Groppi. 2010. Effect of Bleaching on Ethyl Glucuronide in Hair: An *In Vitro* Experiment. *Forensic Science International* 198:23-27.

Morrison, Janet, Lorna Sniegoski, and Wesley Yoo. 1998. Evaluation of Analytical Methodologies for Non-Intrusive Drug Testing: Supercritical Fluid Extraction of Cocaine from Hair. NIJ Report 601-98. Washington, DC: National Institute of Standards and Technology.

Moskowitz, Herbert, and Dary Fiorentino. 2000. A Review of the Literature on the Effects of Low Doses of Alcohol on Driving-Related Skills. Prepared for the U.S. Department of Transportation National Highway Traffic Safety Administration. Washington, DC. Accessed 05.2009 at http://www.nhtsa.dot.gov/people/injury/research/pub/Hs809028/Title.htm.

Mullangi, Ramesh, Shrutidevi Agrawal, and Nuggehally R. Srinivas. 2009. Measurement of Xenobiotics in Saliva: Is Saliva an Attractive Alternative Matrix? Case Studies and Analytical Perspectives. *Biomedical Chromatography* 23:3-25.

Mueller, Melanie, Frank Peters, Marilyn Huestis, George Ricaurte, and Hans Maurer. 2009. Simultaneous Liquid Chromatographic-Electrospray Ionization Mass Spectrometric Quantification of 3.4-Methylenedioxymethamphetamine (MDMA, Ecstasy) and its Metabolites 3.4-Dihydroxymethamphetamine, 4-Hydoxy-3-Methoxymethamphetamine and 3,4-Methylenedioxyamphetamine in Squirrel, Monkey and Human Plasma After Acidic Conjugate Cleavage. *Forensic Science International* 184:64-68.

Mura, Patrick, Pascal Kintz, Véronique Dumestre, Sébastien Raul, and Thierry Hauet. 2005. THC Can Be Detected in Brain while Absent in Blood. Letter to the Editor. *Journal of Analytical Toxicology* 29 (November/December).

Musshoff, Frank, and Burkhard Madea. 2007a. Analytical Pitfalls in Hair Testing. *Analytical and Bioanalytical Chemistry* 388:1475-1494.

Musshoff, Frank, and Burkhard Madea. 2007b. New Trends in Hair Analysis and Scientific Demands on Validation and Technical Notes. *Forensic Science International* 165(203):204-215.

Myrick, Hugh, and Tara Wright. 2008. Clinical Management of Alcohol Abuse and Dependence. Chapter 9 in *The American Psychiatric Publishing Textbook of Substance Abuse Treatment*, 4th Edition. Marc Galanter and Herbert D. Kleber, Editors. Arlington, VA: American Psychiatric Publishers. Pp. 129-142.

Nadulski, Thomas, Frank Sporkert, Martin Schnelle, Andreas Michael Stadelmann, Patrik Roser, Tom Schefter, and Fritz Pragst. 2005. Simultaneous and Sensitive Analysis of THC, 11-OH-THC, THC-COOH, CBD, and CBN by GC-MS in Plasma after Oral Application of Small Doses of THC and Cannabis Extract. *Journal of Analytical Toxicology* 29(8):782-789.

Nakahara, Yuji. 1999. Hair Analysis for Abused and Therapeutic Drugs. *Journal of Chromatography* 733:161-180.

Nakahara, Y., K. Takahashi, and R. Kikura. 1995. Hair Analysis for Drugs of Abuse: Effect of Physicochemical Properties of Drugs on the Incorporation Rates into Hair. *Biological and Pharmaceutical Bulletin* 18(9):1223-1227.

National Institutes of Health (NIH), National Institute of Diabetes and Digestive and Kidney Diseases. 2009. The Kidneys and How They Work. NIH Publication No. 09-3195. Washington, DC. Accessed 11.2010 at http://kidney.niddk.nih.gov/Kudiseases/pubs/yourkidneys.

Navarro, M., S. Pichini, M. Farre, J. Ortuno, P.N. Roset, J. Segura, and R. de La Torre. 2001. Usefulness of Saliva for Measurement of 2,4-Methylenedioxymethamphetamine and its Metabolites: Correlation with Plasma Drug Concentrations and Effect of Salivary pH. *Clinical Chemistry* 47:1788-1795.

Nichols, James H., Robert H. Christenson, William Clarke, Ann Gronowski, Catherine A. Hammett-Stabler, Ellis Jacobs, Steve Kazmierczak, Kent Lewandrowski, Christopher Price, David B. Sacks, Robert L. Sautter, Gregg Shipp, Lori Sokoll, Ian D. Watson, William Winter, and Marcia L. Zucker. 2007. Executive Summary. The National Academy of Clinical Biochemistry Laboratory Medicine Practice Guideline: Evidence-based Practice for Point-of-Care Testing. *Clinica Chimica Acta* 379(1-2): 14-28.

Nielsen, Marie K., Sys S. Johansen, Petur W. Dalsgaard, and Kristian Linnet. 2010. Simultaneous Screening and Quantification of 52 Common Pharmeceuticals and Drugs of Abuse in Hair Using UPL-TOF-MS. *Forensic Science International* 196:85-92.

Nieuw Amerongen, Arie Van, Antoon J.M. Ligtenberg, and Enno C.I. Veerman. 2007. Implications for Diagnostics in the Biochemistry and Physiology of Saliva. *Annals of the New York Academy of Sciences* 1098:1-6.

Nolan, Susan. 2008. Drug-Free Workplace Programmes: New Zealand Perspective. *Forensic Science International* 174:125-132.

Normand, Jacques. 1994. *Under the Influence? Drugs and the American WORKFORCE.* Washington, DC: National Academy Press.

Øiestad, Elisabeth Leere, Unni Johansen, and Asbjørg Solberg Christophersen. 2007. Drug Screening of Preserved Oral fluids by Liquid Chromatography-Tandem Mass Spectrometry. *Clinical Chemistry* 53(2):300-309.

Olof, Beck, Kathinka Leine, Göran Palmskog, and Johan Franck. 2010. Amphetamines Detected in Exhaled Breath from Drug Addicts: A New Possible Method for Drugs-of-Abuse Testing. *Journal of Analytical Toxicology* 34(5):233-237.

Öisjöen, F., J.f. Schneiderman, A.P. Astalan, A. Kalabukhov, and C. Johansson, D. Winkler. 2010. The Need for Stable, Mono-Dispersed, and Biofunctional Magnetic Nanoparticles for One-Step Magnet Immunoassays. *Journal of Physics: Conference Series* 200:122006.

Ojanperä, I., and Ilpo Rasanen. 2008. Forensic Screening by Gas Chromatography. In *Forensic Science Handbook of Analytical Separations, Vol 6.* M.J. Bogusz, Editor. Amsterdam: Elsevier, B.V. Pp. 403-424.

Ojanperä, I., K. Goebel, and E Vuori. 1999. Toxicological Drug Screening by Overpressured Layer Chromatography. *Journal of Liquid Chromatography and Related Technologies* 22(1):161-171.

The Omnibus Transportation Employee Testing Act of 1991. Public Law 102-143. 105 STAT. 917. 28 October 1991.

Online Discussion Forum. Ongoing. Moderated by Robert Swotinsky, MD. Drug and Alcohol Q&As from 2004-2009. Accessed 04.2009 at http://www.occ-doc.net/MRO_information.htm.

Outang, Zheng, and R. Graham Cooks. 2009. Miniature Mass Spectrometers. *Annual Review of Analytical Chemistry* 2:187-214.

Palfrey, Stephen H. 1999. Confirmation of the Presence of Drugs of Abuse in Urine. In *Method in Molecular Medicine. Vol 27. Clinical Applications of Capillary Electrophoresis.* S.M. Palfrey, Editor. Totowa, NJ: Humana Press, Inc.

Dickson S., A. Park, S. Nolan, S. Kenworthy, C. Nicholson, J. Midgley, R. Pinfold, S. Hampton. 2007. The Recovery of Illicit Drugs from Oral fluids Sampling Devices. *Forensic Science International* 165(1):78-84.

Patentstorm US. 2010. Website. Screening Test and Procedure Using Skin Patches. Accessed on 12.2010 at: http://www.patentstorm.us/patents/6585646/description.html.

Paterson, Sue, Sooyeun Lee, and Rosa Cordero. 2010. Analysis of Hair after Contamination with Blood Containing Cocaine and Blood Containing Benzoylecgonine. *Forensic Science International* 194:94-96.

Patient UK website. Salivary Glands. Accessed 1.2011 at http://www.patient.co.uk/diagram/Salivary-Glands.htm. [Picture]

Paul, B.D., and C.S. Dunkley. 2007. Specimen Validity Testing (SVT) – Effects of Oxidizing Agents on Drugs in Urine and Procedures for Detection. *Forensic Science Review* 19:29-47.

Peace, Michelle R., Justin L. Polkis, Lisa D. Tarnai, and Alphonse Poklis. 2002. An Evaluation of the OnTrak Testcup®-er On-Site Urine Testing Device for Drugs Commonly Encountered in Emergency Departments. *Journal of Analytical Toxicology* 26:500-503.

Peace, Michelle R., and Lisa D. Tarnai. 2002. Performance Evaluation of Three On-Site Adulterant Detection Devices for Urine Specimens. *Journal of Analytical Toxicology* 26(7):464-470.

Peace, Michelle R., Lisa D. Tarnai, and Alphonse Poklis. 2000. Performance Evaluation of Four On-Site Drug-Testing Devices for Detection of Drugs of Abuse in Urine. *Journal of Analytical Toxicology* 24(7):589-594.

Peelander, Anna, Johanna Ristimaa, Ilpo Rasanen, Erikki Vuori, and Ikka Ojanperä. 2008. Screening for Basic Drugs in Hair of Drug Addicts by Liquid Chromatography/Time-of-Flight Mass Spectrometry. *Therapeutic Drug Monitoring* 30(6):717-724.

Pehrsson, Anna, Tom Blencowe, Kari Vimpari, Kaarina Langel, Charlotta Engblom, and Pirjo Lillsunde. 2011. An Evaluation of On-Site Oral Fluid Drug Screening Devices DrugWipe® 5+ and Rapid STAT® Using Oral Fluid for Confirmation Analyusis. *Journal of Analytical Toxicology* 35(4):211-218.

Pehrsson, Anna. 2008. Roadside Oral fluids Testing: Comparison of the Results of Drugwipe 5 and Drugwipe Benzodiazepines On-Site Tests with Laboratory Confirmation Results of Oral fluids and Whole Blood. *Forensic Science International* 175:140-148.

Penders, Joris, and Alain Verstraete. 2006. Laboratory Guidelines and Standards in Clinical and Forensic Toxicology. *Accreditation and Quality Assurance* 11:284-290.

Peters, Frank T., Nele Samyn, Thomas Kraemer, Wim J. Riedel, and Hans H. Maurer. 2007. Negative-Ion Chemical Ionization Gas Chromatography–Mass Spectrometry Assay for Enantioselective Measurement of Amphetamines in Oral fluids: Application to a Controlled Study with MDMA and Driving Under the Influence Cases. *Clinical Chemistry* 53:702-710.

Peters, Frank T., Thomas Kraemer, and Hans H. Maurer. 2002. Drug Testing in Blood: Validated Negative-Ion Chemical Ionization Gas Chromatographic-Mass Spectrometric Assay for Determination of Amphetamine and Methamphetamine Enantiomers and Its Application to Toxicology Cases. *Clinical Chemistry* 48(9):1472-1485.

Phillips, Jane Ellen, Stuart Bogema, Paul Fu, Wieslaw Furmaga, Alan H. B. Wu, Vlasta Zic, and Catherine Hammett-Stabler,. 2003. Signify® ER Drug Screen Test Evaluation: Comparison to Triage(R) Drug of Abuse Panel Plus Tricyclic Antidepressants. *Clinica Chimica Acta* 328(1-2):31-38.

Phillips.com. Website. Magnotech: Philips' Magnetic Biosensor Platform Designed for Point-of-Care Testing. Accessed 11.2010 at http://www.newscenter.phillips.com/main/standard/news/backgrounders/2010/20100107_magnetic_biosensor.wpd.

Phinney, Karen W., and Lane C. Sander. 2004. Liquid chromatographic Method for the Determination of Enantiomeric Composition of Amphetamine and Methamphetamine in Hair Samples. *Analytical and Bioanalytical Chemistry* 378(1):144-149.

Phipps, Rebecca J., Jessica J. Smith, William D. Darwin, and Edward J. Cone. 2008. Current Methods for the Separation and Analysis of Cocaine Anaytes. In *Forensic Science Handbook of Analytical Separations, Vol 6.* M.J. Bogusz, Editor. Amsterdam: Elsevier, B.V. Pp. 73-126.

Physorg.com. Website. 2011. New Lab-on-a-Chip Technology Could Improve Accuracy of Lab Tests, Deliver Results Sooner. Accessed 02.2011 at http://www.physorg.com/news/2011-02-lab-on-a-chip-technology-accuracy-lab-results.html.

Pierce, Anya. 2007. Workplace Drug Testing Outside the U.S. In *Drug Abuse Handbook, 2nd Edition.* Steven B. Karch, Editor. Boca Raton, FL: CRC Press. Pp. 765-775.

Pil, K., Francis Esposito, and A. Verstraete. 2010. External Quality Assessment of Multi-Analyte Chromatographic Methods in Oral fluids. *Clinica Chimica Acta* 411:1041-1045.

Pil, K., and A. Verstraete. 2008. Current Developments in Drug Testing in Oral fluids. *Therapeutic Drug Monitoring* 30(2):196-202.

Polettini, A., A. Groppi, and M. Montagna. 1993. Rapid and Highly Selective GC/MS/MS Detection of Heroin and its Metabolites in Hair. *Forensic Science International* 63(1-3):217-225.

Polla, M., C. Stramesi, S. Pichini, I. Palmi, C. Vignali, and G. Dall'Olio. 2009. Hair Testing is Superior to Urine to Disclose Cocaine Consumption in Driver's License Regranting. *Forensic Science International* 189(1-3):e41-e43.

Pötsch, L., and G. Skopp. 1996. Stability of Opiates in Hair Fibers after Exposure to Cosmetic Treatment. *Forensic Science International* 81(2-3):95-102.

Pragst, Fritz, Hans Sachs, and Pascal Kintz. 2010. Hair Analysis for Cocaine Continues to be a Valuable Tool in Forensic and Clinical Toxicology. Letter to the Editor. *Journal of Analytical Toxicology* 34(July/August):354-355.

Pragst, Fritz. 2008. High Performance Liquid Chromatography in Forensic Toxicological Analysis. In *Forensic Science Handbook of Analytical Separations, Vol 6*. M.J. Bogusz, Editor. Amsterdam: Elsevier, B.V. Pp. 447-489.

Pragst, Fritz, and Marie A. Balikova. 2006. State of the Art in Hair Analysis for Detection of Drug and Alcohol Abuse. *Clinica Chimica Acta* 370(1-2):17-49.

Presley, Lance, Michael Lehrer, William Seiter, Dawn Hahn, et al. 2003. High Prevalence of 6-acetylmorphine in Morphine-Positive Oral fluids Specimens. *Forensic Science International* 133:22-25.

Price, Christopher. 2001. Microarrays: The Reincarnation of Multiplexing in Laboratory Medicine, But Now More Relevant? *Clinical Chemistry* 47:1345-1346.

Products. 2007. *LC-GC North America* 25(7):667-672.

Pujadas, Mitona, Simona Pichini, Ester Civit, Elena Santamariña, Katherine Perez, and Rafael de la Torre. 2007. A Simple and Reliable Procedure for the Determination of Psychoactive Drugs in Oral fluids by Gas Chromatography-Mass Spectrometry. *Journal of Pharmaceutical and Biomedical Analysis* 44:594-601.

Pujol, Marie-Laure, Vincent Cirimele, Pierre Julien Tritsch, Marion Villain, Pascal Kintz. 2007. Evaluation of the IDS One-Step[TM] ELISA Kits for the Detection of Illicit Drugs in Hair. *Forensic Science International* 170:189-192.

Quest Diagnostics, Incorporated. 2011. *Drug Testing Index*. Accessed 02.2011 at http://www.questdiagnostics.com/employersolutions/dti/2011_01/dti_index.html.

Quest Diagnostics, Incorporated. 2009. *News from Quest Diagnostics*. Accessed 02.2011 at http://www.questdiagnostics.com/employersolutions/dti/2011_01/dti_index.html.

Quest Diagnostics, Incorporated. 2002. Intercept® Questions and Answers. Accessed 10.2010 at http://www.questdiagnostics.com/employersolutions/files/intercept_faqs.pdf.

Quintela, Oscar, David M. Andrenyak, Archie M. Hoggan, and Dennis J. Crouch. 2007. A Validated Method for the Detection of Δ^9-Tetrahydrocannabinol and 11-nor-Δ^9-Tetrahydrocannabinol-9-Carboxylic Acid in Oral fluids Samples by Liquid Chromatography Coupled with Quadrupole-Time-of-Flight Mass Spectrometry. *Journal of Analytical Toxicology* 31(3):157-164.

Quintela, Oscar, Dennis J. Crouch, and David M. Andrenyak. 2006. Recovery of Drugs of Abuse from the Immunalysis Quantisal ™ Oral fluids Collection Device. *Journal of Analytical Toxicology* 30:614-616.

Quintela, Oscar, David M. Andrenyak, Archie M. Hoggan, and Dennis J. Crouch. 2007. A Validated Method for the Detection of Δ^9-Tetrahydrocannabinol and 11-Nor-9-Carboxy- Δ^9-Tetrahydrocannabinol in Oral fluids Samples by Liquid Chromatography Coupled with Quadrupole-Time-of-Flight Mass Spectrometry. *Journal of Analytical Toxicology* 31:157-164.

Radjenovic, Jelena, Mira Petrovic, and Damiá Barceló. 2007. Analysis of Pharmaceuticals in Wastewater and Removal Using a Membrane Bioreactor. *Analytical and Bioanalytical Chemistry* 387:1365-1377.

Raes, Elke, and Alain G. Verstraete. 2005. Usefulness of Roadside Urine Drug Screening in Drivers Suspected of Driving Under the Influence of Drugs (DUID). *Journal of Analytical Toxicology* 29:632-636.

Raes, Elke, Alain G. Verstraete, and Robert Wennig. 2008. Drugs and Driving. In *Forensic Science Handbook of Analytical Separations, Vol 6*. M.J. Bogusz, Editor. Amsterdam: Elsevier, B.V. Pp. 611-651.

Rai, Balwant. 2007. Oral fluids in Toxicology. *Internet Journal of Toxicology* 3(7).

Ramaekers, J.G., M.R. Moeller, P. van Ruitenbeck, E.L. Theunissen, E. Schneider, and G. Kauert. 2006. Cognition and Motor Control as a Function of Δ^9-THC Concentration in Serum and Oral fluids: Limits of Impairment. *Drug and Alcohol Dependence* 85:114-122.

Rapaka, Rao S., Nora Chiang, and Billy R. Martin. 1997. Pharmacokinetics, Metabolism, and Pharmaceutics of Drugs of Abuse. NIDA Research Monograph 173. Rockville, MD: U.S. Department of Health and Human Services.

Regional Laboratory for Toxicology. 2007. Drugs of Abuse Guidelines. Accessed 06.2009 at http://www.toxlab.co.uk/dasguide.htm.

Reynolds, Lawrence A. 2005. Historical Aspects of Drugs-of-Abuse Testing in the United States. In *Drugs of Abuse: Body Fluid Testing*. Raphael C. Wong and Harley Y Tse, Editors. Totowa, NJ: Humana Press. Pp. 1-10.

Rivier L. 2000. Techniques for Analytical Testing of Unconventional Samples. *Bailliere's Clinical Endocrinology and Metabolism* 14:147-165.

Robinson, Jerome J., and James W. Jones. 2000. Drug Testing in a Drug Court Environment: Common Issues to Address. Drug Courts Resource Series. U.S. Department of Justice Office of Justice Program.

Rocío, M., A. Carrera, Michael M. Meijler, and Kim D. Janda. 2004. Cocaine Pharmacology and Current Pharmacotherapies for its Abuse. *Bioorganic and Medicinal Chemistry* 12:5019-5030.

Röhrich, J., I. Schimmel, S. Zörntlein, J. Becker, and S. Drobnik, T. Kaufmann, V. Kuntz, and R. Urban. 2010. Concentrations of Δ^9-Tetrahydrocannabinol and 11-Nor-9-Carboxy- Δ^9-Tetrahydrocannabinol in Blood and Urine After Passive Exposure to Cannabis Smoke in a Coffee Shop. *Journal of Analytical Toxicology* 34:196-203.

Röhrich, J., S. Zörntlein, J. Becker, and R. Urban. 2010. Detection of Δ^9-Tetrahydrocannabinol and Amphetamine-Type Stimulants in Oral fluids Using the Rapid Stat™ Point-of-Collection Drug-Testing Device. *Journal of Analytical Toxicology* 34(3):155-161.

Rollins D.E., G.G. Wilkins, G.G. Krueger, M.P. Augsburger, A. Mizuno, C. O'Neal, C.R. Borges, M.H. Slawson. 2003. The Effect of Hair Color on the Incorporation of Codeine into Human Hair. *Journal of Analytical Toxicology* 27:545-551.

Romano, Barbara H., G. Spadaro, and V. Valenti. 2003. Determination of Drugs of Abuse in Hair: Evaluation of External Heroin Contamination and Risk of False Positives. *Forensic Science International* 131(2-3):98-102.

Romano, Guido, Nunziata Barbera, and Isabella Lombardo. 2001. Hair Testing for Drugs of Abuse: Evaluation of External Cocaine Contamination and Risk of False Positives. *Forensic Science International* 123:119-129.

Ropero-Miller, Jeri D., Bruce A. Goldberger, and Ray H. Liu, Editors. 2009. *Handbook of Workplace Drug Testing*. Washington, DC: American Association for Clinical Chemistry.

Ropero-Miller, Jeri D., and Bruce A. Goldberger, Editors. 2008. Research and Development in Forensic Toxicology. Final Report. Prepared for the U.S. Department of Justice. Research Triangle Park, NC: RTI International.

Rosay, André, Stacy Najaka, and Denise Herz. 2007. Differences in the Validity of Self-Reported Drug Use Across Five Factors: Gender, Race, Age, Type of Drug, and Offense Seriousness. *Journal of Quantitative Criminology* 23:41-58.

Rowland, Malcolm, and Thomas N. Tozer. 2011. *Clinical Pharmacokinetics: Concepts and Applications, 4rd Edition*. Philadelphia: Wolters Kluwer.

Sachs, H. 2000. Place of Sweat in Drugs of Abuse Testing. Institute of Legal Medicine, Munich. Accessed 09.2010 at www.icadts.org/proceedings/2000/icadts2000-171.pdf.

Samyn, Nele, Gert De Boeck, Michelle Wood, Caroline T.J. Lamers, Dick De Waard, Karel A. Brookhuis, Alain G. Verstraete, and Wim J. Riedel. 2002. Plasma, Oral fluids, and Sweat Wipe Ecstasy Concentrations in Real Life Conditions. *Forensic Science International* 128:90-97.

Samyn, N., and C. van Haeren, C. 2000. On-site Testing of Saliva and Sweat with Drugwipe, and Determination of Concentrations of Drugs of Abuse in Saliva, Plasma and Urine of Suspected Users. *International Journal of Legal Medicine* 113(3):150-154.

Samyn, Nele, Marleen Laloup, and Gert De Boeck. 2007. Bioanalytical Procedures for Determination of Drugs of Abuse. *Analytical and Bioanalytical Chemistry* 388:1437-1453.

Samyn, N., A. Verstraete, A., C. van Haeren, and P. Kintz. 1999. Analysis of Drugs of Abuse in Saliva. *Forensic Science Review* 11(1):2–19.

Sasaki, Tania A. 2008. Drug Screening and Confirmation. *Chromatography Techniques*. July/August: 18-21.

Schaffer, Michael I., and Virginia A. Hill. 2005. Hair Analysis in Drugs-of-Abuse Testing. In *Drugs of Abuse: Body Fluid Testing*. Raphael C. Wong and Harley Y Tse, Editors. Totowa, NJ: Humana Press. Pp. 177-200.

Schepers, Raf J.F., Jonathan M. Oyler, Robert E. Joseph, Jr., Edward J. Cone, Eric T. Moolchan, and Marilyn A. Huestis. 2003. Methamphetamine and Amphetamine Pharmacokinetics in Oral fluids and Plasma after Controlled Oral Methamphetamine Administration to Human Volunteers. *Clinical Chemistry* 49(1):121-132.

Schönberg, Lena, Thomas Grobosch, Dagmar Lampe, and Charlotte Kloft. 2007. Toxicological Screening in Urine: Comparison of Two Automated HPLC Screening Systems, Toxicological Identification System (TOX.I.S.) versus REMEDI™-HS. *Journal of Analytical Toxicology* 31(6):321-327.

Schultz, William B. 1997. Regulation of Over-the-Counter Products to Test for Drugs of Abuse. Testimony to the Subcommittee on Oversight and Investigations, House Committee on Commerce. February 6. Accessed 02.2010 at http://www.fda.gov/NewsEvents/Testimony/ucm114964.htm.

Schwettmann, Lutz, Wolf-Rüdiger Külpmann, and Christian Vidal. 2006. Drug Screening in Urine by Cloned Enzyme Donor Immunoassay (CEDIA) and Kinetic Interaction of Microparticles in Solution (KIMS): A Comparative Study. *Clinical Chemistry Laboratory Medicine* 44(4):479-487.

Schwilke, Eugene, Allan Barnes, Sherri Kacinko, Edward Cone, Eric Moolchan, and Marilyn Huestis. 2006. Opioid Disposition in Human Sweat after Controlled Oral Codeine Administration. *Clinical Chemistry* 52(8):1539-1545.

Schwope, David M., Garry Milman, and Marilyn A. Huestis. 2010. Validation of an Enzyme Immunoassay for Detection and Semiquantification of Cannabinoids in Oral fluids 56:1007-1014.

Security Management. 2006. High on the Job: Drug Dealers and Users are More Savvy in Workplaces Today. 1 February. Accessed 10.2010 at http://goliath.ecnext.com/coms2/gi_0199-5254420/High-on-the-job-drug.html.

Segura, Jordi, Rosa Ventura, and Carmen Jurado. 1998. Derivatization Procedures for Gas Chromatographic-Mass Spectrometric Determination of Xenobiotics in Biological Samples, with Special Attention to Drugs of Abuse and Doping Agents. *Journal of Chromatography B* 713:61-90.

Shankaran, Dhesingh R., K. Vengatajalabathy Gobi, and Norio Miura. 2007. Recent Advancements in Surface Plasmon Resonance Immunosensors for Detection of Small Molecules of Biomedical, Food, and Environmental Interest. *Sensors and Actuators* B 121:158-177.

Shen, Jianzhong, Fei Xu, Haiyang Jiang, Zhanhui Wang, Jing Tong, Pengju Guo, and Shuangyang Ding. 2007. Characterization and Application of Quantum Dot Nanocrystal-Monoclonal Antibody Conjugates for the Determination of Sulfamethazine in Milk by Fluorimmunoassay. *Analytical and Bioanalytical Chemistry* 389:2243-2250.

Shuster, Louis. 1991. Pharmacokinetics, Metabolism, and Disposition of Cocaine. In *Cocaine: Pharmacology, Physiology, and Clinical Strategies*. Joan M. Lakoski, Matthew P. Galloway, Francis J. White, Editors. Boca Raton, FL: CRC Press.

Siek, Theodore J. 2003. Specimen Preparation. In *Principles of Forensic Toxicology, 2nd Edition*. Barry Levine, Editor. Washington, DC: AACC Press. Pp. 67-78.

Skopp, G., L. Pötsch, and M.R. Moeller. 1997. On Cosmetically Treated Hair – Aspects and Pitfalls of Interpretation. *Forensic Science International* 84:43-52.

Smink, B.E., M.P.M Mathijssen, K.J. Lusthof, J.J. de Gier, A.C. G. Egberts, and D.R.A. Uges. 2007. Comparison of Urine and Oral fluids as Matrices for Screening of Thirty-Three Benzodiazepines and Benzodiazepine-like Substances using Immunoassay and LC-MS(-MS). *Journal of Analytical Toxicology* 30(7):478-485.

Smith, Frederick P., and David A. Kidwell. 1996. Cocaine in Hair, Saliva, Skin Swabs, and Urine of Cocaine Users' Children. *Forensic Science International* 83:179-189.

Smith, Michael L., Allan Barnes, and Marilyn Huestis. 2009. Identifying New Cannabis Use with Urine Creatinine-Normalized THCCOOH Concentrations and Time Intervals Between Specimen Collections. *Journal of Analytical Toxicology* 33(May):185-189.

Smith, Michael L., Shawn P. Vorce, Justin M. Holler, Eric Shimomura, Joe Magluilo, Aaron J. Jacobs, and Marilyn A. Huestis. 2007. Review: Modern Instrumental Methods in Forensic Toxicology. *Journal of Analytical Toxicology* 31(5):237-253.

Smith, Michael. L. 2003. Immunoassay. In *Principles of Forensic Toxicology, 2nd Edition*. Barry Levine, Editor. Washington, DC: American Association for Clinical Chemistry (AACC) Press.

Sniegoski, L.T., J.L. Pendergast, M.J. Welch, A.A. Fatah, M. Gackstetter, and R. Thompson. 2009. Evaluation of Oral fluids Testing Devices. NIST Interagency/Internal Report (NISTIR) – 7585. April 1.

Society of Forensic Toxicologists. 2006. Forensic Toxicology Laboratory Guidelines: 2006 Version. SOFT/AAFS. Accessed 10.2010 at http://www.searchpdfengine.com/FORENSIC-TOXICOLOGY-LABORATORY-GUIDELINES.html and http://www.soft-tox.org/files/Guidelines_2006_Final.pdf.

Speedy, T., D. Baldwin, G. Jowett, M. Gallina, and A. Jehanli. 2007. Development and Validation of the Cozart® DDS Oral fluids Collection Device. *Forensic Science International* 170:117-120.

Spiehler, Vina, and Gail Cooper. 2008. Drugs-of-Abuse Testing in Saliva or Oral fluids. In *Forensic Science and Medicine: Drug Testing in Alternate Biological Specimens*. A. J. Jenkins and Yale H. Caplan, Editors. Totowa, NJ: Humana Press.

Spiehler, Vina, and Barry Levine. 2003. Pharmacokinetics and Pharmacodynamics. In *Principles of Forensic Toxicology, 2nd Edition*. Barry Levine, Editor. Washington, DC: AACC Press. Pp. 47-63.

Stout, Peter R. 2007. Hair Testing for Drugs – Challenges for Interpretation. *Forensic Science Review* 19:69-84.

Stout, Peter R., Jeri D. Ropero-Miller, Michael R. Baylor, and John M. Mitchell. 2006. External Contamination of Hair with Cocaine: Evaluation of External Cocaine Contamination and Development of Performance-Testing Materials. *Journal of Analytical Toxicology* 30(8):490-500.

Stowell, A.R., A.R. Gainsford, and R.G. Gullberg. 2008. New Zealand's Breath and Blood Alcohol Testing Programs: Further Data Analysis and Forensic Implications. *Forensic Science International* 178:81-92.

Strano-Rossi, Sabina, Francesco Botrè, Ana Maria Bermejo, and Maria Jesús Tabernero. 2009. A Rapid Method for the Extraction, Enantiomeric Separation, and Quantification of Amphetamines in Hair. *Forensic Science International* 193(1-3):95-100.

Strano-Rossi, Sabina, Christiana Colamonici, and Francesco Botrè. 2008. Parallel Analysis of Stimulants in Saliva and Urine by Gas Chromatography/Mass Spectrometry: Perspectives for "In-Competition" Anti-Doping Analysis. 606:217-222.

Substance Abuse Program Administrators Association (SAPAA). Website homepage: http://www.sapaa.com/about/index.htm.

Sutheimer, Craig A., and John T. Cody. 2009. Subversion of Regulated Workplace Drug Testing – Specimen Adulteration and Substitution. In *Handbook of Workplace Drug Testing, 2nd Edition*. Jeri D. Ropero-Miller, Bruce A. Goldberger, and Ray H. Liu, Editors. Washington, DC: American Association for Clinical Chemistry.

Sweathelp.org. Website. Sweat glands. Accessed 08.2010 at www.sweathelp.org/English/Images/sweat_glands.jpg [Picture]

Swartz, Michael E., and Ira S. Krull, Editors. 1997. Analytical Method Development and Validation. New York: Marcel Dekker, Inc. *Pharmacopeial Forum* 31(2):549.

Tachi, Tomoya, Noritada Kaji, Manabu Tokeshi, and Yoshinobu Baba. 2009. Microchip-Based Homogenoeous Immunoassay Using a Cloned Enzyme Donor. *Analytical Sciences* 25:149-151.

Tagliaro, F., F.P. Smith, Z. De Battisti, G. Manetto, and M. Marigo. 1997. Hair Analysis, a Novel Tool in Forensic and Biomedical Sciences: New Chromatographic and Electrophoretic/Electrokinetic Analytical Strategies. *Journal of Chromatography B* 689:261-271.

Tan, Chongxiao, Nenad Gajovic-Eichelmann, Rainer Polzius, Niko Hildebrandt, and Frank Bier. 2010. Direct Detection of Δ^9-Tetrahydrocannabinol in Aqueous Samples Using a Homogeneous Increasing Florescence Immunoassay (HiFi). 2010. *Analytical and Bioanalytical Chemistry* 398:2133-2130.

Technique That Quickly Identifies Bacteria Has Applications in Food Safety, Health Care, and Homeland Security. *Journal of Environmental Health* 69(10):68-68.

Thierauf, Annette, Heike Gnann, Ariane Wohlfarth, Volker Auwärter, Markus G. Perdekamp, Klaus-Juergen Buttler, Friedrich M. Wurst, and Wolfgang Weinmann. 2010. Urine Tested Positive for Ethyl Glucuronide and Ethyl Sulphate After the Consumption of "Non-Alcoholic" Beer. *Forensic Science International* 202:82-85.

Tsai, Jane S.-C. 2005. Immunoassay Technologies for Drugs-of-Abuse Testing. In *Handbook of Forensic Drug Analysis.* Frederick P. Smith, Editor. Burlington, MA: Elsevier Academic Press. Pp. 13-43.

Tsai, Jane S-C., and Grace L. Lin. 2005. Drug-Testing Technologies and Applications. In *Drugs of Abuse: Body Fluid Testing.* Raphael C. Wong and Harley Y Tse, Editors. Totowa, NJ: Humana Press. Pp. 29-69.

Uemura, Naoto, Rajneesh P. Nath, Martha R. Harkey, Gary L. Henderson, John Mendelson, and Reese T. Jones. 2004. Cocaine Levels in Sweat Collection Patches Vary by Location of Patch Placement and Decline Over Time. *Journal of Analytical Toxicology* 28(4):253-259.

Urinary System. Wikipedia, the Free Encyclopedia. Accessed 01.2011 at http://en.wikipedia.org/wiki/Urinary_system. [Picture]

U.S. Department of Defense (DoD), Assistant Secretary of Defense for Health Affairs. 2009. Status of Drug Use in the Department of Defense Personnel. Accessed 12.2010 at www.tricare.mil/ddrp.

U.S. Department of Defense (DoD). 1994. DoD Instruction 1010.16, DoD Technical Procedures for the Military Drug Abuse Testing Program. Issued December 9. Accessed 10.2010 at www.tricare.mil/tma/ddrp.

U.S. Department of Defense (DoD). 1984. Department of Defense Directive 1010.1, Military Personnel Drug Abuse Testing Program. Issued December 28. Reissued in December 9, 1994. Accessed 10.2010 at http://usmilitary.about.com/library/milinfo/dodreg/bldodreg1010-1.htm.

U.S. Department of Defense (DoD). 1974. Department of Defense Instruction 1010.1. April 4.

U.S. Department of Defense (DoD). nd. Military Drug Program Historical Timeline. Accessed 12.2010 at www.tricare.mil/tma/ddrp.

U.S. Department of Health and Human Services (HHS), Food and Drug Administration (FDA). 2001. *Guidance for Industry: Bioanalytical Method Validation.* Washington, DC: FDA.

U.S. Department of Health and Human Services (HHS), Food and Drug Administration (FDA). 2000. Summary Minutes: Meeting of the Clinical Chemistry and Clinical Toxicology Devices Panel. Open Session. November 13-14. Accessed 10.2010 at www.fda.gov/ohrms/dockets/ac/00/minutes/3666m1.doc.

U.S. Department of Health and Human Services (HHS), National Institutes of Health (NIH) website. Picture of Urinary System. Accessed 12.2010 at www.nlm.nih.gov/.../ency/imagepages/9654.htm and http://www.nlm.nih.gov/medlineplus/ency/images/ency/fullsize/9654.jpg [Picture]

U.S. Department of Health and Human Services (HHS), Substance Abuse and Mental Health Services Administration (SAMHSA). 2010. Current List of Laboratories and Instrumented Initial Testing Facilities Which Meet Minimum Standards To Engage in Urine Drug Testing for Federal Agencies. *Federal Register* 75(202):67749-67751.

U.S. Department of Health and Human Services (HHS), Substance Abuse and Mental Health Services Administration (SAMHSA). 2009. Drug Testing Advisory Board June 3, 2009 Transcript. June 2-3. Accessed 10.2009 at http://nac.samhsa.gov/DTAB/Docs/DTAB6.2.09transcript508.pdf.

U.S. Department of Health and Human Services (HHS), Substance Abuse and Mental Health Services Administration (SAMHSA). 2008. Mandatory Guidelines for Federal Workplace Drug Testing Programs. *Federal Register* 73(228):71858-71907. November 25.

U.S. Department of Health and Human Services (HHS), Substance Abuse and Mental Health Services Administration (SAMHSA). 2006. Lab Tests for Alcohol Abuse: SAMHSA Advisory. *SAMHSA News* 14(6). Accessed 12.2020 at http://www.samhsa.gov/samhsa_news/volumexiv_6/article5.htm.

U.S. Department of Health and Human Services (HHS), Substance Abuse and Mental Health Services Administration (SAMHSA). 2005. Testimony by Robert L. Stephenson, II, Director, Division Workplace Programs, Center For Substance Abuse Prevention, Substance Abuse and Mental Health Services Administration, Department of Health and Human Services before the Subcommittee on Oversight and Investigations, Committee on Energy and Commerce, U. S. House of Representatives, May 17, 2005. Serial No. 109-47, Pp 16-23.

U.S. Department of Health and Human Services (HHS), Substance Abuse and Mental Health Services Administration (SAMHSA). 2004. Proposed Revisions to Mandatory Guidelines for Federal Workplace Drug Testing Programs. *Federal Register* 69(71):19673-19732. April 13.

U.S. Department of Health and Human Services (HHS), Substance Abuse and Mental Health Services Administration (SAMHSA), National Laboratory Certification Program (NLCP). 2000. NCLP: State of the Science – Update # 1. Washington, DC.

U.S. Department of Health and Human Services (HHS), Substance Abuse and Mental Health Services Administration (SAMHSA). 1998. Attachments to Testimony by Joseph H. Autry III, M.D., Acting Deputy Director, Center for Substance Abuse Prevention, Substance Abuse and Mental Health Services Administration, Department of Health and Human Services before the Subcommittee on Oversight and Investigations, Committee on Commerce, U.S. House of Representatives, July 23, 1998. http://workplace.samhsa.gov/DrugTesting/Files_Drug_Testing/Notices_Docs_Resources/Archives/Hearing/Descriptive%20Statements%20for%20Factors%20in%20Table.html.

U.S. Department of Justice, Drug Enforcement Agency (DEA). 2005. *Drugs of Abuse*. Accessed 06.2009 at http://www.usdoj.gov/dea/pubs/abise/1-csa.htm.

U.S. Department of Transportation (DOT). 2010. Procedures for Transportation Workplace Drug and Alcohol Testing Programs. 49 CFR Part 40. *Federal Register* 75(157):49850-49864. August 16.

U.S. Department of Transportation, Federal Transit Administration (FTA). 2010. Random Test Results Remain Stable. *FTA Drug and Alcohol Regulation Updates* 42(Summer):3.

U.S. Department of Transportation, Office of the Secretary. 2008. 49 CFR Part 40 Procedures for Transportation Workplace Drug and Alcohol Testing Programs. Final Rule. *Federal Register*.

U.S. Department of Transportation, Office of the Secretary. 2008. 49 CFR Part 40 Procedures for Transportation Workplace Drug and Alcohol Testing Programs. Final Rule. *Federal Register* 73(123):35961-35975.

U.S. Department of Transportation (DOT). 1988. Interim Final Rule. *Federal Register* 53:47002. November 21.

U.S. Department of Transportation, Office of the Secretary. Nd. Best Practices for DOT Random Drug and Alcohol Testing. Accessed 04.2009 at www.dot.gov/ost/dapc/testingpubs/final_random_brochure.pdf.

U.S. Food and Drug Administration (FDA). 2002 (Dec). Premarket Submission and Labeling Recommendations for Drugs of Abuse Screening Tests. Accessed 04.2009 at http://www.fda.gov/cdrh/oivd/guidance/152.pdf.

U.S. Nuclear Regulatory Commission (NRC). 2008. 10 CFR Part 26 Fitness for Duty Programs; Final Rule. *Federal Register* 73(62):16966-17235.

University of Arizona Department of Chemistry. ND Mass Spectrometry. Accessed 06.2009 at http://www.chem.arizona.edu/massspec/intro_html/intro.html.

Van Eeckhaut, Ann, Katrien Lanckmans, Sophie Sarre, Ilse Smolders, and Yvette Michotte. 2009. Validation of Bioanalytical LC-MS/MS Assays: Evaluation of Matrix Effects. *Journal of Chromatography B* 877:2198-2207.

Van Kasteel, Marcel. 2009. Nanoparticle Technologies for Patient-Centered Diagnosis. *In Vitro Diagnostic Technology*. April. Accessed 06.2009 at http://www.devicelink.com/ivdt/archive/09/04/008.html.

Van Nuijs, A.L., B. Pecceu, L. Theunis, N. Dubois, C. Charlier, P.G. Jorens, L. Bervoets, R. Blust, H. Meulemans, H. Neels, and A. Covaci. 2009. Can Cocaine Use be Evaluated Through Analysis of Wastewater? A Nation-Wide Approach Conducted in Belgium. *Addiction* 104(5):734-741.

Vearrier, D., J.A. Curtis, and M.I. Greenberg. 2010. Biological Testing for Drugs of Abuse. *Molecular, Clinical, and Environmental Toxicology, Experientia Supplementum* 100:489-517.

Ventura, Montse, Simona Pichini, Rosa Ventura, Sonia Leal, Piergiorgio Zaccaro, Roberta Pacifici, and Rafael de la Torre. 2009. Stability of Drugs of Abuse in Oral fluids Collection Devices with Propose of External Quality Assessment Schemes. *Therapeutic Drug Monitoring* 31(2):277-280.

Ventura, Montse, Rosa Ventura, Simona Pichini, Sonia Leal, Piergiorgio Zaccaro, Roberta Pacifici, K. Langohr, and Rafael de la Torre. 2008. ORALVEQ: External Quality Assessment Scheme of Drugs of Abuse in Oral fluids Results Obtained in the First Round Performed in 2007. *Forensic Science International* 182:35-40.

Verstraete, Alain. 2005a. Oral fluids Testing for Driving Under the Influence of Drugs: History, Recent Progress and Remaining Challenges. *Forensic Science International* 150:143-150.

Verstraete, Alain. 2005b. The Results of the Roadside Drug Testing Assessment Project. In *Drugs of Abuse: Body Fluid Testing*. Raphael C. Wong and Harley Y Tse, Editors. Totowa, NJ: Humana Press. Pp. 271-292.

Verstraete, Alain G. 2004. Detection Times of Drugs of Abuse in Blood, Urine, and Oral fluids. *Therapeutic Drug Monitoring* 26(2):200-205.

Verstraete, Alain, and Puddu. 2000. Evaluation of Different Roadside Drug Tests (ROSITA): A European Union Project on Roadside Drug Testing. European Commission. Accessed 09.2010 at http://www.icadts.org/proceedings/2000/icadts2000-174.pdf.

Verstraete, Alain G., and E. Raes. 2006. *ROSITA-2 Project Final Report*. Ghent, Belgium: University of Ghent.

Verstraete, Alain, and J. Michael Walsh. 2007. Point of Collection Testing of Alternative Specimens (Other than Urine). In *Drug Abuse Handbook,* 2nd Edition. Boca Raton, FL: CRC Press. Pp. 898-908.

Villain, Marion, Jean-François Muller, and Pascal Kintz. 2010. Heroin Markers in Hair of a Narcotic Police Officer: Active or Passive Exposure? *Forensic Science International* 196:128-129.

Vogl, W., Editor. 1996. *Urine Specimen Collection Handbook for Federal Workplace Drug Testing Programs.* Rockville, MD: Center for Substance Abuse Prevention, Substance Abuse and Mental Health Services Administration, US Department of Health and Human Services; DHHS publication (SMA) 96-3114.

Walsh, Diana Chapman, Lynn Elison, and Lawrence Gostin. 1992. Worksite Drug Testing. *Annual Review of Public Health* 13:197-221.

Walsh, J. Michael. 2008. New Technology and New Initiatives in U.S. Workplace Testing. *Forensic Science International* 174 (2008):120-124.

Walsh, J. Michael, Dennis J. Crouch, Jonathan P. Danaceau, Leo Canigianelli, Laura Liddicoat, and Randy Adkins. 2007. Technical Note: Evaluation of Ten Oral fluids Point-of-Collection Drug-Testing Devices. *Journal of Analytical Technology* 31(1):44-54.

Wang, W.L. and E.J. Cone. 1995. Testing Human Hair for Drugs of Abuse: Environmental Cocaine Contamination and Washing Effects. *Forensic Science International* 70:39-51.

Watling, Christopher, Gavan R. Palk, James E. Freeman, and Jeremy D. Davey. 2010. Applying Stafford and Warr's Reconceptualization of Deterrence Theory to Drug Driving: Can It Predict Those Likely to Offend? *Accident Analysis and Prevention* 42(2):452-458.

Weis, Brenda K., David Balshaw, John R. Barr,David Brown, Mark Ellisman, Paul Lioy, Gilbert Omenn, Gilbert, et al. 2005. Personalized Exposure Assessment: Promising Approaches for Human Environmental Health Research. *Environmental Health Perspectives* 113(7): 840-848.

Welch, Michael, J., Lorna Sniegoski, and Susan Tai. 2003. Two New Standard Reference Materials for the Determination of Drugs of Abuse in Human Hair. *Analytical and Bioanalytical Chemistry* 376:1205-1211.

Welsh, Eric, J. Jacob Snyder, Kevin L. Klette. 2009. Stabilization of Urinary THC Solutions with a Simple Non-Ionic Surfactant. *Journal of Analytical Toxicology* 33(1):51-55.

Wennig, R. 2000. Potential Problems with the Interpretation of Hair Analysis Results. *Forensic Science International* 107(1-3):5-12.

Westermeyer, Joseph, Ilhan Yargic, and Paul Thuras. 2004. Michigan Assessment-Screening Test for Alcohol and Drugs (MAST/AD): Evaluation in a Clinical Sample. *The American Journal on Addictions* 13:151-162.

Wille, Sarah M.R., Elke Raes, Pirjo Lillsunde, Teemu Gunnar, Marleen Laloup, Nele Samyn, Asbjørb S. Shristophersen, Manfred R. Moeller, Karin P. Hammer, and Alain G. Verstraete. 2009. Relationship Between Oral fluids and Blood Concentrations of Drugs of Abuse in Drivers Suspected of Driving Under the Influence of Drugs. *Therapeutic Drug Monitoring* 31(4):511-519.

Wille, Sarah M.R., Nele Samyna, Maria del Mar Ramírez-Fernándeza, and Gert De Boeck. 2010. Evaluation of On-site Oral fluids Screening Using Drugwipe-5+®, RapidSTAT®, and Drug Test 5000® for the Detection of Drugs of Abuse in Drivers. *Forensic Science International* 198(1-3):2-6.

Wilson, I.D., 2004. Drugs. In *Chromatography, 6th Edition.* Erich Heftmann, Editor. Amsterdam: Elsevier B.V. Pp. 945-986.

Wilson, Lisa, Ahmed Jehanli, Chris Hand, Gail Cooper, and Robert Smith. 2007. Evaluation of a Rapid Oral fluids Point-of-Care Test for MDMA. *Journal of Analytical Toxicology* 31:98-104.

Winter, Michael E. 2010. *Basic Clinical Pharmacokinetics,* 5th *Edition.* Washington, DC: American Association for Clinical Chemistry.

Wisdom, G. Brian. 1976. Enzyme-Immunoassay. *Clinical Chemistry* 22(8):1243-1255.

Wolff, Kim. 2006. Biological Marker of Drug Use. *Psychiatry* 5(12):440-441.

Wong, Raphael, C. 2005. Trends in Drug Testing: Concluding Remarks. In *Drugs of Abuse: Body Fluid Testing.* Raphael C. Wong and Harley Y Tse, Editors. Totowa, NJ: Humana Press. Pp. 293-296.

Wong, Raphael, C. 2002. The Effect of Adulterants on Urine Screen for Drugs of Abuse Detection by an On-Site Dipstick Device. *American Clinical Laboratory* 21(1):21-23.

Wong, Raphael C., and Harley Y. Tse. 2005b. Adulteration Detection by Intect®7. *Forensic Science and Medicine Drugs of Abuse: Body Fluid Testing.* Raphael C. Wong and Harley Y. Tse, Editors. Totowa, NJ: Humana Press. Pp. 233-245.

Wood, Michelle, Marleen Laloup, Maria del Mar Ramirez Fernandez, Kevin M. Jenkins, Michael S. Young, Jan G. Ramaekers, Gert DeBoeck, Nele Samyn. 2005. Quantitative Analysis of Multiple Illicit Drugs in Preserved Oral fluids by Solid-Phase extraction and Liquid Chromatography-Tandem Mass Spectrometry. *Forensic Science International* 150:227-238.

WorkSafe, Inc. Website and Newsletter. http://www.worksafeinc.com.

Wu, Alan, H.B. 2006. A Selected History and Future of Immunoassay Development and Applications in Clinical Chemistry. *Clinica Chimica Acta* 369:119-124.

Wu, Alan H.B. 2002. Testing Urine for Drugs of Abuse. Originally Published in *IVD Technology* July/August. Accessed April 2009 at http://www.ivdtechnology.com

Wylie, F.M., H. Torrance, R.A. Anderson, and J.S. Oliver. 2005. Drugs in Oral fluids: Part I. Validation of an Analytical Procedure for Licit and Illicit Drugs in Oral fluids. *Forensic Science International* 150(2-3):191-198.

Yacoubian, G., Wish, E. D., and Pérez, D. M. 2001. A Comparison of Saliva Testing to Urinalysis in an Arrestee Population. *Journal of Psychoactive Drugs* 33(3):289-294.

Yamada, Hideyuki, Yuji Ishii, and Kazuta Oguri. 2005. Metabolism of Drugs of Abuse: Its Contribution to the Toxicity and the Inter-Individual Differences in Drug Sensitivity. *Journal of Health Science* 51(1):1-7.

Yang, Jane M., and Kent B. Lewandrowski. 2007. Urine Drugs of Abuse Testing at the Point-of-Care: Clinical Interpretation and Programmatic Considerations with Specific Reference to the Syva Rapid Test (SRT). *Clinica Chimica Acta*:27-32.

Yesavage, J.A., V.O. Leirer, M. Denari, and L.E. Hollister. 1985. Carry-over Effects of Marijuana Intoxication on Aircraft Pilot Performance: A Preliminary Report. American Journal of Psychiatry 142:1325-1329.

Zaitsu, Kei, Munehiro Katagi, Hiroe Kamata, Tooru Kamata, Noriaki Shima, Akihiro Miki, Hitoshi Tsuchihashi, and Yasushige Mori. 2009. Determination of the Metabolites of the New Designer Drugs BK-MBDB and BK-MDEA in Human Urine. *Forensic Science International* 188:131-139.

Zuccato, Ettore, Chiara Chiabrando, Sara Castiglioni, Davide Calamari, Renzo Bagnati, Silvia Shiarea, and Roberto Fanelli. 2005. Cocaine in Surface Waters: New Evidence-Based Tool to Monitor Community Drug Abuse. *Environmental Health: A Global Access Science Source 4(14).*

2.7 Glossary

Absorption: Absorption is the process by which a chemical substance moves from the site of administration into the body. It is one of the key pharmacokinetic variables.

Acute effect: The immediate, short-term response to one or a few doses of a drug.

Accuracy: Accuracy refers to the closeness of the measured value to the true value. Accuracy is generally expressed as the percentage difference from the actual value. (Isenschmid and Goldberger 2007:786)

Addiction: Compulsive drug-seeking behavior where the acquisition and use of drugs dominates an individual's life, usually accompanied by physical and psychological dependence.

Adulterated specimen: A specimen that has been altered, as evidenced by test results showing either a substance that is not a normal constituent for that type of specimen or showing an abnormal concentration of an endogenous substance (49 CFR Part 40).

Analyte: The substance or chemical constituent that is measured in an analytical procedure.

Antibody: An antibody is a water-soluble immunoglobin protein produced by the body's immune system that defends the body against antigens by attaching directly to them, coating them to make them recognizable to scavenger cells, producing an antigen-antibody complex that causes the release of enzymes capable of digesting them, or by preventing them from entering cells. Antibodies are manufactured by injecting the target drug (bound to a larger protein molecule) into host laboratory animals.

Antigen: An antigen is a substance the body recognizes as foreign and against which it raises an immune response defense. Antigens are usually high molecular weight proteins or polysaccharides that are chemically complex.

Bioavailability: The availability of a drug to target tissues following administration.

Blood alcohol concentration (BAC): The mass of alcohol in a volume of blood.

Buccal cavity: The space between the inside of the cheek and the teeth.

Capillary electrophoresis: A separation technique, also known as capillary zone electrophoresis, that is used to separate ionic analytes by their charge, frictional forces, and hydrodynamic radius. Introduced in the 1960s, its use in the analysis of drugs grew rapidly in the late 2000s because it requires little sample preparation and very small sample volumes (Cody 2008:154).

Carryover: Carryover means the potential for contamination of a sample by a sample analyzed immediately prior to it. In the urine drug testing laboratory, carryover is used to delineate the concentration of analyte in a sample above which contamination may reasonably be expected to occur. Carryover should be evaluated on each instrument system on which the methods is to be performed (Isenschmid and Goldberger 2007:787).

Chain of custody: Procedures to account for the integrity of each specimen or aliquot by tracking its handling and storage from the point of specimen collection to final disposition of the specimens and its aliquots.

Chronic effect: The long term response to multiple doses or persistent use of a drug.

Confirmatory drug or alcohol test: A second analytical procedure to identify and quantify the presence of alcohol or a specific drug or drug metabolite in a specimen. The purpose of a confirmatory test is to ensure the reliability and accuracy of an initial test result, and is usually performed on a second aliquot of the original specimen.

Controlled substance: A drug or chemical that is regulated under the Federal Controlled Substances Act of 1970, as amended, because of its potential for abuse and dependence.

Cross-reactivity: Cross reactivity is a measure of the response of an antibody in an immunoassay to substances other than and in addition to the target analyte.

Cutoff level: The concentration or decision criteria established for designating and reporting a test result as positive, of questionable validity (referring to validity screening or initial validity test results from a licensee testing facility), or adulterated, substituted, dilute, or invalid (referring to initial or confirmatory test results from an HHS-certified laboratory).

Derivatization: The process by which a substance, especially a chemical compound, is chemically modified to improve its chromatographic resolution. Derivatization is one of the steps in the assay process that needs to be standardized within the laboratory in order to assure comparable results.

Disposition: The kinetic processes of distribution and elimination that occur subsequent to a drug's systemic absorption.

Distribution: Distribution is the movement of the drug and its metabolites throughout the body (internal interstitial and cellular fluids and tissues).

Drug abuse: Use of drugs in a manner or amount inconsistent with the medical or social patterns of a culture. From a legal perspective, the use of substances controlled in Schedules I through V of the Controlled Substances Act (CSA) outside the scope of sound medical practice is drug abuse.

Drugs of abuse: Drugs used outside the scope of sound medical practice; typically including narcotics, depressants, stimulants, hallucinogens, and anabolic steroids.

Elimination: The irreversible loss of drug from the site of measurement, through the processes of excretion and metabolism.

Enzyme immunoassay: A test used to detect and quantify specific antigen-eliciting molecules involved in biological processes.

Extraction: The process of obtaining something from a mixture or compound by chemical, physical, or mechanical means.

False negative: An erroneous result in an analysis that indicates the absence of an analyte that is actually present.

False positive: An erroneous result in an analysis that indicates the presence of an analyte that is actually absent.

Half-life (Elimination half-life): The time interval in which the concentration or amount of drug in the body is reduced by one-half (the starting concentration can be established at any time, it does not need to be when the drug is administered or when the drug is at peak concentration).

Illegal drug: In the NRC rule, any drug that is included in Schedules I to V of Section 202 of the Controlled Substances Act (21 USC 812), but not when used pursuant to a valid prescription or when used as otherwise authorized by law.[73]

Illicit drugs: Five categories of drugs that are unlawful to possess or use in any circumstance or are unlawful to use without a prescription (narcotics, stimulants, depressants (sedatives), hallucinogens, and cannabis).

Initial drug or alcohol test (screening test): The first test used to differentiate a negative specimen from one that requires further testing for drugs or drug metabolites.

Initial specimen validity test: The first test used to determine if a specimen is adulterated, diluted, substituted, or invalid.

Instrumented device: An instrumented device reads the results for the person performing the test.

Insufflation: A mode of drug administration in which a powdered form of the drug is inhaled into the nasal passage.

---

[73] Schedule I **controlled** substances (including cannabis, heroin, 3,4-methylenedioxy-N-methylamphetamine (MDMA), Lysergic acid diethylamide (LSD)) are those that have been found by the Drug Enforcement Administration (DEA) to have a high potential of abuse, no currently accepted medical use, and lack an accepted safety for use even under medical supervision; Schedule II controlled substances (including cocaine, opium, oxycodone, morphine, amphetamines) are available only by prescription and are those that have been found to have a high potential of abuse, have a currently accepted medical use, and potential for abuse that may lead to severe psychological or physical dependence; Schedule III controlled substances (including anabolic steroids, dihydrocodeine) are available only by prescription, have been found to have less potential for abuse than Schedule I and II substances, have a currently accepted medical use, and potential for abuse that may lead to moderate or low physical dependence or high psychological dependence; Schedule IV controlled substances (including benzodiazepines) require a prescription and have been found to have a low potential for abuse relative to substances in Schedule III, have an accepted medical use, and a potential for abuse that may lead to limited physical dependence or psychological dependence relative to Schedule III substances; Schedule V controlled substances (including cough suppressants containing small amounts of codeine) require a prescription and are intended only for medical purposes and have a potential for abuse less than Schedule IV substances.

Interferant: A substance other than the analyte of interest to which the measuring instrument responds to give a falsely elevated result.

Ionized form (of a drug): A drug molecule in which the total number of electrons does not equal the total number of protons, which gives it a net positive or negative electrical charge.

Label: A substance chemically attached to either the antigen or antibody in an immunoassay to convey a measurable property, such as fluorescence or radioactivity.

Limit of detection (LOD): The lowest concentration of an analyte that an analytical procedure can reliably detect, which could be significantly lower than the established cutoff levels.

Limit of quantitation (LOQ): The lowest concentration of an analyte at which the concentration of an analyte can be accurately determined under defined conditions.

Linearity: Linearity is a measure of the procedure's ability (within a given range) to produce results that are directly proportional to the concentration (amount) of the analyte in the sample. Linearity is determined by using a series of calibrators that have been prepared at various known concentrations of analyte.

Matrix: Matrix is the bodily fluid or material used as the specimen to test for an individual's use of drugs or alcohol. Example matrices are breath, urine, blood, and oral fluids.

Medical device: A product that is used for medical purposes in patients in diagnosis, therapy, or surgery.

Melanin: Melanin is a pigment found in the human body. In hair, are two forms of melanin: eumelanin, a dark pigment, which predominates in black and brunette hair; and phaeomelanin, a lighter pigment found in red and blond hair. Hair color reflects the ratio and amount of the two forms of melanin contained in the hair shaft. Some drugs bind more readily to melanin than to other cellular structures in the hair shaft.

Metabolite: The chemical compound(s) produced as the body metabolizes the parent drug.

Metabolism (of drugs): The chemical and physical reactions carried out in the body to prepare a drug for excretion.

Monoclonal antibodies: Monoclonal antibodies are identical antibodies that all bind to the same locations (epitopes) in a single antigen.

Non-instrumented device: A non-instrumented device requires that the person performing the test (e.g., collecting the specimen) interpret the results – for example, by comparing colors on the testing device result indicator with a chart.

Opioid: Synthetic narcotic analgesics that have opiate-like pharmacology but are not derived directly from opium.

Oxidizing adulterant: A substance that acts, alone or in combination with other substances, to oxidize drug or drug metabolites to prevent the detection of the drug or metabolites, or that affects the reagents in either the initial or confirmatory drug test (HHS 2008 Mandatory Guidelines).

Parent drug or parent compound: The original drug substance processed by the body.

Performance test: A test of a laboratory's ability to correctly identify a set of samples designed to ensure accurate and reliable analyte identification and quantification.

Perfusion: The passage of fluid (particularly blood) through a tissue; the passage of fluid into a tissue through transfer from blood.

pH: A measure of hydrogen ion concentration that is used to indicate the acidic/basic properties of an aqueous solution. A solution with a pH of 7 is neutral (neither acidic or basic).

Pharmacodynamics: The study of the relationship of drug concentration to drug effects (Karch 2007b).

Pharmacokinetics: The study of the time course of the processes (absorption, distribution, metabolism, and excretion) a drug undergoes in the body; the study of the quantitative relationship between administered doses of a drug and the observed concentration of the drug and its metabolites in body tissue and fluids.

pKa: The negative logarithm of the substance's acid dissociation constant, Ka, which indicates the tendency of the substance to reversibly dissociate into an ionized form. The pKa is a measure of the acidic or basic properties of a substance. A substance with a pKa less than 2 is a strong acid; a pKa greater than 2 but less than 7 is a weak acid; a pKa greater than 7 but less than 10 is a weak base; a pKa greater than 10 is a strong base. The pKa of a drug or metabolite influences its behavior, particularly how it is affected by the pH of the fluid or tissue in which it is located. The pKa is the pH at which concentrations of ionized and non-ionized forms are equal.

Polarity: Polarity refers to a separation of electric charge leading a molecule having an electric dipole (an asymmetrical arrangement of atoms around the nucleus). This affects the molecule's relationship to other molecules.

Precision: Precision is a measure of exactness and refers to how closely individual measurements agree with one another (i.e., the amount of variability in repeated measurements of a substance, or the degree of scatter).

Psychoactive: Having an effect on the mind or behavior.

Qualitative test: An analysis to identify one or more components of a mixture.

Quality control: Measures taken to ensure than an established standard of quality in results is achieved.

Quality assurance: Measures taken to monitor, verify, and document performance, including, proficiency testing and auditing.

Quantitation: Measurement of the amount or concentration of a material.

Reagent: Generic term for the various commercially-produced compounds used as inputs in the immunoassay process, including the relevant antibodies as well as target drug antigens.

Recovery: A measure of the proportion of the drug present in a specimen at the time of collection that is present in the specimen when it is tested.

Reservoir (drug): Any part of the body (system, fluid, organ) that binds and holds a drug, delaying its elimination from the body.

Retention time: The time taken by a particular compound to elude (i.e., exit) from the chromatographic separation system.

Route of Administration: The path by which a drug is brought into contact with or into the body.

Ruggedness: An attribute of an analytical method, device, or equipment that characterizes its operational stability or sturdiness, measured by such factors as the number of operations or length of time it can perform as designed in field conditions.

Screen: An initial test designed to separate samples containing drugs at or above a particular minimum concentration from samples containing them below that minimum concentration (or not at all).

Sebum: The oily/waxy substance produced by the sebaceous glands (in the skin), to lubricate the skin and hair.

Sensitivity: Sensitivity is the lowest concentration of a drug analyte that can be reliably and reproducibly detected in an analytical solution; the detection limit expressed as a concentration of the analyte in the sample. Analytical sensitivity is the ability of a method or instrument to discriminate between samples having different concentrations or containing different amounts of an analyte (Karch 2007).

Specificity: Specificity is the ability of an analytical method to distinguish the target analyte(s) from other compounds, including those with and without structural similarity. Analytical specificity is the ability of a measurement procedure to determine the analyte it purports to measure and not others (Karch 2007a).

Stability: Stability is a measure of the extent an analyte in a particular matrix remains unchanged during collection, storage, and analysis.

Surface plasmons: Also known as surface plasmon polaritons, surface plasmons are surface electromagnetic waves that occur at the boundary of a metal and the external medium and are very sensitive to changes in this boundary.

Technology: The methods, protocols, matrices, devices, equipment, and combinations thereof used to enhance human analytical and implementation capabilities.

Tolerance: The adaptation of the body to a drug that reduces the dose-response effect that results in a need to increase dose to achieve the same response or effect.

Validation: Validation is the process of documenting or proving that an analytical method is acceptable for its intended purposes.

Vertex posterior: An area on the back of the head identified as the location of least variability in hair growth rate, proportion of hair in the growing phase, and age- and sex-related influences (Kintz et al. 2007:801).

Washing (hair specimen): Washing is the process of cleansing unrelated matter and chemicals from the hair sample prior to testing. This procedure eliminates traces of external contamination from the hair sample. Portions of each wash are saved in case a sample tests positive at which time the ratio of the drug found in the wash will be compared to the ratio of the drug found in the hair.

Window of Detection: The interval during which the drug and/or its metabolites can be detected in a matrix, measured in terms of time since consumption.

3.0 FATIGUE MANAGEMENT

3.1 Introduction

In 2008, the U.S. Nuclear Regulatory Commission (NRC) amended its Fitness for Duty (FFD) rule, 10 CFR Part 26, to impose requirements related to the management of worker fatigue[74] among nuclear power plant (NPP) licensees. In developing the revised rule, however, the NRC recognized that the science and best practices associated with assessing and managing fatigue are constantly evolving. The purpose of this chapter is to provide an update on the research about sleep, wakefulness, and fatigue; the technologies and practices that are emerging for measuring, assessing, and managing fatigue; and the regulatory approaches being taken in other industries to integrate some of the emerging science and technology into regulation and best practice.

This chapter summarizes a large body of scientific and technological information. In addition to the review of fatigue science and technologies, it summarizes advances in the related area of sleep science and describes fatigue management initiatives in a number of key sectors. It presents information obtained from patent and literature searches; attendance at selected conferences; and interviews with key scientists, technology developers, regulators, and industry personnel.

This chapter is organized as follows: Following the introduction, Section 3.2 summarizes recent research on sleep and fatigue, with a focus on information that provides a basis for understanding the purpose, potential, and constraints on newly emerging technologies. Section 3.3 describes efforts under way to develop and deploy technologies to aid fatigue assessment and management. Section 3.4 reviews the status of fatigue management in industries and governmental sectors where fatigue is a significant safety concern. It focuses on support for applied research and actions to require, consider, or adopt technologies and methods to manage fatigue. Finally, Section 3.5 discusses the implications of the earlier sections for the nuclear power industry. Section 3.6 is a bibliography of materials reviewed for this report. Appendix A presents information about specific applied fatigue related technologies. Appendix B presents the National Transportation Safety Board Methodology for Investigating Operator Fatigue in a Transportation Accident. For ease of reference, a glossary of terms and a list of acronyms used in this chapter are included separately in the frontpiece materials, along with a list of the technologies discussed in this chapter.

3.2 Overview of Scientific Literature on Sleep and Fatigue

The summary of the scientific literature is organized according to three main themes:

- Advances in the understanding of fatigue relative to sleep, circadian rhythms, and the sleep-wake cycle;
- Measuring fatigue; and
- Causes and effects of fatigue.

---

[74] This report focuses on fatigue resulting from inadequate sleep and variation in the circadian cycle. It does not address physical fatigue or fatigue resulting from illness (i.e., fibromyalgia).

Together, these themes provide the support for, and the constraints to, the development of technologies and methodologies for assessing and managing fatigue, and for establishing regulatory frameworks addressing fatigue. While this review describes significant advances in the understanding of fatigue and its causes and effects, it is also clear that there is still much to be discovered before there are comprehensive, valid, reliable, and cost-effective technologies and methods available to ensure that workers are continuously fit for duty relative to fatigue concerns.

3.2.1 Sleep, Circadian Rhythms, Homeostatic Pressure, and the Sleep-Wake Cycle

The study of fatigue draws heavily upon research on sleep, wakefulness, and the sleep-wake cycle. The definitions of fatigue, measures of fatigue, concerns about fatigue, and strategies to manage fatigue center around sleep and wakefulness. The technologies used to measure and predict fatigue and its impacts have their origin in research on sleep and the sleep-wake cycle. Since the 1990s, advances in biochemistry and neurophysiology have enabled an increasingly detailed understanding of the physiological mechanisms involved in the sleep-wake cycle, the characteristics of, and the transitions between, wakefulness and sleep (Ashton-Jones 2005; Lavie 2001; McCarley 2007; McCarley and Sinton 2008; Ogilvie 2001; Saper et al. 2005a,b,c). This research has been conducted on many different species of animals and has found strong similarities across species (Cirelli and Tononi 2008).

Daily sleep-wake cycles involve transitions between three distinct states:

- wakefulness;
- non-rapid eye movement (NREM) sleep (also known as slow-wave sleep); and
- rapid eye movement (REM) sleep.

Animals in each of these three states demonstrate neurological profiles with particular:

- electroencephalogram (EEG) - brain wave patterns;
- electrooculogram (EOG) - eye-movement patterns;
- electromyogram (EMG) - electrical activity of muscles; and
- muscle tone characteristics (Somers et al. 1993).

A key feature of the sleep-wake cycle is the consolidated nature of these three distinct states – transition between states occurs relatively rapidly, but once the transition has occurred, the animal tends to remain in the new state for some period of time (i.e., there is neither gradual transition nor frequent fluctuation between states). Animals typically transition from periods of consolidated wakefulness to periods of consolidated sleep. Since the 1960s, two key endogenous biological processes have been postulated as the mechanisms by which these transitions between consolidated states are achieved:

- the circadian pacemaker (tracking time of day and coordinating/synchronizing biological systems); and
- homeostasis (balancing wakefulness and sleep).

Lack of sleep has widespread and well-documented consequences, including impaired cognitive performance, impaired core thermal temperature control, altered dietary metabolism, and impaired immune function (Pace-Shott and Hobson 2002; Saper et al. 2005). Sleep has both quantitative (amount/duration) and qualitative (intensity/continuity) dimensions that together determine the restorative function of sleep and what neurophysiologically constitutes "lack of sleep."

Much of the research on sleep has focused on the relationship between the body's drive for sleep homeostasis, and the circadian pacemaker and circadian rhythms. Homeostasis refers to the body's ability to regulate key functions that establish a dynamic balance and maintain stability in key physiological parameters (e.g., eating, sleeping, and temperature). Homeostatic processes balance wakefulness and sleep. In the sleep literature, this homeostatic process is often represented as sleep pressure or sleep homeostasis (as opposed to wakefulness pressure or wakefulness homeostasis, which is represented by the circadian system). Sleep homeostasis refers to the dynamic balance between sleep and wakefulness in which the body's homeostatic mechanisms "counteract deviations from an average 'reference level' of sleep …[by augmenting] sleep propensity when sleep is curtailed or absent and …[reducing] sleep propensity in response to excess [surplus] sleep" (Borbély and Achermann 2000:377). Sleep pressure, sometimes defined as sleep propensity, builds up over periods of wakefulness and declines during sleep, as shown in Figure 3.1.

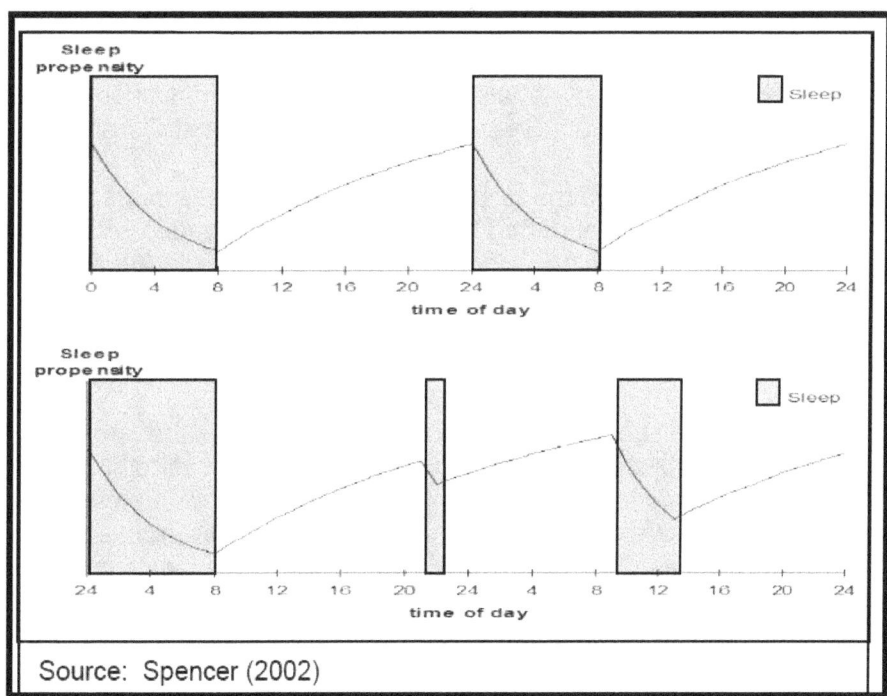

Source: Spencer (2002)

Figure 3.1.   Homeostatic Pressure

A large body of empirical data shows that sleep pressure is reduced by sleeping longer, more deeply (for instance in mammals, NREM sleep rich in slow waves), and/or in a more consolidated manner (less frequently interrupted by brief awakenings) (Belenky 2003; Cirelli and Tononi 2008; Van Dongen et al. 2003).

Inadequate sleep (excessive wakefulness) over successive sleep-wake cycles results in a "sleep debt" that has measurable neurophysiological consequences (Spiegel et al. 1999; Sullinen et al. 2004; Rupp et al. 2009). There is substantial individual variation in the sleep homeostatic process; there are long sleepers and short sleepers. Other factors also affect homeostatic sleep regulation, such as chronic alcohol use, adolescence, and old age (Aeschbach et al. 2001; Di Milia et al. 2011). As discussed below, until recently, the biochemical basis of this homeostatic process was largely unknown and it was treated essentially as a theoretical and empirical "black box" ("the homeostatic process S") in models of the sleep-wake cycle (Borbély and Achermann 2000; Dawson et al. 2011).

Circadian rhythms refer to daily fluctuations in biological processes and physiological functions. "Circadian" is a term from the Latin roots *circa*, meaning "about," and *dies*, meaning "day." For some time a commonly accepted estimate was that between 10 and 15 percent of mammalian genes were regulated by circadian clocks; however, work by Ptitsyn et al., indicates that the majority of mammalian genes exhibit some degree of circadian oscillation (Ptitsyn et al. 2007).

The normal human sleep-wake cycle is timed and coordinated by the central circadian pacemaker, also known as the brain's biological clock, which is located in the suprachiasmatic nucleus of the hypothalamus. The circadian pacemaker synchronizes biological systems on an approximately 24-hour cycle. It does this by regulating temperature, melatonin, and other hormone levels such as the growth hormone (Edery 2000; Mistlberger 2005). Signals from the suprachiasmatic nucleus are involved in promoting wakefulness. As part of the complex feedback system that maintains circadian rhythms and regulates the sleep-wake cycle, melatonin inhibits wakefulness-promoting signals from the suprachiasmatic nucleus (Dijk et al. 1997; Wehr et al. 2001; Aston-Jones 2005; Beersma 2005; Beersma and Gordijin 2007).

During daily awake hours, circadian rhythms lead to predictable changes in alertness, such as the tendency to feel sleepy at some point during the afternoon. This tendency is often referred to as the "post-lunch dip," although the alertness drop has more to do with the point in the circadian cycle and less to do with a person having eaten. Alertness in humans is typically lowest between midnight and 5 a.m., which corresponds to the period when melatonin levels are highest (Czeisler and Dijk 2001; Bes et al. 2009).

The body's circadian rhythms are subject to disruption by external cues that are inconsistent and/or out of phase with the established rhythm. Such disruptions can affect an individual's ability to fall and stay asleep, as well as the quality of the resulting sleep. External cues, particularly light-dark patterns, can reset the circadian pacemaker. This enables individuals to adjust to schedule or time changes. The length of time required for the adjustment depends on how extreme the changes are and individual variability (Groeger et al. 2008). Jet lag, for example, occurs when an individual's circadian rhythm is different from the day-night and activity patterns of the local environment. Most people can adjust their sleep-wake cycle to a full 12-hour time zone change within a few days, although research indicates that disruption of sleep stages may persist past the time an adjusted sleep time has been established. It is more difficult for people to adapt to work schedules that are 12-hours out of phase with their circadian rhythm than to accommodate a 12-hour time-zone change. This is because the light-dark patterns, surrounding activities, and the sleep-wake schedule continue to be in conflict – unlike the case with a time-zone shift where all three shift together. The circadian pacemaker of individuals, such as shift workers, who switch temporarily from one activity-rest pattern to another, as on weekends, can become chronically disrupted and misaligned with external time (Djik and Cajochen 1997; Moore 2007; Lee et al. 2009).

Research by Lee et al. (2009) indicates that disruption of the circadian rhythms that affect sleep and fatigue occurs in two different parts of the suprachiasmatic nucleus that may increase the inability to resynchronize the system. They demonstrate that deep sleep (the initial phase of sleep) and REM sleep are entrained to external signals differently, with REM sleep more resistant to resynchronization to a new schedule.

Research on the neurophysiology involved in the control of sleep-wake behavior has started to delineate more specifically how sleep homeostasis is achieved and to elucidate more clearly the locus and properties of the neuronal populations governing arousal and wakefulness. This research has distinguished two distinct neuronal populations in the brainstem and hypothalamus that are implicated in the production of sleep and wakefulness, respectively.

Saper et al. (2005a,b,c) have proposed a conceptual model in which the sleep- and wake-active neuronal populations have mutually inhibitory connections. Oscillations in the activity of these neuronal populations function as a "flip-flop switch" that governs the transitions between sleep and wakefulness. In the proposed interaction, an increase in activity within one neuronal population inhibits the other, thus creating a positive feedback (i.e., enabling further increase in activity). Hypocretin, a neurotransmitter, stabilizes this process. Malfunction of the hypocretin system is the cause of narcolepsy (Saper et al. 2005c; Sakuri 2007).

Research is focused on articulating the neural network and the neurophysiology of these reciprocal controls (Lu et al. 2006; Datta and MacLean 2007; Nakao et al. 2007). The National Center on Sleep Disorders Research (2003) identified the articulation of the hypocretin (also known as orexin) system and its central role in sleep regulation and behavioral control as one of the greatest achievements of sleep research since the discovery of REM sleep. Decades of research and modeling efforts have established slow-wave activity (SWA) in the EEG during NREM sleep as a marker of the sleep homeostat that is closely associated with cellular changes at the level of thalamic and cortical neurons, and with theta EEG activity during wakefulness (Borbély and Achermann 1999; Johns 1998; Dijk and von Schantz 2005; Datta and MacLean 2007; Saper et al. 2001; Saper et al. 2005a,b,c). The circadian oscillator hypothalamus functions as a self-sustained oscillator that determines the preferred timing of sleep and wakefulness. Established markers of the circadian process include plasma melatonin, cortisol, and core body temperature (Lavie 2001; Siegel 2004). Research indicates that changes in these markers are related to some of the biochemical consequences of sleep and wakefulness (e.g., variation in extracellular adenosine concentration and other sleep regulatory substances) or to variation in connectivity (synaptic strength) in the neuronal networks (Cirelli and Tononi 2008).

Despite extensive research, however, the biological function of sleep remains a mystery and many questions about sleep remain unsolved. What determines the brain's memory for sleep loss, and what neurological deficiency is being regulated by the sleep debt memory remain unknown (Horne 1988; Kalia 2006; Cirelli and Tononi 2008; Vassalli and Dikj 2009). Although it is well established that sleep is a biological necessity, there are no consensus answers as to the fundamental questions of why organisms need to sleep and what function sleep serves. It is known that after some period of sleep deprivation (which varies by species and across individuals within a species) sleep begins to intrude upon wakefulness, and that at some point sleep cannot be avoided. It is also known that sleep deprivation results in a deterioration of performance. However, as Cirelli and Tononi (2008:1607-8) point out:

People may seem superficially awake (moving and with eyes open) even though the EEG slows down or exhibits microsleeps....It is unknown whether the presence of slower activity in the "wake" EEG spectra of sleep-deprived animals or humans is due to "piecemeal" sleep, where some brain regions may be asleep whereas others are awake... to "salt and pepper" sleep-wake, in which within the same brain regions individual neurons may be awake (depolarized) and others may be oscillating between up- and down-states (asleep, ...), or to abnormal cellular activity that is neither wake or sleep. Whatever the underlying cellular events, it seems impossible to completely deprive an animal of sleep for more than 24 hours....Rather, what seems to occur is a kind of "dormiveglia" (sleepwake), a mixed state that is clearly dysfunctional...

An important unresolved question is whether the impairment, cognitive or otherwise, that follows sleep deprivation is merely the consequence of an increased drive for sleep ("sleepiness") or whether brain cells need sleep because they are actually "tired...." It may be that brain cells actually do get tired as a function of waking activities, whether or not the arousal systems are pushing the organization to stay awake.... Pure tiredness can be conceptualized as the inability of brain cells to continue functioning in their normal waking mode, despite the central wake-promoting mechanism telling the brain it should be fully alert...

Altogether, then, while we still do not understand whether sleep deprivation is followed by sleep intrusions and cognitive impairment because we become sleepy, tired, or both, the evidence so far indicates that...lack of sleep has serious consequences, especially for the brain.

The emerging model of the sleep-wake cycle and regulation of sleep is a neurophysiologically refined version of the two-process model that characterizes the dynamic interaction of two oscillatory processes: the sleep homeostat and the circadian pacemaker (Borbély 1982; Achermann and Borbély 1992 and 2003; Åkerstedt and Folkard 1997). Aston-Jones (2005), Behn et al. (2007), Comte et al. (2006), Philips and Robinson (2008), and Rempe et al. (2009) among others, are working on formal models to integrate this new information with the large amount of previously assembled data on the sleep-wake cycle. Dawson et al. (2011) provide a succinct review of the evolution of these models, originally developed to characterize and examine the sleep-wake cycle, into models that predict fatigue and fatigue-caused impairment. These researchers expect that improved understanding of the biochemical and genetic basis of these oscillations will enable further refinement of the biomathematical models that form the conceptual foundation of fatigue research and management.

3.2.2 Defining and Measuring Fatigue

Definitions of Fatigue

There is not a clear consensus on the definition or measurement of fatigue, an attribute of the field noted by virtually all participants (Mallis et al. 2004; Noy et al. 2008). As with many psychosocial variables, definition and measurement are closely intertwined because fatigue, however defined, cannot be directly measured – it has to be characterized by indirect estimates. In addition, because of the variability in definition, there is a not a clear distinction between what constitutes a measure of fatigue and what is a measure of fatigue's effects.

Within the fatigue literature, definitions are influenced by the interests, problem focus, and disciplinary perspective of the researcher. For example, the concept of fatigue, and therefore its definition and measurement, is strongly influenced by the definitions, measurements, and models developed in the research on sleep and circadian rhythms. More generally, three primary interest areas are reflected in the fatigue literature:

- understanding the causes of fatigue;
- understanding how a state of fatigue emerges and what are a fatigue state's characteristics; and
- understanding what are the consequences of fatigue for behavior and performance, and how they can be managed.

To illustrate, the NRC, whose primary concern is the consequences of fatigue on behavior and performance, defines fatigue in Part 26 as "the degradation in an individual's cognitive and motor functioning resulting from inadequate rest."

In addition, definitions of fatigue, and efforts at measuring fatigue vary in terms of whether they emphasize:

- internal biological/neurological factors or external manifestations;
- early internal indicators that precede any external manifestation of alertness deterioration or later indicators of more severe drowsiness; and
- temporally proximate indicators of the state of fatigue versus either causal factors (such as sleep deprivation including time since and amount of last sleep and cumulative sleep debt) or performance degradation effects.

Within this framework, it is interesting to note that some experts define fatigue in terms of multi-dimensional subjective perceptions, while others define it as a simple and clear biological phenomenon, such as the Williamson et al. (2011) definition of fatigue as "a biological drive for recuperative rest." Some researchers' definitions focuses on the ability of the subject to sustain alertness (Ahsberg et al. 2000; Schleicher et al. 2008) while others define fatigue in terms of performance degradation (McCallum et al. 2003).

Rather than putting forward a single definition of fatigue, this review attempts to reflect the state-of-the practice and the multidimensionality and complexity of the concept of fatigue and its management in the workplace. The lack of agreement on definition means, however, that careful attention must be paid to how researchers and studies have defined fatigue, and how they organize variables and measurements into causes of fatigue, states of fatigue, and effects of fatigue.

Measurement of Fatigue

Researchers have used indicators falling into one or more of the following categories to measure fatigue:

- physiological functions;
- physical/behavioral attributes; and
- mental and emotional states.

Each of these is discussed in detail below.
Researchers have used a variety of methods to measure these indicators, including:

- biochemical tests and sensors, often linked to computerized monitoring systems;
- expert ratings; and
- self-assessments.

The types of indicators being measured affect the choice of measurement method(s), and vice versa. Recognizing the indirect nature of their indicators, researchers have expended considerable effort to establish the relationships among the various measures and indicators (see for example, Leproult, et al. 2003; Banks and Dinges 2007).[75]

Physiological Indicators and Their Measurement

<u>**Stage in the Sleep-Wake Cycle**</u>

Circadian rhythms and sleep homeostasis are the key biological processes associated with the sleep-wake cycle and fatigue.[76] The developments described above explicating the biochemistry of the sleep- and wake- neuronal populations may lead to the identification of additional indicators useful for fatigue management. However, the indicators used most frequently to track an individual's stage in the sleep-wake cycle are the "rise and fall" patterns of core body temperature, the levels of melatonin and other hormones including cortisol, and brain wave patterns measured by an EEG.

Core Body Temperature: Core body temperature is the temperature in the part of the body containing the vital organs. It has a distinct circadian variation and is therefore used in sleep and fatigue research to help determine where an individual is in terms of the circadian and sleep homeostatic processes. Core body temperature is measured internally, usually with a thermometer/sensor via rectum or esophagus. When used in fatigue or sleep research, core body temperature is measured either periodically or continuously over time, depending upon whether circadian rhythm is being addressed as a dependent or independent variable in the research. A practical technology that can conveniently measure a person's core temperature over an extended period in a work setting is not available.

Melatonin Levels: An individual's melatonin level can be determined by measuring the concentration of its principal metabolite, 6-sulphatoxy melatonin, in blood, urine, or oral fluids. Standard immunoassay methods are capable of establishing these levels. This review did not identify any researchers using point-of-collection (POC) devices[77] to establish an individual's pattern of melatonin level in a work setting, although the state of the science in immunological testing makes development of such a device feasible should a market for it develop. Because the purpose is to locate the individual within his/her own circadian rhythm and sleep-wake cycle, a number of samples, collected at appropriate intervals, would be needed. In addition, further research is necessary to establish the real-time relationship between melatonin levels and fatigue for a particular individual (Dawson et al. 1996; Roach et al. 2005; Rüger et al. 2005).

---

[75] However, as Aston-Jones (2005:S3) points out, few studies have attempted to link the role of a brain system in sleep-wake regulation with a role in cognitive performance during waking.

[76] The two-process model originated with Borbély (1982). There is also a three-process model that incorporates timing of work and/or sleep as input to the circadian and homeostatic two-process model. It also allows real-time update. The three-process model has also been called the "Three Process Model of Alertness" as well as the "Sleep/Wake Predictor Model" (see Åkerstedt et al. 2004).

[77] As described in Chapter 2, POC devices enable a specimen to be collected and tested outside the laboratory.

Brain Wave Patterns/Brain Activity: Extensive research on arousal and brain activity has generated a large body of empirical data on brain wave patterns, their measurement (EEGs), and the characteristic patterns of wakefulness, alertness, and different stages of sleep. Particular brain wave patterns are markers of the transition from wakefulness to the different stages of sleep, and EEGs are frequently used to monitor individuals' stage in the sleep-wake cycle in laboratory research. As discussed below, EEG data are useful not only in locating individuals within the sleep-wake cycle, but also in characterizing/monitoring the wakefulness-sleep state of the individual.

Brain Activity – Electroencephalography (EEG)

Researchers also use brain wave patterns, as measured by EEG, as an indicator of fatigue. EEG uses a system of sensors (electrodes) to measure specific areas of the brain and to record patterns of brain activity (voltage and frequency). Until recently, EEG systems have been primarily laboratory research tools because the bulky equipment made EEGs impractical as a field technology. However, developments in sensor technology, wireless communications, and portable computers and personal digital assistants (PDAs) have enabled this technology to become more robust, transportable, flexible, and user-friendly. Davis et al. (2009) describe how advances in hardware, software, algorithms, and application design are making EEG classifiers applicable for field deployment – a long-sought goal that has been difficult to achieve.

Some researchers consider EEG to be the gold standard for fatigue measurement (Caldwell et al. 2009) because of its ability to identify the transition to sleep. A study by Larue et al. (2010) found EEG to be one of the most accurate measures of a vehicle driver's vigilance level. However, other recent research has questioned EEG's effectiveness in detecting fatigue and predicting fatigue-related performance impairment (Schleicher et al. 2008), and noted that there is substantial variation in how individuals with similar EEG results perform. A study by Myers et al. (2009) that looked at the predictive validity of various measures concluded that neither EEG nor oculometric technologies that record and assess eye movements (see detailed discussion in "technology" section) reliably predict actual driving behavior.

Researchers are attempting to improve the ability of EEG to detect fatigue, in part by identifying which EEG features provide the best measures of fatigue. For example, Lin (2009) tested six EEG locations and found two locations, the occipito-parietal and the motor areas of the brain, to be the most sensitive to driver drowsiness. Some researchers are attempting to improve the effectiveness of EEG in measuring fatigue by employing a neural net analysis that learns over a number of training samples for each individual. Karrar et al. (2009) developed and tested an advanced pattern-based EEG approach and demonstrated that it had slightly better accuracy than the conventional spectral-based EEG.[78]

Autonomic Nervous System Activity

Research has shown that autonomic nervous system (ANS) activity is associated with fatigue. In particular, there is a strong link between ANS activity and cognitive fatigue.[79] The ANS indicators associated with alertness and sleep/wake patterns include heart rate and heart rate variability, breathing rate and respiratory instability, tidal volume and tidal volume instability

---

[78] However, it failed to detect 30 percent of drowsiness onset events when compared to experts' observations of drivers' drowsiness levels.

[79] In particular, cognitive fatigue is associated with a shift of sympathovagal balance toward sympathetic predominance and reduced vagal tone (Myers et al. 2009).

(which is an index of respiratory instability), skin conductance, and skin temperature. Myers et al. (2009) tested a host of autonomic cardio-respiratory markers, including heart rate and heart rate variability, breathing rate and respiratory instability, skin temperature, as well as motion and postural shifts. They found that heart rate variability and tidal volume instability were the best indicators of cognitive attention on a performance vigilance test.

To date, techniques to measure and analyze autonomic activity of individuals in the workplace on a routine basis are not available. Efforts are currently under way and others are planned for the near future to enhance key physical monitoring technologies (a wrist band actigraph that monitors wrist activity and an eyewear technology that monitors eye activity) and to incorporate measurements of autonomic activity as well. This would expand the range of indicators that could be routinely measured as individuals perform their jobs and go about their daily activities.

Pupillometry

The iris is a dynamic organ whose activity is regulated by the ANS. Physiological reactions of the iris to different sensory stimuli result in variations in pupil size. These reactions are affected by factors that affect the ANS, including fatigue, drugs, and alcohol consumption (McClaren et al. 2002; Monticelli et al. 2009; Morad et al. 2009). Pupil diameter is determined by the relative activity of the nerve fibers that excite the iris muscles.

Analysis of pupil reactivity (pupillometry) is used to evaluate the condition of the ANS and the visual system. Pupillometry is the measurement and recording of pupil diameter as a function of time. Many different diagnostic tools have been developed to assess iris activity, particularly the diameter of the pupil and patterns of pupil response to exposure to light of different wave lengths. Because these response patterns occur very quickly, pupillometry relies on sensor systems that can record pupil diameter at very short intervals (high frequency). Research has demonstrated characteristic patterns that indicate the state of the central and autonomic nervous systems (Kristjansson et al. 2009). Pupil motility in the absence of light exposure has been used to measure arousal state and alertness. Research has demonstrated a distinctive pattern of spontaneous slow pupillary oscillations that occurs in drowsy individuals in conditions of darkness (Lowenstein et al. 1963; Goldwater 1972; Teikari 2007). Capitalizing on the advancements in sensor technology, wireless communications, and portable computers and PDAs, research is under way to incorporate pupillometry into multi-sensor systems to establish the relationship between pupil response, other physiological conditions and performance measurements, and to establish systems to collect data in field settings.

Gene Expression

Burian et al. (2009) have conducted a series of studies to investigate genes whose expression levels[80] vary in response to sleep deprivation. In this research, ribonucleic acid (RNA) was purified from blood samples drawn from study participants at approximately 24-hour intervals before and after a 36-hour sleepless period. The researchers estimated deviance from static expression to identify genes whose expression levels might be used in an assay for fatigue. The goal of this research is to develop real-time assays that could become part of an accident forensics protocol to determine the contribution of fatigue and allow on-the-spot determination of

---

[80] The process of producing a biologically functional molecule of either RNA or protein is called gene expression. Gene expression studies attempt to identify when, under what conditions, and to what extent particular genes are producing proteins (or RNA). Some studies work back from proteins to identify the genes involved.

readiness-to-perform. This effort is benefitting from the extensive research under way in biology, medicine, and pharmacology to develop assays to characterize patterns of gene expression. The growing availability of assay technologies is likely to facilitate the considerable work remaining to establish the relationship of gene expression to fatigue and performance degradation. Consequently, although many field applications of gene testing technologies for fatigue are not likely to be available for several years, this research may well support discoveries that elucidate the physiology of fatigue.[81] The ethical/legal issues associated with such an approach will be significant.

Other researchers are also attempting to develop genetic screening to predict an individual's susceptibility to becoming fatigued (King et al. 2009). For example, Viola (2009) found that the clock gene PER3 plays a role in the generation of different biological rhythms associated with morning, intermediate, or evening types. PER3 affects several aspects of sleep homeostasis but does not appear to have an effect on the circadian phase (Viola et al. 2008; Dijk and Archer 2010). The clock gene effect on homeostasis may explain individual differences in terms of circadian phase misalignment that could result from shift work and/or jet lag and susceptibility to sleep loss. This research is not likely to have applied applications for some time and, as noted in Section 3.3 on applied technologies, the use of genetic-based tools may pose legal issues in the field because of the Genetic Information Non-Discrimination Act (GINA) of 2008 (U.S. Congress 2008).

Physical/Behavioral Indicators and Their Measurement

Body/Wrist Movements, Posture, and Muscle Tone

Wrist movements, measured by wrist actigraphs, can discriminate whether a person lying in bed is asleep or not and if the person is sleeping well or experiencing disrupted sleep. An actigraph is a device that measures and records movement. It typically consist of a piezoelectric accelometer, a low-pass filter (to filter out external vibrations), a timer to start/stop at specific times and to accumulate values over specified periods, a memory to record and store the movement values, and an interface to program the timer and download the data from memory. Research has validated actigraphy as a method for providing data that accurately track sleep/wake cycles and quality of sleep.[82] The American Academy of Sleep Medicine has accepted it for this purpose (Littner et al. 2003).

Although actigraphy can provide accurate information about an individual's sleep/wake cycle, by itself this information does not provide a comprehensive or sufficient means of detecting fatigue or performance degradation. However, it does provide information that can be incorporated into and used to calibrate biomathematical models of fatigue that may be able to provide more accurate detection and prediction of fatigue (Dawson 2009). It also provides a reliable, unobtrusive method for detecting and monitoring when an individual is actually asleep (Russo et al. 2005). Coupling a validated monitor with a validated biomathematical model allows an individual's level of fatigue to be both detected and predicted in real time (Mallis et al. 2004). To enhance the ability of actigraphs to detect fatigue, developers are attempting to expand actigraphs to measure various autonomic activities such as respiration and heart rate as well as

---

[81] Researchers are working to explicate the patterns of gene expression associated with the sleep-wake cycle: Cirelli (2005) and Cirelli and Tononi (1998 and 2009).

[82] A wrist actigraph may be an appropriate monitor for any type of driving task. The only restriction is if an individual sleeps in high-vibration environments. The environmental vibration may "swamp" the actigraph accelerometer readings to the extent that driver sleep cannot be assessed.

wrist activity (Karlen et al. 2007). These enhanced actigraphs are undergoing field testing (Appendix A).

Wrist and body movement patterns, posture, and muscle tone, measured by actigraphs, video monitors, and electromyography (EMG),[83] are also used as indicators of fatigue and sleepiness (Ancoli-Israel et al. 2003). Body posture and movements can also be recorded by video cameras, and the resulting images analyzed and correlated with other data to establish patterns that indicate fatigue. In addition, as an individual falls asleep, muscle tone and the electrical activity of muscles change. In general muscle activity (and muscle tone) decreases as an individual falls asleep. These changes can be reliably measured by EMG (Fridlund and Cacioppo 1986; Tassinary et al. 2007). However, muscle tone measures are limited in that they distinguish only between asleep and awake; they have not been shown to be reliable measures of level of alertness.

Eye Movements and Eyelid Closure

Monitoring and assessing physical eye movements and eyelid closure have become an increasingly common method for measuring fatigue. Research has confirmed that percentage of eye closure (PERCLOS) is correlated with EEG (i.e., concurrent validity).[84] PERCLOS also has validity in terms of predicting performance degradation (i.e., predictive validity). Predictive validity is improved when PERCLOS is supplemented with other oculometrics, such as average length of eyelid closure (AVCLOS), microsleep events (MSEs),[85] eye gaze (the range of peripheral vision), eye focus (whether the person attends to instruments, rear view mirrors, etc), and pupil dilation (see pupillometry discussion above). Researchers have demonstrated substantial inter-individual oculometric variability – for example, subjects differ in blink duration independent of sleepiness levels (Van Dongen et al. 2005; Ingre et al. 2006a; and Schleicher et al. 2008). Therefore, to increase the accuracy of these technologies in detecting fatigue, researchers have been focusing on methods to calibrate the oculometrics to account for this individual variability. Algorithms using multiple oculometrics have demonstrated stronger correlations with other physical manifestations of fatigue and performance degradation than those using only a single measure (Myers et al. 2009).

Oculometrics have typically been captured using video camera technology. However, the reliability of video monitoring as a means of capturing data both in the laboratory and in the field has been called into question (Johns et al. 2007; Hartley et al. 2000). In spite of these findings, developers of new oculometric video technologies, claim their technologies are more robust and capable of overcoming previous reliability issues. To improve reliability, developers such as Eye-Com Corporation, have produced eyewear monitoring technologies that they claim can more accurately and reliably measure multiple eye metrics (see Appendix A). Independent evaluations of the newest oculometric technologies are not yet available, although the Eye-Com technology is being used by the U.S. military (Ruppert 2009).

Oculometric approaches focus on detecting rather than predicting fatigue. However, advances in oculometrics have improved its ability to detect the onset of drowsiness in addition to detecting acute sleepiness characterized by the occurrence of microsleeps. To date,

---

[83] Electromyography measures the electrical signals generated by muscles. Muscle activity patterns vary during different stages of the sleep-wake cycle.

[84] The significance of this finding, however, depends on the extent to which EEG is a valid measure of fatigue (see the discussion of EEG and brain activity).

[85] An MSE is based on eyelid closure of some duration beyond that of the typical blink of a non-drowsy individual.

oculometric data, unlike wrist activity, have not been integrated into biomathematical models to predict a future state of fatigue. Moreover, oculometric monitoring technologies, unlike actigraphs, are not applicable for continuous use. Although eyewear microchip technology, as opposed to video monitoring technology, could monitor eye movements when the eyes are closed during sleep, wearing eyewear while sleeping is not considered practical.

Developers of eyewear microchip technology, like developers of actigraphs, report that it may be possible to expand the capacity of the microchip to include other measures in addition to oculometrics and to integrate information from other sensors using wireless communication. Research is under way to determine which autonomic activities can be effectively monitored by or integrated into the eyewear microchips (Appendix A).

Verbal Behavior

Developing indicators based on verbal behavior may be another potential means of estimating fatigue. Speech involves interaction between the sensory and motor systems with voice control accomplished through a feedback process between these two systems. As fatigue increases, this feedback system is disrupted and speech sounds and patterns are affected. Indicators of fatigue in speech include changes in pitch and word duration, slurring, stammering, pauses, slowness, and inability to recall words (Greeley at al. 2007). In addition to psychologists who are researching the mechanisms involved in this process, a few researchers are designing and assessing a computerized approach for continuous speech monitoring to estimate fatigue (see Shahidi et al. 2009; Krajewski et al. 2009). This work is benefiting from advancements in speech recognition software for other purposes and rapidly advancing computing power. However, this NUREG/CR does not identify any technologies using verbal behavior that had reached the prototype, field-test stage.

Cognitive and Affective Indicators and Their Measurement

Cognitive indicators of fatigue can include degraded alertness/attention, problems with sustained concentration, tendency to be easily distracted, confusion, forgetfulness, memory problems, and performance worries. Psychomotor and cognitive speed, vigilant and executive attention, working memory, and higher cognitive abilities are particularly affected by sleep loss. These cognitive decrements can accumulate to severe levels over periods of chronic sleep restriction without the full awareness of the affected individual (Goel et al. 2009b). Affect indicators can include demotivation (such as boredom, lack of desire and enthusiasm, temporary feelings of depression) and coping or emotional/interactional fatigue (such as anxiety, avoidance, comfort-seeking, irritability, feeling stressed) (Luna et al. 1997; Kamdar et al. 2004; Wijesuriya et al. 2007; Williamson and Friswell 2009). These effects show considerable variability across individuals. Microsleeps, sleep attacks, and lapses in cognition are considered to be an indication of the instability between the sleep and wake states (Banks and Dinges 2007; Goel et al. 2009a). These indicators of fatigue are also addressed as effects of fatigue; the methods used to measure them are discussed below.

3.2.3 Types of Measurement and Estimation Methods

The type of measurement method used determines the types of indicators that are measured. Therefore, this discussion of types of measurement methods has some inevitable overlap with the discussion above regarding types of indicators. However, it is useful to understand which types of indicators can and cannot be captured by different measurement methods.

Biochemical Tests and Physiological and Behavioral Sensor Systems

Biochemical tests and sensor systems that measure physiological and behavioral attributes, often linked to computerized monitoring systems, are used to track and assess objective indicators of fatigue (as opposed to relying on expert observation or subjective self-assessment). These types of tools, including the software systems used to integrate and interpret the data generated by them, have been the recipient of extensive research and development efforts over the past decade or so, benefiting greatly from breakthroughs in the computer, communications, and biological sciences. Immunoassay tests, thermometers, EEGs, and EMGs tend to remain primarily laboratory-based research tools, although as discussed below, some have been adapted for use in the field.[86]

Two computerized sensor system approaches that have field applications are video monitoring systems and actigraphy. Increasingly miniaturized and sophisticated video monitoring systems are capable of coordinating multiple high-frequency video recorders and sensing systems, including EOGs, and feeding high-quality images into computers with algorithms for analyzing oculometrics, facial tone, posture, and other physical features. A new eyewear technology that has a microchip in the frame is now available. The microchip captures a wide range of oculometrics. In the future, it might be able to capture other physical and physiological (autonomic) metrics.

Actigraphy involves a person wearing a device, typically resembling a wristwatch, that measures wrist activity. Advances in actigraphy and sensor systems are providing devices that are small enough to wear and capable of measuring a wider range of physical attributes (such as posture shifts and movement) and autonomic physiological attributes (such as electrodermal activity [skin conductance], skin temperature, heart rate and heart rate variability, and breathing rate and respiratory instability). Actigraphs are capable of providing reliable sleep-wake data for individuals over intervals long enough to address cumulative fatigue and sleep debt because they can be worn continuously over an extended period of time, as opposed to video systems and eyewear technology.

The algorithms used to analyze biochemical test results and sensor systems data can estimate fatigue based on either pre-set thresholds or evolving criteria. The thresholds can be either standardized or individualized. The trend is to develop individualized algorithms and/or algorithms that learn and evolve over time. Another trend is to develop sensor platforms that can record and/or receive measures from multiple monitoring systems. The challenge is to determine how to best select and integrate these metrics into an algorithm that effectively detects and predicts fatigue. One strategy is to link sensor data with biomathematical models to predict fatigue and performance more effectively.

Expert Observer Ratings (Behavioral Observation)

Some studies have used expert observers to rate the fatigue level of individuals. A study by Schleicher et al. (2008) investigated the relationship between expert observer ratings, self-assessments, and computerized monitoring and assessment of oculometrics. This study did not involve sleep deprivation but rather had subjects perform a 2.75-hour monotonous task to produce the onset of fatigue (as defined in the study). Video recordings of subjects' faces at

---

[86] Advances in miniaturization, wireless communications, and durability have increased the field use potential of EEG and immunoassay testing. See the discussion of specific technologies.

various intervals during the task were observed and scored by expert observers using a method similar to that originally used in the study by Wierwille and Ellsworth (1994). The fatigue ratings assigned by the expert observers were:

- Stage 1 = not drowsy: fast blinks and saccades,[87] normal facial tonus;
- Stage 2 = drowsy: frequent blinks, limp face, yawning, scratching;
- Stage 3 = very drowsy: clearly prolonged eyelid closures, rare blinks, staring or drifting eyes;
- Stage 4 = extremely drowsy: overlong lid closures > 2 seconds, micro sleeps and abrupt arousals.

The study found substantial correlations between observer ratings, self-assessments, and oculometrics. In addition, Karrar et al. (2009) and Lal and Craig (2002) have investigated the association between expert ratings of video tapes of subjects and other types of indicators and have found reasonable concurrent validity.

Expert observers are often used to validate other fatigue detection techniques, to diagnose fatigue management conditions and issues, and to provide periodic monitoring or assessment of fatigue in particular settings or among particular populations. It is likely that such expert observers will increasingly use some of the sensor systems and technologies discussed above to enhance their capabilities. Expert observers are not often used for long-term, routine monitoring and detection assignments. However, many organizations that perform safety- and security-sensitive activities implement behavioral observation programs that train colleagues and supervisors to recognize the indications of fatigue and give them responsibility to notice and take action if they observe behaviors that indicate fatigue impairment.

Self-Assessments

Self-assessments of fatigue are used in basic scientific research, self-management in health care, and in some work settings where fatigue has been identified as a threat to safety and security. Research has shown a strong correlation between subjective perceptions and objective indicators of some aspects of fatigue and alertness. Self-report fatigue scales have strong concurrent validity with some objective measures of fatigue, such as oculometrics and video-recorded physical manifestations (Lal and Craig 2002; Schleicher et al. 2008; Karrar et al. 2009). The research of Leproult et al. (2003) highlights the importance of appropriately disaggregating the assessment of fatigue into its different aspects or dimensions. They found that the decrement in an individual's subjective alertness from a given amount of sleep deprivation was unrelated to the decrement it caused in that individual's performance on alertness tests. In addition, self-report measures have less individual variability than performance-based measures. According to Schleicher et al. (2008), subjective rating scales may be even more sensitive than other types of indicators to fatigue caused by monotonous or low-workload tasks in non-sleep deprived subjects. They found that drivers are quite aware of increasing sleepiness but have a tendency to underestimate the impact their drowsiness has on their performance abilities. This is especially true with individuals experiencing cumulative fatigue.

---

[87] Saccades are small, rapid, jerky movements of the eye especially as it jumps from fixation on one point to another, as in reading.

Van Dongen et al. (2003) examined how sleep-deprived subjects performing a driving test assessed their own driving ability, and how they judged the driving ability of a similarly sleep-deprived "hypothetical other." Self-ratings of their own driving ability declined substantially with increasing sleep deprivation and *correlated 0.70 – 0.76 with performance. Self-ratings of their own ability to drive were somewhat higher than their ratings for the hypothetical other, suggesting that subjects had more confidence in themselves than someone else in the same state. Also, in controlled settings, sleep deprivation appeared to have a greater and more enduring effect on performance than on self ratings; self ratings of sleepiness increased initially then leveled off after the first or second day. Therefore, although self-perceptions are fairly accurate, objective indicators with graded warning levels might be useful in supplementing subjective perceptions. In addition, Schmidt et al. (2009b) found that drivers misjudged their vigilance state after approximately 3 hours of continuous monotonous daytime driving, reporting a subjectively improved vigilance state when objective measures (auditory reaction time, EEG, ECG, heart rate) indicated a continuous degradation in vigilance over this period.

The use of self-report measures has a long history in studies of fatigue and sleep deprivation. One of the first self-report scales, the Stanford Sleepiness Scale (SSS), has demonstrated strong correlations with other objective indicators of fatigue, such as the amount of time a person is kept awake, EEG, and various performance tests. Other scales have been developed, including the Karolinska Sleepiness Scale (KSS), the Epworth Sleepiness Scale (ESS), the Retrospective Alertness Inventory, the Shiftwork Survey Index, and the Swedish Occupational Fatigue Inventory (SOFI). Some researchers use simple visual analogue scales anchored to extremes of alertness or sleepiness (for examples see Shen et al. 2006). These self-report rating scales often do not explicitly differentiate dimensions of sleepiness/alertness. A more comprehensive representation of the fatigue construct might address a broader range of factors and attempt to differentiate between key dimensions. Differences between both tasks and workers may affect vulnerability to and type of fatigue experienced (Folkard et al. 1979; Harrison and Horne 2000).

Hitchcock and Matthews (2005) attempted to include all relevant subjective perceptions of fatigue into a comprehensive self-assessment tool to be used by drivers. This tool includes subjective perceptions of:

- physical indicators (muscular fatigue, exhaustion, visual fatigue, sleepiness);
- cognitive indicators (degraded alertness/attention, problems with sustained concentration, easily distracted, confusion, forgetfulness, memory problems, performance worries); and
- affect indicators
  - demotivation (aversion to effort, boredom, lack of desire and enthusiasm, temporary feelings of depression); and
  - coping and interactional capabilities (feeling stressed, being distressed, anxiousness, avoidance, comfort-seeking, irritability).

Verbal performance, an area being examined by researchers to develop behavioral indicators of fatigue, was not considered in their model. A more complete self-assessment tool might incorporate subjective perceptions of verbal fatigue, including verbal slurring, stammering, pauses, slowness, mispronunciations, and inability to recall words.

A Swedish group (Ahsberg et al. 1997) developed SOFI, a multi-dimensional scale which allows ratings on 95 possible fatigue issues associated with a broad range of occupational tasks. This scale was derived from factor analysis of questionnaires from a large sample (705 persons in 16 different occupations). Factor analysis identified five dimensions of self-perceptions of fatigue: (1) lack of energy; (2) physical exertion issues; (3) physical discomfort; (4) lack of motivation; and (5) sleepiness. Their research found that lack of energy was a general underlying factor, while the others are more situation-specific. Validation samples suggest that these dimensions are differentially sensitive to different types of work and to when the work occurs. Night-shift NPP supervisors showed the highest ratings on the sleepiness factor, whereas factory workers showed the highest ratings on fatigue-related physical discomfort. Ahsberg et al. (2000) conducted subsequent testing of this instrument on tasks specifically designed to induce monotony and boredom to assess the sensitivity of these dimensions to this aspect of the task.

Åkerstedt et al. (2008d) found that other task characteristics can influence self-report ratings of sleepiness. For example, a study in which subjects were given the opportunity to walk around versus to simply relax and then perform a reaction time test showed that walking around resulted in slightly reduced sleepiness ratings.

Although self-assessment scales and questionnaires have been useful for understanding how different task environments and schedules affect fatigue, these tools may have limited value in the field because people appear to be able to assess some dimensions of their fatigue levels with reasonable accuracy without the assistance of self-assessment instruments. Instead, helping the individual grasp the probability of degraded performance and associated consequences may be more important and useful by increasing the likelihood that appropriate actions or interventions are taken when a person does experience fatigue. These tools typically address current manifestations of fatigue. Researchers have not focused on the ability of self-assessments to predict the probability of becoming fatigued or detect early onset of fatigue prior to obvious fatigue manifestation. Although guides for assessing one's current state of fatigue seem to add little to the individual's understanding, similar self-assessment tools to help individuals assess life, work, and sleep patterns in terms of their contribution to fatigue may be informative and useful to the individual. An instrument to help assess fatigue retrospectively (i.e., the Retrospective Alertness Inventory [Folkard et al. 1995]), could provide a starting point. The review found no information about the perceived utility of self-assessment guides or instruments to supervisors for use in conducting fatigue assessments.

Biomathematical Models

Considerable effort has been devoted to adapting and refining biomathematical models of the sleep-wake cycle to estimate the prevalence, extent, and performance consequences of fatigue for both work groups and individuals. Although biomathematical models do not directly measure an individual's biological processes or states, or estimate impact on actual job performance, they provide an analytic framework and algorithms for incorporating information about past or projected work hours, and schedules with the timing of prior sleep and waking (and sometimes the nature of the work tasks) to estimate "average" fatigue levels over time. These models, referred to as two-step models, are based on how work-rest schedules, would affect sleep/wake processes on average. Because of significant individual variability[88], these tools typically are not designed to predict the fatigue level of any particular individual but rather to estimate group-

---

[88] A wide range of personality variables have been evaluated in relation to fatigue and vulnerability to shiftwork circadian disruptions. These include tests of morningness-eveningness (Horne and Osterberg 1976), evaluations of anxiety and mood

average fatigue levels. These models are being used to assess which work schedules are better or worse in terms of contributing to the potential for fatigue across the work group as a whole. An active community of biomathematical modelers has been working on the development and refinement models for many years (see for example, Borbély and Achermann 2003; Dawson 2009; Dinges 2004 and Dinges et al. 2004; Gundel et al. 2007; Gunzelmann and Gluck 2008; Hursh 2009a,b; Ji et al. 2006; Mallis et al. 2004; McCauley et al. 2009a,b; Rajaraman et al. 2009; Van Dongen 2009). Consequently, the models discussed in the literature have often been further refined and updated by the time the article is published.

Research is under way to improve the specificity of the models by including additional variables (e.g., naps and the nature of the task) and to improve their ability to estimate and/or predict individual-level fatigue, for example by including genotype information (Bes et al. 2009). Complementary efforts are under way to develop models that are more reliable, whose reliability parameters are better understood, and to estimate and/or predict fatigue-caused performance impairment (Dinges 2004; Dawson 2009; Gunzelmann et al. 2009; Dawson et al. 2011). Progress on understanding the neurophysiology of the sleep-wake cycle, as discussed above, is resulting in revision of the underlying structure of the biomathematical models (see for example McCauley et al. 2009; Rempe et al. 2009). Individual variability, data needs, and the indirect relationship between impairment measures and impact on job performance pose major challenges to the goal of developing individualized models that can predict fatigue hazard for a particular shift of work. There is an entire literature on biomathematical modeling of fatigue and the relative merits and shortcomings of modeling approaches and particular modeling tools. A comprehensive review of this literature is beyond the scope of this report.

3.2.4 Assessing Causes and Effects of Fatigue

In addition to understanding the concept of fatigue and estimating a person's state of fatigue, managing fatigue requires assessing and addressing the causes and effects of fatigue. Basic fatigue research has typically focused on the individual level. However, addressing workplace fatigue requires examining the fatigue issue at the individual, work group, job category, and organizational levels.

Figure 3.4 depicts a cause-state-effect model of fatigue derived from the literature reviewed for this chapter. Starting in the middle of this figure, the *state of fatigue* is shown as a continuum beginning with the earliest detectable stage (possibly preceding any outward sign of degraded alertness) to a stage of extreme drowsiness (immediately preceding the onset of sleep). The physiological, behavioral, and affective indicators, as addressed in the previous section, are shown in the center diamond-shaped box. This figure illustrates that:

- Lifestyle and work related factors, in combination with mediating conditions, can cause a future state of fatigue; and
- The state of fatigue can cause performance degradation.

Cause, state, and effect indicators can be assessed in terms of their predictive validity, but it is important to differentiate between the ability of causal factors to predict a state of fatigue versus

---

traits (Craig et al. 2006), and assessments of characteristics such as flexibility, vigor, languor, and rigidity (Folkard et al. 1979; Di Milia et al. 2005). The basic concept underlying this research is that individual traits will affect physiological variables, such as sleep need and vulnerability to the disrupting influences of shiftwork. To the extent that workers can be selected or assigned based on their "circadian type," it is presumed that fewer negative consequences of shiftwork will result.

the ability of indicators of the state of fatigue to predict performance effects. Different levels of predictive validity are associated with the first versus the second cause/effect phase.

Causes of Fatigue

The causes shown in Figure 3.2 (shown below) were extracted from the literature on fatigue. Causes can be more or less immediate and can interact in complex ways. The literature also indicates that causes of degraded alertness may differ from causes of severe drowsiness. Distractions or monotony can cause inattention but may not necessarily produce severe drowsiness while sustained monotony might eventually cause severe drowsiness. Other causes of drowsiness can be compounded by monotony. Causes include both job and lifestyle factors, as well as conditions or attributes that mediate the effect of these factors on fatigue. The key workplace factors of concern have been shift work, shift duration, cumulative work hours, and regularity of work schedules but other factors are increasingly being researched and addressed, such as workload, monotony, and environmental stressors/enhancers. Lifestyle issues have not been addressed to the same extent. Other conditions that can aggravate lifestyle and job factors, including stress, cumulative sleep debt, sleep disorders (particularly sleep apnea and insomnia), and obesity, are receiving greater research attention. Workplace fatigue management programs are starting to incorporate some of these causal factors (Matthews et al. 2009).

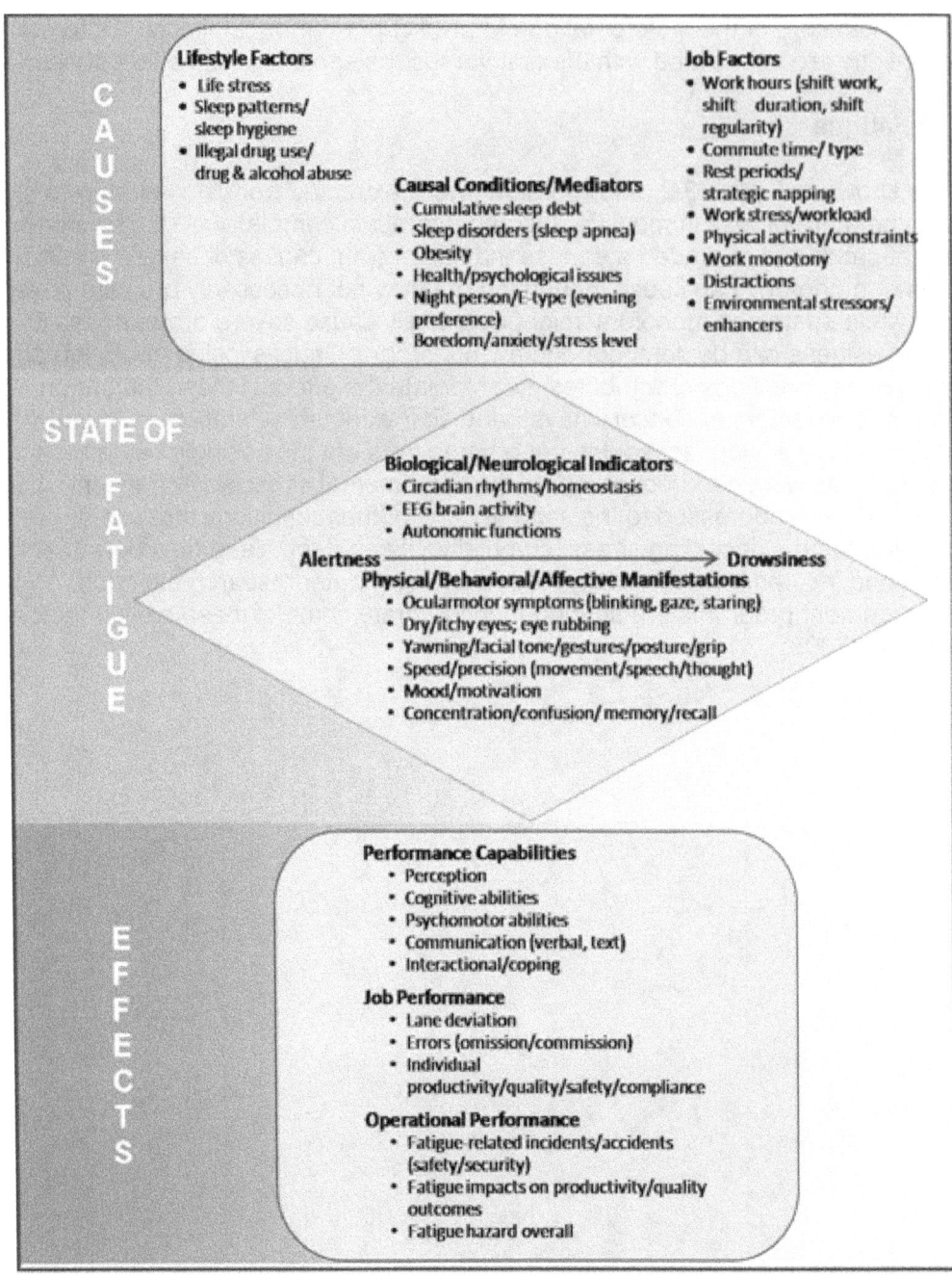

Figure 3.2.   Conceptual Cause-State-Effect Model of Fatigue

Research has established that individual variation, in terms of need for sleep, vulnerability to circadian rhythm disruptions, and ability to recover from sleep deprivation, is considerable. Some of this variation is associated with well-understood individual differences such as obesity and/or having a sleep disorder. Obesity and sleep apnea are becoming critical concerns in the transportation sector. A study of 193 truck drivers found 82 percent to be overweight and 53.4 percent to be obese. The overweight/obese drivers were 8.95 times more likely to be rated by expert observers as fatigued and 1.69 times more likely to be identified as fatigued using a PERCLOS measure. Overweight/obese individuals are more likely to become fatigued for many reasons, sleep apnea being one of these reasons. Tregear et al. (2009) estimated that the probability of a motor vehicle crash is 30 to 42 percent greater for an individual with obstructive sleep apnea (OSA) than individuals without this disorder, and that the probability increases with the severity of the disorder. Paim et al. (2008) investigated sleep disorders among NPP shift workers and found that 27 percent of participants in the study met the diagnostic criteria for a clinical sleep disorder.

Less obvious personal conditions and predispositions have also been shown to contribute to individual variation. Predispositions include morningness/eveningness (Horne and Ostberg 1976), anxiety and mood traits (Craig et al. 2006), flexibility, vigor, languor and rigidity (Folkard et al. 1979; Di Milia et al. 2005), and introversion/extroversion (Killgore et al. 2007).[89]

Research efforts tend to investigate causal factors. Efforts are under way to develop biomathematical models that include how the two main endogenous processes associated with the sleep/wake cycle (i.e., circadian rhythms and sleep homeostasis), in conjunction with other causal factors, predict the state of fatigue. Generally, the greater the number of relevant causal factors taken into account, the greater the predictive validity of the model. Also, just as individual variation exists with respect to the key endogenous processes used in biomathematical fatigue modeling, individuals respond differently to many of the causal contributors identified in Figure 3.2. Fatigue models that incorporate additional causal factors and take individual variation into account are still in the development stage and are not yet sufficiently tested and validated for application in the field. In addition, it is important to note that all of the models stop short of attempting to predict actual job performance. Even the most sophisticated models only attempt to estimate impact on specific constrained tasks, such as reaction time on a vigilance test, rather than on the integrated and diverse capabilities required for job performance.

Effects of Fatigue

In this discussion, the effects of fatigue experienced by the person as the state of being fatigued are distinguished from the effects of fatigue on performance. As indicated in Figure 3.2, the causes of fatigue affect the state of the individual, for example by altering the individual's circadian rhythm, EEG patterns, mood, and a variety of other variables, such that the individual enters a state of fatigue. The causes of fatigue also affect the performance of the individual. For this discussion, the subset of the effects of fatigue on performance are grouped into three levels: individual performance capabilities; job performance; and operational safety performance.

<u>Indications of a Fatigue State</u>

In general, the variables identified in Figure 3.2 as indicators of an individual's state of fatigue are not treated or discussed consistently in the fatigue literature. Neurological studies, for example, provide convincing evidence that performance decrements associated with sleep loss are the consequence of actual changes in cerebral function (Williamson et al. 2011). The review did not find a body of work that integrates the considerable amount of empirical

---

[89] Killgore et al. (2007) studied the effects of introversion/extraversion on psychomotor vigilance test (PVT) performance. They found that after a single night of sleep deprivation, introverts showed faster reaction times than extraverts did. This effect was not observed on subsequent nights. The authors interpret the results as indicating differential levels of baseline cortical activation manifested as the trait differences, which confer a relative performance advantage for introverts after a single night of sleep deprivation.

information about these physiological variables into a coherent description of fatigue states and their relationship to causal factors nor performance effects.[90] Information about the state of fatigue could be of great help in understanding the observed variability between individuals regarding their susceptibility to the effects of sleep deprivation. Information about fatigue states could also help in understanding the finding that sleep-deprived subjects sometimes demonstrate little impaired performance on one task, while showing great impairment on another (King et al. 2009). This is an area that warrants additional attention.

Performance Capabilities

Performance testing associated with fatigue research initially focused on measuring attention, vigilance, reaction time, and coordination using a performance vigilance test (PVT).[91] It typically included consideration of sleep homeostasis factors, circadian influences, and the nature of task effects (Williamson et al. 2011). In order to determine what constitutes a "dangerous" level of fatigue, a few studies have used PVTs to compare the effect of sleep deprivation to the effect of an elevated blood alcohol concentration (BAC). Australian researchers Williamson and Fiswell (2009) demonstrated psychomotor performance degradation at around 18 hours of wakefulness (commencing from 06:00 hrs) that was equivalent to performance decrements observed at 0.05 percent BAC (the legal BAC limit for drivers in Australia). Significant performance deficits were only found when high levels of sleep deprivation coincided with the low point of the circadian rhythm. Performance on simple, monotonous, and unstimulating tasks is particularly vulnerable to the impairment from fatigue and can result in slowed or incorrect responses, or failure to respond. Williamson et al. (2011) hypothesize that the lack of clarity about the effects of fatigue on performance of complex tasks may result from the difficulty of determining what aspect of an individual's performance is affected.

There is little clear evidence as to what constitutes a "dangerous" level of fatigue (Lamond and Dawson 1999). Some researchers suggest that the effect of sustained wakefulness on performance has been over-estimated (Maruff et al. 2005). Others suggest that it has been underestimated because of the use of simple PVT-type tests; noting that the impact of fatigue on performance depends on the type of performance capabilities being examined. Lack of sleep may not measurably degrade psychomotor performance until the person is extremely sleepy. Automated or highly over-learned tasks such as tracking are more resistant to fatigue than mental tasks, such as decision-making and other higher cognitive tasks involving short-term memory and attention (Miller 1996).[92] EEG data indicate that one can steer a vehicle successfully on a straight highway while the brain's cortex is asleep (O'Hanlon and Kelley 1977).

A less severe state of fatigue has, however, been shown to affect a driver's ability to perceive hazards (Smith et al. 2009). Alertness, concentration, and the ability to maintain focus may be even more readily affected than higher-level cognitive processing. Fatigue and other causes of

---

[90] It is poss ble this literature exists and was not identified as a consequence of the focus on technologies and methods. A number of the measurement methods described in Section 3.2.2. would provide information about these state indicators and evidence about the effects of fatigue on them.

[91] These tests are sometimes referred to as stimulus response reaction tests (SRRT).

[92] A study by Landrigan et al. (2004) provides some evidence for the relationship of sleep deprivation to on-the-job safety in cognitive work. They showed that reducing work hours for on-call physician interns reduced serious errors by 50 percent. McKenna et al. (2007) evaluated performance on a choice task involving risk and how it is framed (i.e., as a potential gain or loss). The results suggest that sleep deprivation decreases risk taking when a potential loss is presented, but increases risk taking when the problem is framed as a potential gain. These results suggest that in work situations where sleep deprivation and safety may be an issue, framing decisions in terms of conservative approaches may be a reasonable strategy. Also, educating persons as to how fatigue may affect their decision making may be of value.

impairment are likely to affect the ability to sustain alertness and concentration first, then perception and higher-level cognitive functioning, and motor functions last. Conversely, alertness may recover from cumulative sleep deprivation more quickly than performance on PVT tests. A study examining severe sleep deprivation followed by a 9-hour sleep recovery period found alertness recovered after a single 9-hour sleep period, physiological measures recovered after two 9-hour sleep periods, but PVT performance remained below baseline for the entire recovery period (Van Dongen et al. 2002; Van Dongen et al. 2004).[93]

A similar study was conducted by Axelsson et al. (2008) in which subjects were restricted to four hours of sleep for five days, followed by seven recovery days of eight hours of sleep. As expected, both self-reported sleepiness scores increased and PVT scores deteriorated over the restricted sleep days. However, during the recovery days, self-reports of alertness rebounded faster than the person's PVT scores. After three recovery days, sleepiness scores returned to baseline, but the performance scores did not. The researchers also found substantial individual variation in the extent to which perception of alertness corresponded to improvement in PVT performance scores. The implication of this latter finding is that two people may rate themselves as equally alert but one might display greater performance impairment than the other.

The fatigue research distinguishes between two distinct but related concepts: *vulnerability* and *resilience*. Vulnerability refers to how easily one becomes fatigued in the sense of showing indications of a state of fatigue, and resilience refers to the extent to which being fatigued affects the individual's performance ability or ability to recuperate. Just as research has demonstrated that *vulnerability* to fatigue varies considerably among individuals, studies have also shown individual variation in terms of *resilience* to fatigue. Studies have demonstrated substantial individual variation in terms of performance effects resulting from seemingly similar states of fatigue that are a result of sleep loss. Kecklund et al. (2009) found fatigue resilience is as important as the amount of sleep loss in predicting performance degradation. However, this review found no research that investigated factors that might explain variability in the effect of fatigue on an individual's ability to perform particular tasks, as was reported by King et al. (2009). Research has addressed the differential impact of fatigue on different aspects of performance (psychomotor vs. higher levels of cognitive functioning) but has not succeeded in delineating the source of observed *individual* variation (see Di Milia et al. 2011 for a discussion of the literature on demographic factors and fatigue). A better understanding of what causes individual variation with respect to being able to sustain performance while fatigued could provide guidance for the development of prevention and mitigation strategies.

Job Performance

There are a number of studies investigating the effect of fatigue on job performance, a large proportion of which focus on vehicle driver performance. Many of these studies were conducted in simulated driving conditions. Several simulator studies have shown that lane deviation is associated with length of time driving and other attributes of the driving task that were expected to contribute to fatigue. Smith et al. (2009) found that a technology designed to assess a driver's ability to perceive hazards could detect fatigue at a lower level (i.e., was more sensitive) compared to lane deviation technologies. Technologies to measure lane deviation in the field are commercially available and have been pilot tested in some driving work situations (see

---

[93] It is noteworthy that performance on all the measures improved, however, at different rates and that performance on none of the measures continued to deteriorate, as would be suggested by the studies of cumulative sleep debt.

Section on 3.3 on applied technologies). Williamson et al. (2011) reviewed research on the effect of fatigue on job performance in terms of accidents and injuries, errors, and slowed responding. They concluded that there is strong evidence that fatigue-inducing conditions produce impairment in job performance that affects safety.

Operational Performance

Effects on performance capabilities or on actual job performance do not necessarily translate into operational safety, security, quality or productivity impacts, or a requirement to stop work. Organizations involved in safety-related work activities implement defense-in-depth strategies to decrease the probability that human performance errors will result in a consequential safety incident. Alternatively, the need for and value of the service provided (for example by physicians, public safety personnel, military personnel) may outweigh the risk from errors because of fatigue (Dawson and Zee 2005). Investigating the role of fatigue in incidents/accidents provides some evidence that fatigue has an impact on operational performance. However, studies of this type are not common outside of a few commercial transportation sectors. Also, the data used in these studies may not be highly accurate because of inadequate methods and processes for investigating the role of fatigue in incidents and accidents, and the inability to control for confounding variables.

One of the best ways to assess operational impact may be to have relevant experts evaluate the potential impact on safety, security, and/or other operational performance dimensions of interest. This assessment would need to consider normal as well as possible off-normal conditions. It should also address the extent to which managing fatigue versus increasing the resiliency of operations from human error would contribute to a more effective and efficient solution.

3.2.5 Literature Summary

There is a strong scientific basis for concluding that the use of multiple indicators, including multiple categories of indicators, improves detection and prediction of fatigue and fatigue-related impairment. For example:

- multiple indicators across various categories of indicators improve the ability to *detect* and *characterize* the *state of fatigue*;
- multiple causal indicators improve the ability to *predict* the *state of fatigue*;
- multiple indicators of the state of fatigue are better able to *predict performance effects*; and
- multiple indicators can improve the ability to assess the role of fatigue in accident/incident investigation.

Capturing multiple types of indicators is likely to require multiple methods. Researchers are coupling monitoring technologies with biomathematical fatigue models to improve both the detection and prediction of fatigue and degraded performance. Combining data from different technologies is the future direction of fatigue research.

The sensitivity and accuracy of different categories of indicators and particular indicators within categories may depend on characteristics of the individual or the work situation. The following considerations affect which types of indictors and methods to use:

- whether the focus is directed at assessing the fatigue hazard for a particular work situation or estimating a particular individual's likelihood of being fatigued;
- the need for interventions to address causes and/or the effects of fatigue; and
- the kinds of indicators that can feasibly be measured, given the situation.

Researchers are replacing simple thresholds with complex algorithms that combine multiple indicators within and across categories of indicators, such as head movement and EEG threshold measures (Hussain et al. 2008). In addition, researchers are developing algorithms that learn and improve over time using Markov models, neural nets, Bayesian techniques, and fuzzy logic combined with scalable monotonic chaining.[94] The basic feature of these advanced approaches is the use of a relatively large "training" data set to develop and fine-tune feature extraction and classification procedures that increase concurrent and predictive validity. These techniques are in preliminary stages of development but may eventually provide a more accurate discrimination of fatigue states and improved ability to predict performance effects.

Although the literature clearly supports the use of multiple indictors and/or technologies, it provides only limited insights regarding the particular combinations of indicators and measurement techniques that would provide the most accurate estimate of individual-level fatigue and/or the fatigue hazard for a particular work situation. Current scientific knowledge is also far from conclusive as to how to weigh or resolve contradictions among indicators.

3.3 Overview of Applied Fatigue-Related Technologies

A number of the technologies used by fatigue researchers have been adapted for use in applied work settings. A high proportion of these applied technologies are designed primarily for use in commercial trucks in the transportation sector, but are increasingly in use in other transportation sectors and in non-commercial automobiles and trucks. In addition, the Australian mining sector has piloted and/or adopted some of these technologies. Although the technologies have generally not been piloted extensively or adopted widely in other work settings, many have potential utility beyond work situations involving driving and heavy machine operation.

This overview differentiates technologies in terms of their intended uses, implementation context, and requirements. It summarizes their strengths and weaknesses, with particular attention to attributes that would make them more or less suitable for application in the nuclear power industry. The technologies reviewed perform different functions, address different aspects of fatigue, and have typically been designed for application in specific work settings. Every new technology must be field tested in the particular context in which it will be deployed. The common consensus is that no currently available technology or combination of technologies provides an adequate and sufficient means of detecting and/or addressing fatigue, even in the transportation sector. The utility of any particular type of technology or combination of technologies will depend on the particular work situation, including its role in the system of technologies already in place.

---

[94] Markov models are being used to incorporate learning with respect to facial feature recognition (Dinges et al. 2005). Neural nets are being used to improve individual fatigue estimates based on EEG (Lin 2009). Circadian Technologies Inc. has developed the Fatigue Accident Causation Testing System (FACTS) consisting of a diagnostic survey instrument, along with a weighted risk model based on "fuzzy scalable monotonic chaining," to help investigators determine the role of fatigue as a causal factor in driving accidents. Biomathematical fatigue modelers are attempting to improve the prediction of individual performance using Bayesian estimation involving traits learned from reference data (Mott et al. 2009).

3.3.1 Categorization of Technologies

The review found numerous technologies that are currently available, as well as some that are fairly close to commercial viability. Greater detail on the technologies can be found in Appendix A.

A few categorizations of fatigue-related technologies exist in the literature, none of which provided an adequate basis for organizing the technologies developed to address fatigue and fatigue-related impairment. A categorization system based on how the different technologies fit into a larger fatigue management system is proposed here.

This approach builds on the key distinctions used in other categorization schemes and demonstrates how the technologies contribute to an overall fatigue management system. The fatigue management perspective suggests a fundamental distinction among the technologies: most of the technologies have the intended function of providing some type of intervention or countermeasure and a smaller number are oriented toward assessment of the fatigue hazard.

3.3.2 Review of Intervention Technologies

Fatigue Prediction/Prevention Technologies

Applied fatigue prediction/prevention technologies look at causes of fatigue in order to predict and prevent the probability of fatigue on the part of individual workers and/or the work group as a whole. The following types of technologies directed at predicting and/or preventing causes of fatigue were identified:

- biomathematical scheduling and roster management tools;
- biomathematical tools to predict/detect individual-level fatigue;
- employee assistance tools; and
- employee screening and selection tools.

<u>Biomathematical Scheduling and Roster Management Tools</u>

Until recently, biomathematical models of the sleep-wake cycle and fatigue have primarily been tools for basic research, seeking to integrate the empirical data obtained from studies by researchers from a wide range of disciplines.[95] However, applied biomathematical fatigue technologies are now being used to predict the prevalence and extent of fatigue for work groups and to evaluate alternative work and non-work schedules. These biomathematical tools do not directly measure individuals' biological processes; instead, they use projected or past work hours, and work schedules to predict workforce fatigue levels based on how these factors, on average, are expected to affect the sleep/wake cycle. Torgovitsky et al. (2009) are building a model specifically designed to assess an individual's response, rather than a group average response, to work schedules.[96] Another proximate research goal is to develop a technology

---

[95] As discussed above, these models were developed to illustrate the complex interaction between sleep-wake and circadian systems (Dawson et al. 2011) and to explain the growing body of empirical data, but with little understanding of the neurophysiological processes that constitute either the drive for wakefulness or sleep. With the explication of the neuronal populations and their interactions at a molecular and anatomical level, this is just now beginning to change.

[96] Torgovitsky et al. (2009) report that they conduct neurobehavioral testing at different combinations of circadian phase and homeostatic sleep drive to estimate subject-specific effects using smoothing spline ANOVA models with restricted number of knots, a method used to study potential covariate effects.

that can reflect fatigue patterns of workers based on their "circadian type" in order to guide worker selection of different types of work schedules and to improve the design and evaluation of work-rest schedules.

Researchers have historically used biomathematical fatigue models to investigate how endogenous sleep/wake processes correlate with indications of fatigue and/or performance impairment. A key goal was to improve the validity of these models with respect to predicting, detecting, and estimating fatigue. Tools now exist that use biomathematical fatigue models to evaluate alternative work schedules and optimize schedule solutions by balancing fatigue-informed practices and business needs. These biomathematical scheduling technologies provide feedback regarding "time-at-risk" for particular schedules and predict the likelihood of fatigue for the work group. Individual variability on multiple parameters has thus far thwarted efforts to develop models capable of predicting fatigue for a particular individual. Advanced scheduling tools incorporate additional factors besides work schedules, such as workload, nature of the task (especially monotony), environmental conditions, and naps.

In addition, developers are attempting to individualize these tools by incorporating individual input regarding sleep time and quality, relevant individual characteristics, operator feedback, and system learning (Van Dongen 2009; Dawson et al. 2011). Van Dongen et al. (2007) applied Bayesian forecasting to the basic two-process model using available data to optimize model parameters for individuals. They were able to predict future performance within the 95 percent confidence intervals established by the model. Predictive accuracy increased and confidence intervals narrowed over time as more data became available. However, this individualized approach would require workers to provide accurate sleep/wake history data, which may not be in their interest to do.

The United Kingdom Civil Aviation Authority is using System for Aircrew Fatigue Evaluation (SAFE) (Kinetic, UK) to predict aircrew fatigue. This tool incorporates crew input that allows the tool to learn and improve over time. Individual-level input allows the tool to address variability and become a better individual-level fatigue prediction tool as it advances.[97] Individual-level information can be used to improve the scheduling tool and could be used to apportion scheduling to the extent possible for the particular work situation, such as assigning driving routes that are most compatible with a person's specific traits. At a minimum, more individualized scheduling tools could help in making decisions regarding overtime. Another method for individualizing scheduling tools is to input individual-level data from actigraph, oculometric, and/or PVT technologies.

Developers anticipate that biomathematical scheduling tools will bring about major advances in fatigue management as they incorporate advances in fatigue science. Currently, best practices for schedule design include:

- hour limits (day, week, month, and longer periods of time);
- schedule length, schedule rotation (forward rotation and reduced rotation frequency); and
- schedule regularity.

These practices do not address all commonly occurring sources of fatigue, in large part because they do not incorporate information about what individuals do with their non-work time.

---

[97] Data on the utility and effectiveness of the SAFE model are not publicly available.

Therefore they do not take into account the actual amount and quality of rest/sleep individuals have had when they come to work. In addition, they do not reflect research indicating that a shift worked at night contributes more to fatigue than a shift of the same length worked during the daytime[98] (see Gundel et al. 2007; Dawson 2009; and Dawson et al. 2011 for reviews of the current state of biomathematical models).

Biomathematical models are reaching a degree of refinement and validation that make them feasible to use in optimizing schedule and roster management within the parameters established by regulations designed to prevent excessive work hours (McColgan and Nash 2009). However, no country currently mandates the use of biomathematical scheduling tools. Australia encourages their use in various transportation sectors but has withheld regulatory endorsement until guidelines governing their use are completed. The experience indicates that early adopters often did not adequately understand the tools, scheduling personnel sometimes abused the process, and the process often conveyed a false sense of accuracy and certainty about fitness (Dawson 2009; Dawson et al. 2011). These tools, for example, became unpopular with airline staff. Some flight staff reported that they were told they could not be fatigued because the model said they were not, despite acknowledgement by model developers that many factors affect workers' fatigue, not just the work schedule and that there is significant individual variability even in terms of how work schedules affect different people. In fact, group-level estimates of fatigue do not predict individual-level fatigue nor do they represent *readiness-to-perform* measures.

Australian regulators have expressed the view that these scheduling tools could be useful if used properly and not as the only strategy to manage fatigue. Their emphasis is now on developing a fatigue management approach that draws upon biomathematical models but uses multiple controls. Biomathematical model developers acknowledge that factors unrelated to hours of work and schedule design contribute to fatigue and suggest that these models only be used as guidelines for assessing and selecting schedule solutions.

Even in this limited capacity, these tools will require the development of organizational policies governing their use (or nonuse), delineating the factors other than model output that will be considered in scheduling and roster design, and addressing issues of liability and the allocation of responsibility for decision making, data collection, and monitoring (Rosekind 2009). Policies will also be needed regarding the collection, protection and use of information from employees (e.g.,, what aspects of the model solutions and organizational actions will be "public domain" versus "confidential" versus. "discoverable" if there is a negative safety outcome) (Rosekind 2009). To the extent that biomathematical models require individual-level data, confidentiality issues and data reliability could become greater concerns.

In addition, as emphasized by all reviewers, more evaluation is needed to determine how well these tools can help optimize schedule design and roster management in terms of balancing fatigue concerns with operational constraints and goals. Nevertheless, despite their limitations, there is general agreement that biomathematical models can be useful for highlighting fatigue "hotspots," particularly when a large number of duties and associated rosters need to be analyzed and multiple different schedules coordinated.

---

[98] For example, work hour regulations have at their basis the assumption that rest and recovery time (hence the amount of sleep workers obtain) depends on how much time people are at work. Analysis of extensive survey data obtained in the American Time Use Survey suggests that the only variable showing a reciprocal relationship with amount of time reported sleeping is the amount of time reported working, but the correlation (-0.36 for weekdays), while significant, is quite small (Basner et al. 2007). Advancements in biomathematical models are attempting to take this type of individual-level information into account.

The following biomathematical tools are designed to evaluate and optimize schedule design (only the first two are commercially available):[99]

- SAFTE (Sleep, Activity, Fatigue, and Task Effectiveness model, based on the University of Pennsylvania Medical School model, available from NTI Inc., Science Applications International Corporation, and Archinoetics);
- CAS (Circadian Alertness Simulator model, available from Circadian Technologies);
- Sleep/Wake Predictor Model (also called the Three Process Model of Alertness or TPMA, associated with Åkerstedt and colleagues).

Biomathematical Models to Predict/Detect Individual-Level Fatigue

Rajaraman et al. (2009) are developing a biomathematical model to predict performance, as opposed to fatigue. This model employs the two-process model of sleep regulation as the underlying template but incorporates previously collected PVT response-time data for individuals subjected to sleep deprivation to customize the parameters for an individual.[100] A major limitation of their initial effort was that a number of past PVT measures had to be available from an individual before model-parameter customization and prediction could commence. They have improved on their method for individualized performance prediction to enable estimation of an individual's performance as early as the first PVT observation. This approach uses Bayesian inference that combines *a priori* information about the model parameters with information obtained from the individual's performance measurements. Using simulated performance data for known model parameter values, they found their newer method yielded parameter estimates that asymptotically converged to their true values as the number of performance observations for an individual increased and the amount of uncertainty in the data decreased. Dawson et al. (2011) note that these models are essentially estimates derived from estimates and that caution is required in interpreting their statistical characteristics.

The applied use of biomathematical models is moving beyond fatigue prediction based on schedule and roster evaluation. Applied individual-level biomathematical fatigue models that directly *detect*, as opposed to *predict*, individual-level fatigue in the field are not feasible for reasons previously discussed,[101] but developers are combining predictive biomathematical models with fatigue detection technologies in order to enhance the ability to both predict and detect individual-level fatigue in real time. For example, developers at Archinoetics are coupling biomathematical models with data obtained from actigraphs worn by the workers. Developers of ocularmetric technologies have not yet incorporated oculometric data into biomathematical fatigue models but this is likely to occur in the near future.

By coupling these technologies, it may be possible to both:

- improve biomathematical prediction technologies by incorporating detection technology data; and

---

[99] Other biomathematical models under development and/or commercially available but not designed to evaluate and optimize work schedules include the two-process model, SAFE, interactive neurobehavioral model, and FAID (see Barr et al. 2009 and Aviation, Space, and Environmental Medicine, 2004).

[100] Performance vigilance tests and psychomotor vigilance tests are used in the literature to mean the same thing.

[101] There are issues involving measuring these endogenous processes on a continuous or sufficiently frequent basis in the field to produce estimates of individual fatigue based on the biomathematical fatigue model. There is also the problem that in the field there are many other factors that contribute to fatigue. Just using biomathematical fatigue modeling associated with sleep/wake cycles and patterns is not likely to have a high level of validity in terms of detecting or predicting a worker's state of fatigue.

- improve the sensitivity and accuracy of individual-level fatigue detection technologies by incorporating biomathematical fatigue models.

These coupled technologies could improve and individualize biomathematical scheduling and roster management tools, improve fatigue detection, improve worker and control room fatigue alerting, and contribute to informed real-time decisions about rest breaks and overtime waivers.

It is not clear how well these systems will work because they are just being introduced into the workplace. However, it seems that eventually the marriage of real-time physiological data with biomathematical fatigue models will inevitably improve both fatigue prediction and detection at the group and individual levels. According to Archinoetics' marketing information, a growing number of companies, especially in Australia, are considering piloting the combined biomathematical model/actigraph system (see Appendix A).

Employee Assistance Tools

Although awareness of fatigue as a risk factor has increased in the workplace, behavioral changes to better manage fatigue have been much more difficult to achieve. This review identified only a few tools, beyond education on the basics of fatigue management, that are directed at aiding workers to make fatigue-informed behavioral changes. There are also self-assessment tools that help individuals assess their life, work, and sleep patterns to determine the likelihood of future fatigue or more serious health impacts. In addition, employee assistance tools, such as screening for sleep apnea and stress management counseling, can be useful to employees with life stress or sleep disturbances even if fatigue is not found to be a major hazard for their particular work situation.

Employee Screening and Selection Tools

The most common screening technology applied in work environments involves screening individuals for sleep disorders. Sleep disorders and obstructive sleep apnea are becoming an increasing concern for several types of workers, including commercial truckers, aircrews, and law enforcement officers. There have also been efforts to screen individuals on other dimensions to determine their propensity to become fatigued, such as their tolerance for shift work.

Sleep Screening

Typically, screening for sleep apnea requires an overnight polysomnography in a sleep center. However, Gerson et al. (2009) demonstrated that a two-step method of screening involving a questionnaire and confirmation by a self-applied ambulatory device (such as actigraphy) to measure physiological indicators is a reliable alternative. Research has demonstrated that actigraphy is a highly accurate and reliable means of detecting sleep disorders. It provides basic information on quantity and quality of sleep. However, if actigraphy finds disrupted sleep, it cannot diagnose the source of the problem. Some actigraph systems are being marketed for home use.

Shiftwork Tolerance Screening

Circadian Technologies is developing a technique called Shiftwork Adaptation Testing System (SATS) to predict how well an individual can tolerate shift work (Trutschel et al. 2009). In the future, it may be possible to genetically screen individuals. Research has found gene PER3 plays a role in the generation of biological rhythms that differentiate morning, intermediate, or evening types (see Section 3.2.2. on Gene Expression Analysis). A major issue confronting genetic-based screening technologies is Public Law 110-233, the Genetic Information Non-Discrimination Act (GINA), enacted in 2008 (U.S. Congress 2008) to prevent discrimination against individuals with regard to health insurance and employment.

Some model development efforts that are not based on genetic information may enable future technologies that can be used to screen workers for adjustment to shift work. Measuring an individual's melatonin level over a period of time has been used to determine how well a person might adapt to particular work shift schedules (Dawson et al. 1997). In addition, Lehrer (2005, 2009) is developing a theoretical model that integrates stress and coping research with shiftwork research to enhance understanding of shiftwork stress. Specifically, the model examines how control and support mechanisms influence shiftwork adjustment. By applying demand, control, and support to shiftwork, this model applies Karaseks' (1979) expanded demand-control model of stress to shiftwork. The results were validated using survey instruments.[102]

Research using this predictive, integrative model has investigated the roles of individual variability, adaptive coping strategies, schedule preference, and their relationships in predicting bio-psycho-social adaption under stress. This research has found that work schedule preference is significantly related to shiftwork adjustment estimates. It also attempts to show how to moderate maladaptive response patterns through a better understanding of the nature of coping and coping malleability in response to moderating effects of support and control. The goal of this research is to identify and encourage interventions aimed at facilitating adaptive lifestyle changes and workplace strategies. It might further provide new biological/psychological insights regarding stress and fatigue coping mechanisms and abilities.

Fatigue Detection/Measurement Technologies

Fatigue detection/measurement technologies include biochemical assays, monitoring systems, and self-assessments.

Biochemical Assays

Although not currently in use as a fatigue management tool, the technology exists (immunoassay and mass spectrometry) to test blood, urine, and/or oral fluids for the metabolite of melatonin, 6-sulfatoxymelatonin, to determine the level of melatonin in an individual's system (Bojkowski et al. 1987). Though a simple melatonin test cannot currently be used to estimate an individual's state of fatigue, measuring an individual's melatonin secretion over a time interval has been used to determine how well a person might adapt, or has adapted, to particular work shift schedules (Dawson and Armstrong 1996; Roach et al. 2005). Testing for melatonin levels could prove useful in assessing the role of fatigue for post-accident testing. However, substantially more research to validate melatonin level as a reliable and useful indicator would be needed before this could occur.

---

[102] In this model, control and support operate interactively to predict adjustment, with higher levels of internal control buffering the relationship between support and adjustment, strengthening the demand-control-support conceptualization.

Monitoring Systems

Actigraph Monitoring Systems

Actigraph technology has traditionally measured wrist movements by having the person wear a device resembling a wristwatch. The wrist device is intended to be worn 24/7. Research has validated its accuracy in tracking sleep/wake cycles and the American Academy of Sleep Medicine has endorsed it for this purpose. Applied systems have combined wrist movement tracking with biomathematical models of fatigue to both detect and predict fatigue.

The validity of measures relying on actigraphy, like the eyewear technology, will be compromised if the individual does not wear the device. Some pilot test results have indicated poor acceptance by the workers who report the wrist device is bulky and uncomfortable to wear.[103] The Sleep Band software is designed to ensure that an individual cannot subvert the system by having someone else wear the band while the employee stays up. The Federal Motor Carrier Safety Administration (FMCSA) pilot study (Dinges et al. 2005) included an actigraph. However, the actigraph was used along with other technologies and, consequently, it is not clear whether actigraph feedback to the driver was responsible for changes in driver performance or alertness. Drivers did sleep more on their non-work days. Actigraph feedback alone might encourage the wearer to get more sleep, especially if he or she knew that the data would be available to management either on a regular or for-cause basis.

Advances in actigraphy are directed at capturing a wider range of physical activity, such as posture shifts and movement. Advancements may also capture physiological/ autonomic activity, such as electrodermal activity (skin conductance), skin temperature, heart rate and heart rate variability, and breathing rate and respiratory instability.

EEG Monitoring Systems

Systems to measure and analyze EEG signals are being incorporated into real-time fatigue detection and countermeasure devices. According to Caldwell et al. (2009), the B-Alert system, developed by Advanced Brain Monitoring, Inc., has been validated as an alertness indicator for driver fatigue and fatigue vulnerability. It relies on a wireless headset to enhance usability. In addition, Caldwell et al. (2009) and Lin et al. (2005) identify EEG-based drowsiness estimation systems that are based on computed correlations between EEG signals and fluctuations in driving performance to develop individualized models for real-time monitoring. Operational conditions involving high electric fields may limit the use of EEG monitoring systems.

Video Monitoring Systems

Video monitoring systems are fixed-station devices. They can be mounted on a vehicle dashboard to provide continuous monitoring during the driving period or at a computer workstation. If the work situation is not constrained enough for fixed-station video to capture, alternatives are fixed-station video oculometric instruments that give on-the-spot readings, as opposed to continuous readings. These are less accurate because they are not continuous. They also require a baseline and may be subject to subversion or error if the person is fatigued during the baseline reading.

---

[103] A new version of Sleep Band, ReadiBand™, is now available and apparently has addressed some of the comfort-related criticisms of SleepBand.

Video monitoring systems have focused on measuring oculometrics, sometimes combined with other physical features that can be captured by video recording, such as facial tone and posture. They have been pilot tested in vehicles and are being used in some instances on a regular basis in the transportation sector. FMCSA sponsored a pilot test of a video monitoring system called CoPilot. The results indicated that oculometric feedback showed small but reliable effects in reducing the occurrence of eyelid closures but the feedback to the driver did not significantly improve driver performance as measured by a lane tracking technology (Dinges et al. 2005).

There are several issues involving oculometrics for managing fatigue. Oculometrics may alert the driver to the fact that his or her fatigue is becoming increasingly severe, but it does not help the driver prevent performance impairment unless the driver elects to stop driving. Also, oculometric measures may not alert the driver in sufficient time to take action. Advances in oculometric research are improving early detection of fatigue onset and predictive accuracy in terms of performance impairment. These refinements capture a greater range of eye movements, such as breadth of eye gaze and extent to which the driver attends to instruments and rear-view mirrors, rather than focusing exclusively on PERCLOS and microsleeps. The goal is to improve early detection and the identification of levels of fatigue severity to enhance the effectiveness of warning messages to drivers. A key drawback of oculometrics is that this technique may not be transferrable to work situations where workers are not as physically constrained as drivers.

Eyewear Monitoring Systems

The newest monitoring technology involves placing a microchip in eyewear frames. Current eyewear microchip technology can capture a wide range of oculometrics. Eyewear oculometric technology has some advantages over video monitoring in that it is more reliable and applicable in more types of work environments.
Eyewear technologies tested in the field have not been well-accepted by workers because they have found them to be bulky and uncomfortable. A new eyewear technology, Eye-Com, is more streamlined. Though not yet commercially available, the eyewear will be developed by Oakley™, and the glasses will be very similar to popular Oakley™ sport glasses. Several additional features will be included in this new eyewear technology, such as enabling the wearer to send messages or control actions using their eyes. More advanced eyewear technology might be able to capture physiological metrics as well as oculometrics, but these capabilities are currently at the conceptual stage.

Self-Assessments

Self-assessment tools have been shown to have fairly strong validity in experimental studies. Three types of self-assessment instruments have been developed. They focus on:

- assessing current level of sleepiness;
- identifying and assessing current symptoms of fatigue; and
- helping individuals assess the likelihood they may become fatigued and identifying ways to manage their work, given that likelihood.

Most of the self-assessment tools have been developed to provide information to researchers, often to validate other more objective indicators of fatigue. The various sleepiness scales (see

Appendix A.5) have been used in this way. Some self-assessment tools have been used to enhance awareness of fatigue by both the individuals and those setting their schedules, and to provide a standardized way for individuals to report their fatigue or sleepiness level. Validated self-assessment instruments can also be used in the workplace to assess the prevalence of fatigue (see fatigue hazard assessment below) and to raise awareness of fatigue. The Retrospective Alertness Inventory is intended for this purpose (Folkard et al. 1995). Instruments that help assess and manage the likelihood of becoming fatigued may also be of potential value to the individual. Among the tools are a diary for documenting sleep history and quality (Shen et al. 2006; Kecklund and Åkerstedt 1997). If used as part of a workplace fatigue management effort (as opposed to a research project), issues of privacy, information protection, accuracy/consistency, and potential for subversion would need to be addressed. The ability to obtain valid self-assessment information is important to the ongoing efforts to calibrate and improve biomathematical fatigue scheduling tools.

Performance Impairment Detection/Mitigation Technologies

Performance Testing Technologies

Performance tests can range from simple reaction time tests (stimulus-response-reaction tests, SRRTs) to tests of higher level cognitive functioning. SRRTs, also called PVTs, use mean reaction times (lapses) to detect performance degradation indicative of reduced alertness. Cognitive tests evaluate a range from lower level perceptual to higher-level analytical and decision-making skills. A perceptual test might require the individual to accurately compare patterns, identify colors, or recognize tones as high, medium or low pitch (Miller 1996). Examples of higher cognitive tests include:

- simple mathematical processing (such as adding numbers up to or beyond three-digit sums with reasonable speed and accuracy);
- code substitution (where the individual is called upon to substitute numbers for letters in a simple code-solving process); and
- short-term memory (for example, a set of letters is briefly presented, followed by a series of letters presented one at a time and the subject's task is to determine if the letters in both sets match).

Some performance testing technologies involve a simple psychomotor test while others include a battery of tests that cover various performance dimensions, allowing testing to be customized to the particular job situation. Palm-PVT is a handheld device that has reduced the time required to complete the simple psychomotor test to 5 minutes.[104] A technology called Act-React-Test system (ART90) incorporates a battery of 8-10 tests covering visual perception, reaction time, concentration, cognitive processing, and personality. It has been used in Europe and an evaluation study found that test scores could predict 66 percent of driving mistakes (see Charlton and Ashton 1998; Hartley et al. 2000).

Performance testing can be conducted at the start of and/or at key times during a work shift when fatigue is more likely to be an issue (i.e., after lunch or other times when circadian rhythms are known to be low). The utility of these tests depends on the extent to which they can accurately and reliably detect and measure performance impairment.

---

[104] The original PVT required a 10-minute data collection period and was delivered on a microcomputer that measured 21x11x6 cm and weighed 658 g.

It should be noted that performance tests do not detect fatigue per se – their intent is to detect impairment regardless of the cause. These performance testing technologies do not measure and detect an impaired state-of-being but, rather, performance degradation resulting from an impaired state. Psychomotor tests generally pick up impairment that is comparable to the impairment of an individual with approximately a 0.1 percent BAC level, which is a fairly severe level of impairment (Toquam and Bittner, Jr. 1996, 1994).

Because higher-level cognitive functioning may be more susceptible to impairment, hazard perception tests for drivers might be expected to be a more effective means of detecting a driver's state of impairment. However, Smith et al. (2009) developed and tested a 90-minute PC-based simulator task, called the Queensland Hazard Perception Test (QHPT), to test a driver's hazard perception ability. Although a significant relationship between sleepiness scores and QHPT hazard perception was found, this test was not more sensitive than the simple psychomotor reaction time test (correlation of 0.36 versus 0.38). Heitmann et al., (2009) evaluated a different higher-level cognitive test that involved shape recognition in a medical work setting and similarly found that the current version of the test was not very sensitive in that it only detected severe impairment and did not reflect gradual alertness changes.[105] Even if advanced cognitive tests improve sensitivity, higher-level cognitive testing is likely to be more time demanding and perhaps more intimidating to workers.

In addition to sensitivity, "recruiting" is another big issue in terms of the validity of performance test technologies. The issue is whether test takers can recruit (i.e., motivate or "psych themselves up") to give the performance test energy and attention for the short period of time necessary. Persons with high levels of fatigue (and moderate levels of illegal drugs) may be able to recruit sufficient attention to pass many of these tests. Impaired employees may be able to perform adequately well on a familiar, well-learned performance test even though they may not be able to adequately perform non-routine tasks, such as coping effectively with abnormal work conditions or accident situations. Recruiting is only slightly less of a problem for many of the higher-level cognitive tests, such as shape recognition and simple mathematical processing tests. As the research indicates, impairment affects one's ability to sustain alertness before it is likely to impact short-term performance involving either psychomotor or higher-level cognitive functioning.

Determining how to measure and interpret performance results is another issue. If an absolute performance threshold is established, the more sensitive this threshold is with respect to being able to detect all impaired persons and, the greater the chance of false positives (i.e., low specificity in terms of a higher likelihood of falsely identifying non-impaired persons). The reason for this is individual variation. For example, requiring a fairly high level of performance may mean that not all persons who fail are actually in an impaired state, such as fatigued, intoxicated, or under the influence of drugs. To address this issue, performance testing technologies often use an established individual baseline rather than using a population norm. This solution presents the following dilemma (Gilliland and Schlegel 1995, 1993; Comer 1994): "Suppose Jack meets his baseline but performs less well than Jill who fails to meet hers. Is it equitable or sensible to allow Jack to work while preventing Jill who actually performed better on the test?"

---

[105] Bowles-Langley Technology, Inc. has developed the BLT impairment test that is an inexpensive, computerized shape recognition text that requires the user to make a Yes/No decision about whether all items in a given screen are the same.

Job Performance Monitoring/Alerting Systems

Rather than testing a person's performance on various dimensions assumed to be relevant to a particular job, job performance monitoring/alerting systems evaluate actual job performance. Most of these technologies have been directed at driving and some at operating dangerous machinery. Driving performance technologies monitor lane deviation variability (LDV) and/or heading error variability (HEV). They typically use video camera technology to detect swerving, crossing lane dividers, unnecessary lane changes, and driver reaction time to traffic situations.

Job performance monitoring systems are typically in-vehicle and on-going, rather than periodic. In-vehicle job performance monitoring systems have been pilot tested and adopted on a trial basis in commercial trucking. Mercedes-Benz and Volvo have introduced in-vehicle job performance monitoring systems in high-end car models.[106] The State of Arizona is piloting a technology that monitors and evaluates simulated driving performance. Driving simulators equipped with a steering wheel and driving pedals are located at weigh stations throughout the state. Drivers perform a brief driving simulation task on a computer screen to assess their driving performance. This technology attempts to test job performance as opposed to monitoring job performance. An evaluation of the utility of the Arizona pilot test is not yet available.

The strength of these technologies is that they are passive and do not require workers to engage in tasks in addition to their job tasks (this is not true for simulated performance testing technologies). Because these technologies assess actual job performance, they have high face validity. In addition, they can function as a job performance aid. For these reasons, workers may accept this type of technology more readily and be less likely to perceive it as an invasion of their privacy. Ongoing job performance monitoring might be more reliable than attempting to test the subject's driving ability because, in the latter case, the employee can recruit his or her energies to do well on a brief performance test.

There are several potential weaknesses of job monitoring technologies. One limitation is that performance monitoring does not equate to monitoring FFD or readiness to perform (RTP) because these technologies might not be sufficiently sensitive. Furthermore, a lack of performance errors does not mean that impairment is not present. Performance monitoring typically focuses on lower-level psychomotor aspects of job performance. Laboratory tests comparing technologies that monitor physical attributes to technologies that monitor job performance found the latter to be somewhat less sensitive in reflecting the level of sleep deprivation of subjects.

Face validity, in terms of measuring actual job performance, does not necessarily translate into effectiveness in detecting fatigue. If the objective is to detect a state of impairment in order to intervene and avoid potential consequences, these technologies might not detect impairment in a timely enough fashion to prevent adverse consequences. Job performance monitoring technologies might have some utility as one of several methods for ensuring FFD but they may be insufficiently sensitive to be a stand-alone technology.

---

[106] A performance monitoring drowsiness detection/alert system called Attention Assist is standard equipment on the Mercedes-Benz 2010 E-Class (Automotive Fleet 2009). Attention Assist system includes highly sensitive sensors that continuously monitor and observe the driver's steering behavior across 70 different parameters, among which is a steering angle sensor that recognizes patterns of minor steering corrections. Once the system detects a drowsy driving pattern, it emits both an audible and visible warning (the latter being an espresso cup icon in the instrument cluster).

Further, if job performance degradation is detected, it may be difficult to interpret. Will an absolute performance threshold be used and, if so, on what basis will this threshold be decided? Alternatively, will the determination of degraded performance be based on an individual's baseline performance and, if so, how much of a decrement must be recorded before the individual is determined unfit to perform his or her job? Moreover, using thresholds established from baseline performance data might result in persons with similar absolute performance results being treated differently. Finally, monitoring and assessing actual job performance in non-driving work situations might be difficult.

Performance monitoring systems might be used as a job aid to help the driver be more aware of his or her level of fatigue rather than by management to make decisions regarding whether the person can or cannot work. The FMCSA pilot study included a lane tracking technology, SafeTRAC®, as the measure of driving performance (Dinges et al. 2005). Feedback from the group of four technologies used in the FMCSA pilot slightly reduced drowsiness (as measured by PERCLOS measures from Co-Pilot) but did not improve driving performance (as measured by SafeTRAC®). Fewer than half of the drivers considered the feedback useful (Dinges et al. 2005). No other field evaluations of job performance monitoring technologies were identified in this review.

Job Controls/Job Aids

The simplest job control instrument is an alerter. An alerter does not attempt to measure fatigue but rather tries to verify that the operator is alert. An alerter requires the operator to hit a button periodically or an alert will sound. Virtually all U.S. main-line passenger and freight locomotives are now equipped with some type of alerter system (sometimes called a deadman system). Canada has used alerters in locomotives for almost 15 years. Studies indicate alerters have had limited success. Oman et al. (2009) found that between 1996 and 2002, accidents involving fatigue and alertness occurred on average three times per year, and approximately 70 percent of these accidents involved alerter-equipped locomotives. They suggest that changes in alerters could greatly reduce preemptive automatic resetting behavior and make the alerter function more reliably as a psychomotor vigilance probe, visual distraction detector, and job aid. An alerter technology would not be appropriate for work situations that require concentration as opposed to just staying alert.

New locomotives in the U.S. will implement a positive train control (PTC) system that perceives and warns the operator of hazards. The objective of this system is to enhance operator performance as opposed to serving only as a control. Hazard detection and warning systems are also being implemented in trucks and cars. Advanced job aids to mitigate operator error and enhance operator performance are being designed into advanced NPP control rooms.

Defense-in-Depth Systems

Job aids are only one means of protecting against operator errors. Defense-in-depth technologies are evolving. The nuclear industry may well be a leader in designing defense-in-depth systems. This technological area goes beyond the scope of this report. The analyses done to develop defense-in-depth will provide information pertinent to determining the potential impact operator error can have on system operation and the potential safety and security consequences. This information is a core input to the assessment of a fatigue hazard.

3.3.3 Assessment Technologies

Fatigue Hazard Assessment

It is easy to assume that assessing the fatigue hazard for a work situation is a simple extension of assessing the prevalence of individual-level fatigue for key categories of workers. Assessing the fatigue hazard for a work situation, however, must go beyond determining the extent to which workers are fatigued. It must even go beyond assessing the potential impact of fatigue on worker performance. It is also necessary to determine how impaired worker performance could affect operational or organizational level performance. A serious fatigue hazard exists only if it is determined that impaired individual-level performance can significantly affect operational safety, security, or quality.

Assessing the fatigue hazard is fundamental to designing an effective fatigue management system for the particular work situation. Fatigue hazard assessments include:

- assessing the presence/prevalence of fatigue – detecting incidences of fatigue and determining the extent and pervasiveness of worker fatigue; and
- impact assessment – assessing the extent to which fatigue, either a single fatigued person or a fatigued group or crew, could affect operational or organizational performance.

<u>Assessing the Presence/Prevalence of Fatigue</u>

If performance testing and monitoring are used in the workplace, the resulting information can be analyzed over time to determine the prevalence of fatigue. Even if these technologies are not normally used, strategic introduction of one or more of these technologies during key periods in the schedule could be used to assess the extent, prevalence, and pattern of fatigue in the workplace.

Alternatively, periodic self-assessments or retrospective alertness assessments can used to assess the prevalence of fatigue, as well as when and where fatigue is most evident. For example, a relatively simple assessment tool, such as the Retrospective Alertness Inventory, can be applied over a wide range of individuals and conditions on a repeated basis (Folkard et al. 1995). One challenge is to obtain honest responses, given that individuals may fear negative consequences, such as loss of work or income, for reporting fatigue. In addition, multiple self-assessments at key times across many shifts may be necessary to assess a fatigue hazard.

<u>Impact Assessment</u>

There are two ways of assessing the impact of fatigue. One strategy is to analyze reported incidents and accidents in order to estimate the percent of accidents that may have involved fatigue. However, this requires that incidents and accidents are systematically investigated to determine whether fatigue was a contributing or primary causal factor (see section on incident/accident investigation technologies below). The second approach is to conduct risk and vulnerability assessments using experts to determine the probability that fatigue-related human error could affect operational safety, security, quality, or productivity, and the extent of the possible consequences. Systems with built-in protections to reduce the consequences of human error would decrease the fatigue hazard.

The transportation sector has relied heavily on the first strategy, analyzing accidents, and has not focused on fatigue-oriented risk and vulnerability assessments to the same extent. This reflects the assumption that fatigue on the part of drivers equates to a fatigue hazard and that degraded driving performance can have serious, often fatal, outcomes that have both public and worker safety, and business consequences. The transportation sector is pursuing strategies to limit the impact of fatigue and degraded individual performance on operational performance. For example, rather than relying solely on the driver to perceive hazards, automated sensors, hazard alerts, and automatic overrides of driver actions are being introduced. As defense-in-depth technologies progress, the fatigue hazard posed by individual-level fatigue is likely to be reduced.

A number of airlines are incorporating the impact assessment approach in their fatigue management strategies. For example, a British airline, easyJet, has developed a framework for risk assessment as part of their Fatigue Risk Management System (FRMS). A project funded by the European Union, called Human Interaction in the Lifestyle of Aviation Systems, is enhancing this risk assessment system (Stewart et al. 2009).

Incident/Accident Investigation Tools

There are a number of tools that have been developed in the transportation sector to assess the contribution of fatigue to incidents and accidents. These include accident investigation protocols developed by the National Transportation Safety Board (NTSB) and Circadian Technologies, as well as a tool based on biomathematical modeling.

The NTSB's protocol to investigate whether fatigue was a contributing factor to transportation accidents may be found in Appendix B to this report. Other sectors have adapted and applied this investigation protocol.

Circadian Technologies has developed an accident investigation tool called the Fatigue Accident Causation Testing System (FACTS). FACTS consists of a diagnostic survey instrument, along with a weighted risk model based on "fuzzy[107] scalable monotonic chaining," to help investigators determine the role of fatigue as a causal factor in driving accidents. The purpose of using "fuzzy" methods is to develop a robust algorithm for probability calculations involving numerical and verbal attributes, particularly with the possibility of missing data and relatively uncertain circumstances of an accident event. Sirios et al. (2007) compared the FACTS technology with the 1995 NTSB analysis of 107 heavy truck accidents and found a high degree of correlation.

In addition, developers of one of the biomathematical tools, Fatigue Audit InterDyne (FAID), claim that it can be used to predict accident probability and to investigate whether fatigue was a contributing factor to an incident or accident. Combining a biomathematical model with continuous fatigue monitoring technologies would provide more direct data to assess whether fatigue may have been a contributing or primary factor. Continuous monitoring technologies alone would provide useful direct information about the individual's fatigue level.

---

[107] Fuzzy math is a branch of mathematics that developed in the 1960s and has shown utility in modeling.

Fatigue Management Evaluation Technologies

This category captures technologies that can be used to evaluate the fatigue management system as a whole and/or the effectiveness of specific fatigue-oriented interventions. Some technologies included in other categories are also included in this category. For example, actigraphy can be used not only as a fatigue intervention technology in terms of detecting fatigue, but it can also be used to evaluate other fatigue-oriented interventions by measuring the extent and prevalence of worker fatigue before and after the intervention. One vendor, Archinoetics, uses actigraphy precisely in this way to assess the effectiveness of introducing a new optimized schedule solution generated by their biomathematical scheduling tool. Actigraphy or oculometrics could also be used to help assess the overall fatigue management system by analyzing a baseline period prior to implementing a fatigue management system, and periodically after implementation to determine whether the system was effective in decreasing worker fatigue. In addition, aggregate actigraphy analysis for work groups over equivalent periods in the operational cycle could provide the basis for comparing the effectiveness of fatigue management across different NPPs. Other types of technologies, such as periodic retrospective fatigue assessments, could also provide some basis for evaluating fatigue management over time and/or across NPPs.

3.3.4 Summary of Applied Fatigue-Related Technologies

Most applied tools are derived from technologies originally developed for scientific research. Several have demonstrated usefulness in applied settings. The review found no field evaluations demonstrating the effectiveness of most of these tools. Some tools are in an early development stage and have not been field-tested. Other tools have been in use for only a short time and lack data to assess their effectiveness. In many cases, fatigue interventions were adopted without a strategy to evaluate their effectiveness. A growing number of pilot studies have been conducted. While informative, an important limitation for the nuclear industry is that most field tests and adoptions have occurred in the transportation sector, particularly in commercial trucking. Another important limitation is that few of the technologies have been in use long enough to determine their effectiveness in reducing accident rates and to ensure that issues related to incorporating the technology into normal operating procedures have been identified and addressed.

The FMCSA study (Dinges et al. 2005) looked at several technologies simultaneously because it would be too difficult to pilot test each technology separately. Their findings are ambiguous as to whether these technologies were useful and are not very conclusive because of the pilot study design. The oculometric technology, CoPilot, appeared to have some effect in promoting driver alertness. Perhaps it prompted drivers to engage in some activity to counter their drowsiness, such as eating, talking on their radios, or even taking a short rest. Drivers increased the amount they slept on the non-work days as a result of feedback from the technologies (but it cannot be determined to which of the technologies this effect can be attributed).[108] Driver performance, as measured by SafeTRAC®, showed no significant change in response to feedback from the technologies. From the research, one might expect that driving performance would not be changed by the use of these technologies because normal

---

[108] Drivers increased their sleep on non-work days but not on work days. Sleep on non-work days increased on average 26 minutes per non-work day. Drivers averaged only 5-6 1/4 hours of sleep per day on work days. Dinges et al. (2005) claim that research shows severe sleep debt and decreased alertness can develop within a few days at these sleep durations. Whether or not research conclusively supports this claim, the interesting finding is that fatigue feedback encouraged persons to sleep longer, at least on their days off.

driving performance is fairly resilient to fatigue. In addition, an unusual driving situation where the driver must quickly respond to avoid an accident is not likely to occur with sufficient frequency to be captured in the pilot studies. Moreover, the drivers were not required to pull over to rest if the fatigue feedback was above a certain level. Thus, the effects of these technologies may have been more evident if the technologies were implemented within an overall fatigue management system and over a longer period of time.[109]

This pilot study also captured information on worker acceptance and perceptions as to the utility of the technology. Drivers tended to have positive attitudes toward fatigue management technologies. They preferred the job performance monitoring technology (SafeTRAC®) to the oculometric monitoring technology to detect operator fatigue (CoPilot). However, feedback from CoPilot seemed to increase driver alertness, whereas SafeTRAC® feedback had no significant performance effect. Drivers were also favorable to PVT if it was used only as a personal aid and the time required to perform the PVT could be shortened.[110] Drivers were most favorable toward fatigue training.

There is a general consensus that none of the available technologies are valid enough to be used as a stand-alone approach to fatigue management. However, these technologies do not necessarily have to meet some pre-established threshold of accuracy to be useful, especially if the technology is designed to aid learning and improve performance in the particular work context and/or there is reason to believe that a technology can contribute to the overall fatigue management program. None of these technologies should be seen as a stand-alone intervention that is expected to eliminate the fatigue hazard. The various technologies are more or less appropriate and effective in different work situations and, in most cases, using more than one type of technology will be the most appropriate and effective approach.[111]

Most researchers also advocate the use of multiple metrics to increase detection and prediction accuracy. This can involve using multiple technologies or having a technology that effectively monitors and integrates multiple metrics. Using multiple technologies to detect an individual's level of fatigue can compensate for technology reliability issues but can also create an issue in terms of how to integrate or prioritize the information provided from the different technologies. Researchers are developing algorithms to combine multiple drowsiness metrics from different technologies as well as developing single technologies that can integrate multiple metrics. The Driver Drowsiness Monitoring System (DDMS), developed and tested by the Virginia Tech Transportation Institute, is a prototype system that uses an algorithm to combine multiple drowsiness metrics across two different technologies: PERCLOS and lane position metrics (Wierwille et al. 1996; Olson 2006; Baker et al. 2007). The researchers expect this system to have increased reliability compared to technologies using only one type of metric. Further research is required to test the system's algorithm. In addition, an effort sponsored by the U.S. Air Force is developing a technology called Enhanced Psychomotor Vigilance Task (EPVT) that combines PVT and oculometrics (see Stern and Brown 2005). Future versions of wrist actigraphs and oculometric eye wear may be able to capture various types of autonomic activity. Also, data from monitoring technologies may increasingly be integrated into biomathematical models to further increase the accuracy of detection.

---

[109] The pilot was a 28-day study.

[110] A 10-minute PVT was used in this study. The Palm-PVT has a 5-minute test version. Reducing PVT times may not be effective given these tests are not highly sensitive in detecting fatigue. There is on-going work by NASA to reduce PVT duration to 3 minutes.

[111] It should be kept in mind that fatigue-oriented technologies, even if multiple types are employed in an overall fatigue management system, will have limits because fatigue is only one potential source of performance impairment.

3.4   Integrated Approach to Fatigue Management

There is wide consensus that effective fatigue management requires an integrated, holistic approach. The Federal Aviation Administration (FAA) adopted the term, FRMS, now being used by many working in this area, to refer to this integrated, holistic fatigue management approach. [112] According to the FAA, this approach should optimally be a component of a larger Safety Management System (SMS). [113] However, even without a larger SMS, an FRMS could be implemented and would likely have a positive impact (Gander et al. 2011). In collaboration with the International Civil Aviation Organization and other civil aviation authorities, the FAA is in the process of developing guidance, but as of yet there is no accepted guidance to define what an FRMS should entail (Clark 2009).

The U.K. Department for Transport commissioned a world-wide study to explore operators', regulators', and researchers' experience and views pertaining to an FRMS. This study found that a number of commercial trucking operators, especially in Australia, have implemented an FRMS (Fourie et al. 2010). [114] The study found views toward the system were mainly positive across all entities (operators, regulators, and researchers). However, data do not yet exist to demonstrate its effectiveness (Jackson et al. 2009). The main criticism was that standardized, consensus guidelines regarding what an FRMS should entail did not yet exist. Jackson et al. (2009) found that only one Australian regulator reported providing guidelines for operators on the use of biomathematical models of fatigue to inform schedule design and staffing levels. Starting in 2009, a number of British road transport operators will begin to define, implement, and assess an FRMS in coordination with the U.K. Department for Transport. Other sectors are also implementing an FRMS. For example, the Philadelphia Police Department established a Comprehensive Police Fatigue Management Program with participation from the National Institute of Justice and Centers of Disease Control (Vila 2006).

Hersman (2009), a member of the NTSB, employs a "house" analogy to describe a structure for such an integrated fatigue management system. She sees the fatigue management system as consisting of a foundation, a framework, wiring and plumbing, and a roof. Just as constructing a house begins with the foundation, so does a fatigue management system. Fatigue management started with what she refers to as the foundation, work hour limits, but has gradually expanded to include a more comprehensive set of components as follows.

- Foundation: The foundation of a fatigue management system is work hour limits. Work hour limits can address shift hour limits as well as weekly, monthly, and yearly hour

---

[112] Integrated approaches to fatigue management trace their origin to the NASA Ames Fatigue/Jet Lag program, later renamed the Fatigue Countermeasures Program (http://human-factors.arc.nasa.gov/zteam/). This nearly 20-year research and outreach program performed ground-breaking research on the causes and consequences of fatigue in aviation operations, and established best research practices that are in use today. Additionally, the program developed an education and training module that was presented to over 2500 individuals. The material in the program served as a basis for individual airlines to develop alertness management systems. Many aspects of an integrated approach, such as education and training, countermeasures, and healthy sleep habits, were developed and articulated in this program. Currently this program is not supported within NASA, but the referenced website documents the various research studies and interventions evaluated. The program was influential in helping establish similar activities across the FAA and other parts of the U.S. Department of Transportation (DOT).

[113] Safety Management System is a term used by the FAA to refer to an integrated approach to safety.

[114] There is a study by Rosekind et al. (2006) evaluating the effectiveness of an alertness management program for a major commercial airline. The program included education involving the basics of sleep and fatigue, alertness strategies, and the assessment of alternative schedules. Pilots were measured on a variety of variables before and after the alertness program intervention (knowledge, sleep duration, PVT performance). The results indicated that all measures improved significantly following the implementation of the program. In particular, following the program intervention, pilots slept an average of 1 hour and 9 minutes longer during layover flight shifts.

limits and, to the extent possible, should be informed by scientific information and research.[115]

- Framework: The framework consists of scheduling practices. Scheduling practices need to ensure that workers can get 8 hours of uninterrupted sleep by taking into account not only hours of actual service but also duty hours and commute hours. In addition, shift rotation, roster management, and staffing levels should be scientifically informed.
- Wiring, Plumbing, and Key Features: Wiring and plumbing include other critical fatigue management practices, such as education and training, screening for sleep disorders, and identifying medications that need to be reported. Other features that might add to the effectiveness and acceptance of the system include implementing fatigue detection technologies, instituting fatigue call-in options, incorporating strategies to promote on-duty fatigue mitigation and off-duty sleep, and other desirable fatigue countermeasures.
- Roof: The roof is comprised of overarching organizational policies (especially written policies); the oversight, maintenance, integration, assessment, and improvement of the many fatigue-related practices; a strong safety culture; and management support.

Several have attempted to list the types of interventions that would make up a comprehensive, holistic, and integrated fatigue management system (see Caldwell et al. 2008). Some warn that this sort of toolbox approach should be informed by an overarching fatigue management plan to help ensure the right tools are selected and that these tools both support and are supported by a larger fatigue management system (Fourie et al. 2010).

The components of a fatigue management system might include:

- Management Commitment - Management commitment and support is essential to all safety programs.
- Culture - The safety culture should explicitly counter tendencies to dismiss fatigue and over-estimate one's abilities while fatigued.
- Policies - There should be clear written policies covering many aspects of the comprehensive program, including:
  - work hour limits and scheduling that address actual time (such as driving time or flight time), limbo time, and commute time;
  - leave options (sick leave and other);
  - fatigue call-in options and repercussions;
  - appropriate data protection and confidentiality;
  - disciplinary measures related to fatigue as needed;
  - drug use restrictions, reporting requirements, and guidelines regarding pharmacological countermeasures; and
  - role of fatigue-oriented intervention technologies and guidelines regarding interpreting and using results of these technologies.
- Awareness - Awareness has been increasingly achieved but changing behavior to prevent fatigue is harder to realize (Fourie et al. 2010). Customized education and training are needed along with greater assistance to help persons prevent and mitigate fatigue. Education and training programs are attempting to provide more usable

---

[115] There is some evidence that mistakes begin to increase after 8 hours, and that performance decrements are even more notable after 12 hours of work, but the evidence is not altogether consistent. In the transportation sector, this phenomenon has been referred to as "get-there-it is." The driver becomes increasingly anxious to get to the end of the driving shift and to the day's destination.

information to workers and are encouraging the whole family to participate in these programs to increase the likelihood that changes in behavior will occur.

- Education - Research indicates that employee education and training should be sufficiently in-depth to address:
  - sleep and health effects on fatigue;
  - recognition of individual differences in need for sleep and fatigue vulnerability;
  - sleep disorders;
  - sleep hygiene (good and bad sleep-related habits) and sleep optimization strategies;
  - strategic napping and other restorative techniques;
  - sleep self-assessment (with consultation and confidentiality);
  - effects of caffeine consumption and effective use of caffeine given shift-type;[116] and
  - the use and effects of pharmaceuticals and other substances to adjust to circadian disruptions and short-term fatigue.[117]
- Training key staff (schedule/roster designers and managers/supervisors/Medical Review Officers [MROs])
  - *Education.* Education for roster designers is extremely important. Most roster designers have little understanding of the underlying science of fatigue and fatigue-informed scheduling practices (McColgan and Nash 2009). Currently, no uniform training exists in the U.S. for individuals who are responsible for managing work schedules. Stentz et al. (2009) suggest that there should be training and a credential for work schedule managers.
  - *Fatigue Intervention Technologies.* Intervention technologies should be selected and used carefully – strengths and weaknesses of tools should be evaluated, and field testing conducted to ensure that the technologies are appropriate to the particular work situation. Also, clear guidelines as to how results are to be interpreted and acted upon are important for many of these technologies.
  - *Napping.* Some studies show napping improves psychomotor performance (Bonnefond et al. 2001; Purnell et al. 2002; Schweitzer et al. 2006). Organizations might provide training about and promote strategic napping during long shifts.
  - *Workload and Monotony.* Fatigue may be most hazardous when workloads are low and monotony sets in (Matthews et al. 2009). Fatigue countermeasures might be directed at optimizing effort over the course of the shift.
  - *Incident investigation to include systematic examination of fatigue.* Setting up systematic processes to examine whether fatigue may have been a contributing factor to an event or incident would be useful in helping to assess whether fatigue is a hazard. There may be tools that can facilitate fatigue investigation, such as the Fatigue Accident/Incident Causation Testing System (FACTS), developed by Circadian Technologies, Inc.

---

[116] Moore-Ede et al. (2009) examined whether a commonly used, legitimate fatigue countermeasure might have a similar effect to that of illicit pharmaceuticals (e.g., cocaine, amphetamines, marjuana) such that the initial stimulation phase is followed by a depression phase with excessive fatigue. Micro-sleep events were suppressed using caffeinated chewing gum but a rebound effect in the seventh session was observed, suggesting that alertness stimulants, while having short-term value as a fatigue countermeasure, should be used with caution.

[117] Researchers argue that guidelines and training are needed regarding the effects of a myriad of chemical substances on driver performance and health for both drivers and medical providers responsible for qualifying drivers (Krueger and Leaman 2009; Leaman and Krueger 2009). Pharmacological countermeasures may offer the greatest promise for overcoming natural biological/physiological limitations. Evidence suggests that at least some of the currently available compounds for sleep, particularly short-acting hypnotics, should be more strongly considered for operational use (Caldwell 2009). There is, however, significant resistance to the use of pharmacological agents in spite of substantial evidence that artificially enhancing sleep and/or alertness poses less of an immediate safety threat than untreated fatigue. Caldwell argues that the use of select medications is a better alternative than forcing personnel to fight their basic physiological propensities unaided and that these medications should be included in a comprehensive fatigue management system as needed.

- *Performance measures and operational feedback regarding fatigue management effectiveness.* Assessing the effectiveness of fatigue management requires having performance measures and operational feedback.
- As the work force ages, and obesity and diabetes in the worker population becomes a greater problem, sleep apnea has become a major concern in many sectors, particularly transportation and other sedentary types of work. Many comprehensive sleep management programs include:
  - screening for sleep apnea (a commonly used tool is the Berlin questionnaire for sleep apnea)[118];
  - providing treatment for those with severe obstructive sleep apnea;
  - less accepted types of employee screening, such as screening tolerance for shift work.
- *Appropriate data control protocols and confidentiality guidelines.* There are aspects of a fatigue management program that would require confidentiality and adequate data protection. For example, confidentiality and data protection would be necessary if an organization uses performance testing, collects information on individuals' health and medication use, encourages self-reports or self-evaluations in terms of sleep and fatigue, or screens persons for sleep disorders.
- *Supply chain outreach.* The Australian Department of Transport has adopted the position that for fatigue management to be effective, all the parties (the operator, consigner, etc.) must be committed to managing driver fatigue and must take reasonable steps to ensure that their actions do not cause a driver to drive while fatigued.
- *Science-informed technical basis.* A good fatigue management system should keep up with science and adopt new science-informed practices as appropriate for the work situation. Regulators and/or industrial associations may fund appropriate research and help conduct field testing of potentially useful technologies.

A sector-based approach may facilitate the adoption of fatigue management practices. Short-term financial pressures, as well as pressures to be competitive, may inhibit adoption of fatigue management strategies and technologies unless competitors are doing the same. As occurred in the airline industry, industrial associations may have the combination of sector knowledge, resources, and access that would enable them to develop guidelines for the sector as a whole and promote collective compliance. In regulated industries, it may be necessary for the regulator to be engaged in this process. For sectors with an international market, it may be important to consider existing standards in other countries and impacts on competitiveness. Some aspects of this approach are discussed in the *Proceedings of the Aviation Fatigue Management Symposium: Partnerships for Solutions* (U.S. DOT/FAA 2008).

In addition to addressing fatigue management for an industrial sector as a whole, on a national and perhaps international level, the Australian Department of Transport looks at fatigue management from a supply chain perspective involving entities outside the particular industrial sector. They consider addressing the transport supply chain to be essential for effective fatigue management because it reduces pressure on shipping companies and drivers to compete in terms of decreasing transit times (Jackson 2009).

---

[118] Note that confidentiality is very important for these questionnaires.

Designing a fatigue management system requires more than identifying the components of this system. Effective design of a fatigue management system for a particular work situation requires a systematic process that includes:

- Assessing the fatigue hazard to provide:
  - a baseline for evaluating the fatigue management system;
- Designing the fatigue management system to identify key intervention strategies and policies that make up an overall integrated approach; and
- Selecting appropriate technologies that are integrated into an overall systems approach, support the key strategies, and are guided by clearly specified policies and practices.

3.5   Regulatory and Fatigue Management Trends

Fatigue has been a long-term concern in several sectors and is becoming a greater concern due to workforces that are aging, becoming increasingly obese, and/or making long commutes that extend their workdays. In addition, economic pressures that have reduced staffing levels are further contributing to a fatigue hazard. Notable accidents in which fatigue was implicated as a contributor have increased the salience of fatigue as a key public safety concern. Consequently, regulators in many sectors have been directing greater attention to the problem of fatigue.

Although research on fatigue has increased understanding of the causes and consequences of this complex phenomenon, there is not yet scientific consensus regarding its implications for regulatory standards, guidelines, and requirements. Similarly, although there is consensus that technologies to address fatigue could be very useful, there is little consensus as to which technologies should be adopted in which operational contexts, or whether any of the technologies are sufficiently effective and reliable at this time.

In this section we briefly review initiatives to understand and address fatigue in a number of sectors, including thetransportation (including commercial trucking, air, rail, and marine), military, mining, law enforcement/security, and medical sectors. Fatigue research and technology development have focused disproportionately on the transportation sector, primarily the commercial trucking industry. Fatigue-related strategies, practices, and technologies developed with drivers in mind may not be as useful for addressing fatigue in other work situations. Therefore, this overview concludes by identifying key differences that might affect the effectiveness of various fatigue strategies and technologies, particularly noting how the work situation of drivers may differ from other work situations.

3.5.1 Transportation Sector

The fatigue hazard is very apparent in the transportation sector, with a large number of accidents across the various transportation modes resulting, in part, from fatigue.[119] Consequently, the DOT has directed more attention to the issue of fatigue than most other federal agencies. Further, there is reason to believe the fatigue problem in the transportation sector may be increasing because of the combined effects of an aging workforce and increasing worker obesity. These factors also apply to the NPP workforce (see Paim et al.

---

[119] Researchers have attempted to estimate the fatigue hazard in various transportation sectors but data limitations restrict the accuracy of these estimates.

2008) and workforces in various other industries that have the potential to affect public health and safety.

A survey of 2280 truck drivers indicated that more than 50 percent reported experiencing drowsiness and 13 percent reported experiencing serious drowsiness (e.g., nodding off/falling asleep) on half or more of their trips (Dinges and Maislin 2009). Other studies indicate that fatigue may be a larger problem in the general driving population. Based on surveying a sample of drivers, the Institute of Medicine (2006) estimated that on average, 7.5 million drivers in the U.S. have fallen asleep at the wheel within the past month, and another 7.5 million drivers have done so during the prior 2-6 months, indicating a large number of persons whose state of fatigue while driving poses a hazard.[120] A study by Czeisler et al. (2009) found that 97 percent of drivers in sleep-related crashes admit to having driven drowsy during the year before the fall-asleep crash.

In addition, various analyses of accident databases estimate that fatigue contributes to between 15 to 30 percent of ground transport mishaps (Caldwell 2009; Kecklund et al. 2009; Sirois et al. 2007). The Large Truck Crash Causation Study conducted jointly by the FMCSA and the National Highway Traffic Safety Administration (NHTSA) analyzed a representative national sample of 963 large truck crashes from 2001-2003 and found 13 percent of the truck drivers were fatigued at the time of the crash. The incidence of fatigue was slightly higher (15 percent) among passenger vehicle drivers who collided with large trucks.[121] An analysis of an earlier accident database conducted nearly two decades ago examined 182 heavy truck accidents that resulted in truck driver fatalities. This study estimated that fatigue was a probable causal factor in 31 percent of these fatal accidents (NTSB 1990). A study conducted in Australia estimated that fatigue, inattention, distraction, and monotony contribute to 40 percent of fatal crashes and 34 percent of all crashes (Queensland Transport 2003 – see Michael and Meuter 2009).[122]

Although the methodologies used in the analyses of accidents are not sufficient to specify a definitive quantitative estimate of the influence of fatigue on accident rates, they clearly indicate that fatigue is a major hazard in commercial transportation. In addition, some analyses have used an occupational epidemiology approach and case-control studies to estimate the relative risk and odds ratios of drivers who have crashed compared to those who have not. These analyses have consistently found that sleep of less than 9 hours in 48 hours, or less than 4 hours in 24 hours raises the relative risk of a crash, as does the feeling of "imminent sleep onset."

Data regarding the extent to which fatigue has contributed to incidents is less available in the other transportation sectors. However, an analysis of NASA's Aviation Safety Reporting System (ASRS), a confidential self-reporting system for flight crews and others to report difficulties and incidents, suggests that 21 percent of incidents reported were fatigue-related. A 1996 U.S. Coast Guard study analyzing 297 commercial marine casualty investigations estimated that

---

[120] The DOT registered 196,165,666 licensed drivers in 2003 – 7.5 million drivers falling asleep at the wheel in the last month represents almost 4 percent of the total driving population; 15 million drivers falling asleep at the wheel in the last 2-6 months represents almost 8 percent of the total driving population.

[121] Another study using the same database found that the amount of sleep the driver reported was inversely correlated with the likelihood that drivers' actions were responsible for crashes; however, long hours of driving in a day or working overtime the previous week did not increase crash risk (see the *Driver Fatigue and Alertness Study*, mentioned in Cohen et al. 2009). This suggests that if a person has had adequate sleep, they may be able to work long hours on occasion without major consequences. Recent naturalistic driving experiments support this finding (Jovanis et al. 2009). In contrast, other studies have found a positive relationship between continuous hours of driving and the odds of a crash.

[122] The authors considered fatigue, inattention, distraction, and monotony to be too overlapping to separate.

fatigue was a contributing factor in 16 percent of vessel casualties and in 33 percent of the personnel injuries (McCallum et al. 1996).

In addition to sponsoring studies, the DOT has pilot-tested and assessed various fatigue countermeasure technologies, particularly in the commercial trucking sector.

Commercial Trucking

The DOT has directed considerable attention to fatigue in the commercial trucking sector. The FMCSA is currently revising hours-of-service (HOS) regulations for drivers of commercial vehicles, including the maximum number of hours of driving allowed in a shift, the minimum number of hours off-duty between shifts, and the maximum number of hours of work allowed in multi-day periods (e.g., calendar weeks). The HOS regulations enacted in 2003 specify both driving time limits, as well as an hours-off component; after 34 consecutive hours of off-duty time drivers can begin a new 7-day period during which they can drive or be on-duty[123] for a cumulative total of 70 hours. This means the seven-day clock restarts after a 34-hour off-duty period. The off-duty regulatory specification provided drivers an additional two hours of off-duty time with the intent of increasing drivers' sleep time. The current effort to revise these regulations has prompted several studies, some of which the FMCSA has sponsored or helped conduct.

A study by Olson et al. (2009) found that HOS regulations accomplished the objective of increasing drivers' sleep time. However, this study did not find a relationship between the small increase in the average amount of sleep, or individual differences in amount of sleep, and safety-critical events. An interesting finding of this study was that the frequency of safety-critical events is highest on the first day after returning to work following days off. Jovanis et al. (2009) analyzed data from both the 1980s and 2004, and found that crash odds increase with continuous hours of driving and that rest breaks tend to have the greatest impact on reducing the risk of an accident. However, they also found that in both time periods crash odds increased after more than 34 hours off-duty. Consequently, they note that the 34-hour "restart" policy may need additional examination.

Aviation

For over 10 years, the NTSB has included fatigue on the list of "Most Wanted" challenges to aviation safety. In 2008, the NTSB recommended that the FAA develop guidance, based on empirical studies and scientific research findings, to manage fatigue in aviation operations. This recommendation followed key fatigue-related mishaps, such as the crash of Corporate Airline flight 5966, the runway overrun by Delta connection flight 6488, the off-runway excursion of Pinnacle Airlines flight 4712, and the airport overshoot by Go1 Flight 1002 (Caldwell 2009). The crash of Colan Air flight 3407 in February, in which pilot fatigue was implicated, increased the sense of urgency for stricter fatigue management regulations. The FAA published the proposed rule in September 2010, and was scheduled to issue the final rule by mid-summer 2011. The proposed rule addresses transient fatigue (i.e., the immediate, short-term fatigue that can be addressed by a recuperative rest opportunity), cumulative fatigue, impacts of changing time zones and nighttime flying, and training on symptoms and mitigation of fatigue (FAA 2010).

---

[123] On-duty time can include non-driving time, such as rest breaks and time loading/unloading, spent at weigh stations, etc.

Although these mishaps implicated fatigue as a contributor, systematic data on the number of fatigue-related aviation accidents do not exist. An NTSB analysis of aviation accidents, based on the Aviation Accident Database that dates back to 1962, found that only 0.45 percent noted fatigue as a cause (Price 2009). Price (2009) concluded that analysis of this database underestimates the role of fatigue because the accident investigations did not include adequate methods for assessing whether fatigue was a contributor. Caldwell (2009) conducted a search of the scientific literature and other resources to explore the impact of fatigue on aviation safety; he found no other studies that estimated the percent of aviation accidents that involved fatigue.

Although pilots and crews have been the primary focus of concern in the aviation sector, fatigue may be a concern for aviation maintenance personnel as well. Systematic data to assess the extent to which aviation maintenance poses a fatigue hazard are not available. One study estimated that approximately 12 percent of major aircraft accidents and 50 percent of engine-related flight delays and cancellations worldwide result from maintenance deficiencies (Marx and Graeber 1994). This study points to the fact that aviation maintenance may have an operational impact but it does not necessarily indicate that fatigue is responsible for maintenance issues. However, aviation maintenance personnel have reported fatigue to be one of the most prevalent causes of accidents and deficiencies (Hobbs and Williamson 2003).

FAA regulations primarily address flight time limits and minimum pre- and post-duty pilot rest periods. As Caldwell et al. (2009) point out, neither the duty time nor the rest break requirements in FAA regulations account for the timing of the work or the rest break (and supposed sleep period) with respect to the time of day or placement relative to the circadian phase of the individual.[124] However, they note that this is beginning to change. The Flight Safety Foundation, Airbus, and Boeing are collaborating on identifying ways to maintain "maximal alertness levels" of the crew during ultra-long range (ULR) flights (flights that extend the longest flight duty times to more than 20 hours). The steering committee advising this effort is basing its recommendations largely on the FRMS approach that does reflect these considerations. The FAA, in turn, is using the Sleep, Activity, Fatigue and Task Effectiveness Model (SAFTE) to predict crew performance levels in its case-by-case review of the carriers' route-specific plans.

The carriers are developing and submitting a plan for each proposed ULR flight city pairs and routes, and are required to collect and evaluate subjective and objective data during these flights to validate the effectiveness of fatigue mitigation strategies. The FAA also requires carriers to educate all ULR crewmembers and provide them with city-pair specific route guides that include suggested practices for maximizing total sleep times (Caldwell et al. 2009). In addition to the ULR process, the FAA is now focusing attention on two new areas. One involves exploring the utility of biomathematical fatigue models for predicting the relative fatigue hazard associated with alternative schedule designs and roster management solutions.

Advocates of science-informed biomathematical models suggest that the prescriptive rest break requirements and work hour limits not only provide inadequate safeguards, but they might be more restrictive than the optimal schedule and roster solutions suggested by biomathematical models. Several airlines are working with Boeing and key biomathematical fatigue modelers to apply these models to flight crew scheduling (Klemets and Romig 2009).

---

[124] They emphasize that rest breaks that are out of synchronization with an individual's circadian cycle, especially when the sleep period occurs during subjective daytime, result in shorter, poorer quality, and less restorative sleep.

The other key emphasis in aviation is to develop a systematic FRMS approach. In concert with the International Civil Aviation Organization and other civil aviation authorities, the FAA is currently identifying the elements of and guidance for an FRMS. Although it is possible that the FAA might require implementation of an FRMS as a standalone system, it seems more likely that it will encourage aviation service providers to incorporate it in their SMSs.[125]

Rail

The rail sector has long recognized fatigue to be a safety concern. In fact, the first HOS regulations in the U.S. were established in 1907 for the rail industry. Since that time, HOS have been further restricted and, in 2008, the U.S. Congress mandated fatigue management be adopted in the railroad industry.

A 2009 study found that between 1996 and 2002, rail accidents involving fatigue and/or alertness occurred on average three times per year (Oman et al. 2009). Gerson et al. (2009) screened a sample of railroad workers using the Epworth Sleepiness Scale. Over 40 percent of the respondents reported that they experienced excessive daytime sleepiness (EDS), which is significantly higher than the estimates (ranging from two to eight percent) for the general population. Rail workers experienced irregular schedules and sometimes backward shift rotations. Research has demonstrated that both these practices aggravate fatigue by hindering an individual's ability to biologically adjust to disruptions of their circadian rhythms.

The Federal Railroad Administration (FRA) applied a biomathematical fatigue model (SAFTE) to assess fatigue in the rail industry (Hursh et al. 2008). Examining the 30-day work histories of locomotive crews prior to 400 human factors accidents and 1000 non-human factors accidents, this assessment found a strong relationship between crew fatigue scores and the probability of a human factors accident.[126]

Alerter systems have existed in locomotives for approximately 15 years but these systems have proven to be inadequate. The alerter system requires the locomotive engineer to hit a button every few minutes (the frequency increasing or decreasing depending on the extent of the potential safety hazards given the time and place) or an alert is sounded. Oman et al. (2009) found that approximately 70 percent of fatigue-related accidents involved alerter equipped locomotives. This finding suggests that fatigued persons are able to hit a button every few minutes and that this action is insufficient to keep them alert. Locomotive engineers have reported that this motion becomes so habitual that they sometimes move their arms to push the button when they are sleeping at home in their beds (Oman et al. 2009). New locomotives will have a positive train control (PTC) system installed. These systems are designed to prevent train-to-train collisions, enforce speed restrictions, and provide protection for roadway workers. The rail industry is investigating whether other technologies might also be beneficial in new locomotives or for scheduling purposes (Oman et al. 2009).

Various rail carriers are considering the implementation of fatigue management systems. For example, the Union Pacific Railroad has developed and is implementing an FRMS (Holland

---

[125] The SMS is essentially a quality management approach to controlling risk that provides an organizational framework to support a sound safety culture. The FAA initiated it in 2006 as a voluntary program described in its Advisory Circular No. 120-92 (FAA 2006).

[126] Biomathematical fatigue modeling tools that assess schedule design and roster management have only been used to assess the prevalence of fatigue in the rail industry; they have not been adopted as tools to optimize schedule design and roster management (see section on applied technologies).

2008). Starting in 1990 with initial research and education, the Union Pacific Railroad implemented a risk management model in 2005 that specifically addressed the challenges of 24/7 operations, an unsupervised workforce, workforce aging and turnover, and the implications of an increasingly 24/7 society. The initiative emphasizes the scientific basis of the approach that includes the following elements:

- policy (both corporate and local);
- training and education;
- ensuring adequate average sleep opportunity (supported by software analysis using the FAID model to plan and monitor schedules);
- ensuring employee preparedness;
- ongoing research (in collaboration with the DOT and the Department of Labor); and
- additional countermeasures (including a planned nap program, sleep disorder screening, and measurements).

Transportation Sector in Other Countries

The Department of Transport in Australia and the Ministry of Transport in Germany have been particularly active in addressing the fatigue hazard. The Australian Department of Transport has developed model legislation entitled "Heavy Vehicle Driver Fatigue Reform."[127] The reform includes three different options for operators. The Standard Hours (default option) is a set of prescriptive work/rest requirements to which everyone must adhere. This option allows up to 12 hours of driving in a 24-hour period. If operators want additional driving hours for their drivers, they can enter an accreditation module (either the Basic option or Advanced Fatigue Management option) that requires them to meet established standards for fatigue hazard analysis and management. The model legislation requires an audit of the operators' fatigue management program on a regular basis, and drivers are required to complete mandatory health and training modules. Basic Fatigue Management accredited operators can allow drivers to drive up to 14 hours in a 24-hour period; for operators with Advanced Fatigue Management accreditation, the allowance extends to 15-16 hours (subject to approval, which includes approval from a fatigue expert).

Another component unique to this legislative regime involves the chain of responsibility requirements for all parties in the transport supply chain. All parties in the supply chain have a general duty to manage driver fatigue and must take reasonable steps to ensure that their actions do not cause a driver to drive while fatigued.[128]

A survey of 10 aviation and road transport regulators from the U.S., Australia, and New Zealand found that only one Australian regulator reported providing guidelines for operators on the use of biomathematical models of fatigue to inform schedule design and staffing levels (Fourie et al. 2010).

3.5.2 Military and Space

Fatigue has been a long-term concern for the military and space sectors in the U.S. and other countries. Each of the U.S. military branches has a division that conducts, sponsors, and

---

[127] This model legislation can be accessed at http://www.ntc.gov.au/viewpage.aspx?documentid=1495 and is discussed on their website http://www.ntc.gov.au/viewpage.aspx?documentid=1409.
[128] The Australian Department of Transport website dedicates an extensive component to this fatigue reform.

reviews research on fatigue, and tests technologies and methods to manage fatigue. These efforts are typically part of broader human factors initiatives. Military research and assessment efforts have focused particularly on fatigue of the warfighter during sustained operations and, in collaboration with the non-military sector, on fatigue in transportation modes (land, air, and sea).[129] The military has focused particularly on methods for predicting fatigue and its impacts, countermeasures, including pharmaceuticals, and on incorporating fatigue in operations planning and management (Caldwell et al. 2009; King 2005; Storm 2008; Kronauer and Stone 2004). The military has also emphasized examination of the effects of fatigue on team performance (see for example, Darlington et al. 2006) and on performance in complex, multi-system environments (Lawton et al. 2005).

The U.S. military has invested heavily in the development of human performance modeling. As part of this effort, the U.S. Department of Defense funded a project by the Sandia National Laboratories to develop a model of soldier fatigue and its potential impacts on a "system of systems" that better reflects the operational context than previous efforts that focused only on a single system (e.g., cockpit operations) (Lawton et al. 2005). The military branches have monitored or collaborated in many of the efforts to develop predictive models that better represent the complex operational environment and individual variability, including the Sleep, Activity, Fatigue, and Task Effectiveness (SAFTE) and the Fatigue Avoidance Scheduling Tool (FAST) models. They have also applied these models to examine how actual work hours compare to the established schedules and the implications for fatigue (Mason 2009). The Air Force Research Laboratory and the Air Force School of Aerospace Medicine periodically conduct a "Military Aviation Fatigue Countermeasures Workshop" that introduces participants to advancements in methods and technologies. The 2004 workshop, for example, included instruction on the use of the FAST model. In addition to using these models to evaluate alternative schedules (Beshany 2009), the military branches also participate in applied research using biomathematical fatigue models to evaluate the impacts of fatigue on warfighters, evaluate potential countermeasure strategies for warfighting, conduct post-incident investigations, and develop mitigation strategies (see Hursh et al. 2004; Kronauer and Stone 2004; and Hursh and Balkin 2004).

In 2005, the Naval Safety Center teamed up with the Human Performance Center to explore how to increase awareness and inform intervention strategies concerning impaired and fatigued driving by military personnel. More than most other sectors, the military and NASA have focused on understanding and developing strategies for dealing with personnel who will be fatigued, as well as participating in the development, evaluation, and testing of scheduling strategies to minimize fatigue and maximize recovery. These initiatives have included evaluations of restorative strategies such as breaks, sleep (both naps and longer sleep periods), stimulants/performance boosters, and sedatives/sleep aids (Miller et al. 2007).

3.5.3 Other Sectors

Regulators for other sectors have not addressed fatigue to the same extent as those discussed above, although the issue of fatigue and how to manage it has been the focus of considerable attention in the medical and public safety/security sectors. Fatigue is also a potential concern for mining, as well as law enforcement and security, however, our investigation did not find much information on fatigue-related regulations or indications of organized efforts to develop,

---

[129] Motor vehicle crashes are the leading cause of death of military personnel. An estimated 13 percent of resulting fatalities are identified as fatigue-related; 31 percent are identified as alcohol-related (Naval Safety Center 2005).

pilot test, evaluate, or adopt technologies or fatigue management tools in these sectors. We found a few mentions of pilot tests or adoption of fatigue-related practices and/or technologies, including some for the off-shore oil and gas industry and the health care sector:

Mining

Caterpillar, a manufacturer of heavy machinery, and Delphi, a supplier, worked together to evaluate an oculomotor technology in large mining equipment such as Caterpillar's large off-highway haul trucks. The evaluation of this technology concluded that it was useful but not a sufficient means of detecting and managing fatigue in this sector (Dufour et al. 2009).

Law Enforcement/Security

The Philadelphia Police Department established a Comprehensive Police Fatigue Management Program with participation from the National Institute of Justice and Centers for Disease Control (CDC). Evaluation results for this program were not yet available at the time this review was completed (Vila 2006).

The Division of Sleep Medicine at Harvard and Brigham Medical Schools, in collaboration with an evaluation firm, implemented and evaluated a screening and treatment program for obstructive sleep apnea in a city police department (Lockley et al. 2009). The evaluation found the results of the sleep screening and treatment program to be positive.

Medical

Fatigue is a growing safety concern in the medical sector, for both physicians and nurses. In 1993, the European Union established a "Working Time Directive" for physicians and other safety and health workers (excluding physicians in training). The directive established maximum average working week hours (48), minimum rest periods (11 consecutive hours/day; one rest day/week), rest breaks for working days longer than 6 hours; minimum paid holidays per year (four weeks/year); and maximum average night shift length (8 hours). Requirements for physicians in training in the European Union were phased in later (Simoens and Hurst 2006).

In 2008, the American Institute of Medicine issued *Resident Duty Hours: Enhancing Sleep, Supervision and Safety*, which provided a rationale and policy recommendations concerning fatigue management for physicians in training in the U.S. (Ulmer et al. 2008). Aside from work hour controls, including rest break requirements, this report did not recommend the use of predictive or fatigue monitoring technologies. Heitmann et al. (2009) pilot tested a cognitive performance test involving shape recognition in a medical work setting and found that the current version of the test was not very sensitive in that it only detected severe impairment and did not reflect gradual alertness changes. There is growing attention to fatigue among health professionals and its effects on patient safety, with much of the attention focused on scheduling.

3.5.4 Concluding Points

Attention to fatigue management in safety- and security-sensitive workers is growing throughout the world. A review of the literature and discussions with experts confirm that the transportation sector, particularly commercial trucking, followed by aviation, continues to be the leader in developing, testing, and deploying technologies to monitor and manage fatigue. Other sectors, such as health care; law enforcement; mining, oil and gas development; the military; and the nuclear industry are assessing the causes and incidence of fatigue, exploring ways to reduce fatigue, and its impacts on safety, security, and performance.

The trend across the various transportation sectors is to move beyond prescriptive regulations addressing hours of service and rest periods towards a more comprehensive approach. Many transportation regulators are requiring regulated entities to adopt an FRMS and are working with key transportation industries to establish FRMS guidelines and standards and address the scheduling and personnel management challenges associated with this complex approach.

Fatigue management practices and technologies directed at drivers tend to be the most developed. Most applied fatigue technologies have been and continue to be developed with drivers of vehicles in mind. Only a few of these technologies have been adapted to, pilot tested, and/or deployed in other work contexts. However, breakthroughs in imaging, communications, data processing, and systems integration are expanding the capability base rapidly. While it is important to be aware of what is being done to address driver fatigue, it is critical to take into account differences between work situations involving drivers and workers in other settings and activities. These differences will affect both the nature of the problem (the extent to which fatigue is a potential hazard) and the requirements for a solution (the way the fatigue hazard can most effectively and optimally be addressed). One expert noted that the marine and rail sectors have control operations that might be more similar to an NPP control room than to a driver work situation. There are also key differences across the categories of workers at NPPs that are subject to FFD and fatigue regulation. Some key factors that may affect the applicability of technologies designed with drivers in mind for NPP and other types of workers include:

- *Type of Task.* Drivers are continuously performing the same task and must constantly attend to potential hazards and make corrections to the vehicle's direction and speed. In contrast, nuclear control room operators perform a monitoring task with infrequent interaction with the controls of the plant; security personnel perform monitoring and surveillance tasks; and maintenance personnel perform a wide variety of tasks in different locations. Job monitoring may be more difficult when job tasks are more diverse.
- *Work Posture.* While operating a vehicle, a person's position is restricted. The driver is generally seated and facing forward, looking out the windshield. NPP control room operators, security, and maintenance personnel change position from sitting to standing and location. Fatigue detection technologies that require restricted movement would not work in these work situations.
- *Type of Staff.* Vehicles normally have one operator. NPP control rooms have operator crews; security and maintenance personnel work both alone and in teams. Difficulties, costs, and the applicability of methods for detecting fatigue for multiple operators and different categories of workers need to be considered in the development and selection of technologies and fatigue management systems.
- *Staff Arrangements.* The presence or absence of other people influences individuals' opportunity for social interaction and the opportunity for intervention by others in the case of observable fatigue. Staff arrangements vary, with operators typically working in groups, and maintenance and security personnel often working alone. Impacts on team interactions and the potential to disrupt co-workers are considerations for technologies and fatigue management systems.
- *Environmental Conditions.* The environmental conditions within a truck or other vehicle are relatively constant and well defined. The environmental conditions experienced by

some categories of NPP staff subject to FFD regulations are more varied. Protective clothing may also be a considerationd, such as gloves, face shields, fire-fighting clothing, anti-contamination clothing, and steam suits.

3.6 Conclusion: Relevance to the Nuclear Industry

This overview provides an update on recent developments and emerging trends in:

- fatigue research;
- fatigue-related technology development and deployment;
- fatigue management systems; and
- fatigue management and regulatory practices in various sectors.

Its purpose is to highlight the existence and status of these developments and trends, and to identify resources that can be used to obtain more detailed information about them. It provides a framework for understanding the drivers, locus of activity, and stage of development of fatigue management strategies and technologies based on a review of the literature, identification and review of technologies and methods, and selected interviews with industry experts relative to:

- types of fatigue intervention technologies;
- types of fatigue assessment technologies;
- potential uses and utility of specific types of technologies and combinations of technologies; and
- implementation issues.

The 2008 Part 26 rule requires licensees to *establish a fatigue management policy* and to *evaluate* and *audit the management of worker fatigue*. This report provides information concerning the state of the practice in fatigue assessment and management, including the status of technologies that might be useful in monitoring and/or aiding in the management of fatigue.

Analysis of this information suggests:

- The policy and practice of fatigue management should reflect and support an overall approach to fatigue, with an emphasis on prevention;
- Fatigue management should be part of an overall FFD management system;
- Tools to evaluate schedules in terms of their impact on fatigue are now commercially available, and can provide information to inform workforce and crew scheduling approaches and strategies;
- Technologies, such as actigraphs, have been tested and used in a sufficient number of contexts that they can be considered potential resources to address particular fatigue management issues, including verification of the effectiveness of fatigue management practices; and
- Audits of fatigue management programs should evaluate the effectiveness of the fatigue management system in preventing fatigue and consider using some of the analytic tools to help identify where and why remaining fatigue hazards are occurring.

Part 26 also requires licensees to train and test personnel on their knowledge concerning the contributors to fatigue and various risk factors, and their ability to identify symptoms of worker fatigue. The resources identified in this report, including the self-assessment tools, may provide a guide to information to be conveyed in this training. Review of the literature on training is outside the scope of this report.

This overview also provides information about emerging methods and tools that are applicable to some of the specific requirements in Part 26, such as the conduct of fatigue assessments for specific purposes and for supervisory assessments relative to issuance of a waiver of work hour controls.

Although none of the technologies on the market now have been validated for widespread deployment in the context of an NPP, the years of effort in the military and transportation sectors, and rapid advancements in microelectronics and video technologies, have resulted in a number of technologies with solid theoretical and empirical foundations that have demonstrated performance in the field. The technologies that use actigraphy and oculometrics, including pupillometry, to measure fatigue and detect sleep may be of greatest interest to the industry for the reasons discussed in this report. Some of the oculometric technologies may become more pertinent as the industry shifts to greater use of digital instruments and controls.

Given the complexity of workforce schedules at many nuclear facilities, biomathematical models may be a useful tool for evaluating alternative schedules and crew structures. These tools are becoming more user-friendly. By virtue of their relatively widespread use in the airline and transportation industries, many of the complex policy, personal preference, and other implementation issues associated with scheduling have been identified. A growing number of the modeling tools provide both schedule evaluation and schedule design.

Unlike some other sectors and work environments, NPPs do not face constraints on shift timing, duration, and travel across time zones that create some of the challenges for fatigue management in other sectors. However, many safety- and security- sensitive jobs in NPPs do have the compounding effects of night work, monotonous/vigilance tasks, and monotonous environments that create alertness and sleepiness challenges. Some of the computerized vision monitoring technologies appear to be sufficiently accurate, flexible, and unobtrusive to perform effectively in the NPP environment, if the consequences of inattention are sufficient to warrant monitoring and intervention, or the conditions promote sleepiness. A clear lesson learned from the transportation industry is that the technology must be adapted to the particular situation, which requires both analysis to characterize the problem and the context, and field tests to demonstrate performance and the absence of unintended consequences.

As this review indicates, work has been under way developing tools and methods to predict, detect, and mitigate fatigue for many years. An indication of the maturity of some of these technologies is their incorporation into consumer products, such as automobiles and commercial vehicles (Automotive Fleet 2009). Others, such as actigraphs, have been widely deployed within the medical community to address problems related to fatigue, such as sleep apnea and other sleep-disrupting conditions. Others have been adapted for use in interactive video games and virtual reality tools. As a result, several of the basic technologies and systems that may have applicability to the nuclear industry have both validated scientific bases and extensive empirical performance documentation, and may be ready for the next step of pilot-testing in a nuclear facility.

3.7   References

Achermann, P., and A. Borbély. (2003). Mathematical Models of Sleep Regulation. *Frontiers in Bioscience* 8:683-693.

Achermann, P., and A. Borbély. (1992). Combining Different Models of Sleep Regulation. *Journal of Sleep Research* 1:144-147.

Aeschbach D., T. Postolachea, L. Shera, J. Matthewsa, M. Jackson, and T. Wehra. (2001). Evidence from the Waking Electroencephalogram that Short Sleepers Live under Higher Homeostatic Sleep Pressure than Long Sleepers. *Neuroscience* 102(3):493-714.

Ahsberg, E., F. Gamberale, and K. Gustafsson. (2000). Perceived Fatigue After Mental Work: An Experimental Evaluation of a Fatigue Inventory. *Ergonomics* 43(2):252-268.

Ahsberg, E., F. Gamerale, and A. Kjellberg. 1997. Perceived Quality of Fatigue During Different Occupational Tasks: Development of a Questionnaire. *International Journal of Industrial Ergonomics* 20:121-35.

Åkerstedt, T., J. Connor, A. Gray, and G. Kecklund. 2008a. Predicting Road Crashes from a Mathematical Model of Alertness Regulation – The Sleep/Wake Predictor. *Accident Analysis and Prevention.* 40:1480-1485.

Åkerstedt, T., M. Ingre, J. Broman, and G. Kecklund. 2008b. Disturbed Sleep in Shift Workers, Day Workers, and Insomniacs. *Chronobiology International* 25(2/3):333-348.

Åkerstedt, T., M. Ingre, G. Kecklund, S. Folkard, and J. Axelsson. 2008c. Accounting for Partial Sleep Deprivation and Cumulative Sleepiness in the Three-Process Model of Alertness Regulation. *Chronobiology International.* 25(2/3):209-319.

Åkerstedt, T., G. Kecklund, and J. Axelsson. 2008d. Effects of Context on Sleepiness Self-Ratings During Repeated Partial Sleep Deprivation. *Chronobiology International* 25(2/3):271-278.

Åkerstedt, T. 2007. Altered Sleep/Wake Patterns and Mental Performance. *Physiology and Behavior* 90:209-218.

Åkerstedt, T., J. Axelsson, and G. Kecklund. 2007. Individual Validation of Model Predictions of Sleepiness and Sleep Hours. *Somnologie* 11:169-174.

Åkerstedt, T., S. Folkard, and C. Portin. 2004. Predictions from the Three-Process Model of Alertness. *Aviation, Space, and Environmental Medicine* 75 (3 Suppl):A75-A83.

Åkerstedt, T., G. Kecklund, M. Gillberg, A. Lowden, and J. Axelsson. 2000. Sleepiness *and Days of Recovery.* Transportation Research Part F. 3:251-261.

Åkerstedt, T., and M. Gillberg. 1990. Subjective and Objective Sleepiness in the Active Individual. *International Journal of Neuroscience* 52:29-37.

Åkerstedt, T. 1998. Shift Work and Disturbed Sleep/Wakefulness. *Sleep Medicine Reviews* 2(2):117-128.

Åkerstedt, T., and S. Folkard. 1997. The Three-Process Model of Alertness and Its Extension to Performance, Sleep Latency, and Sleep Length. *Chronobiology International* 14(2):115-123.

Åkerstedt, T. 1996. *Wide Awake at Odd Hours*. Stockholm: Swedish Council for Work Life Research.

Åkerstedt, T., and S. Folkard. 1995. Validation of the S and C Components of the Three-Process Model of Alertness Regulation. *Sleep* 18:1-6.

Allen, P., E. Wadsworth, and A. Smith. 2007. The Prevention and Management of Seafarers' Fatigue: A Review. *International Maritime Health* 58:1-4.

Ancoli-Israel, S., R. Cole, C. Alessi, M. Chambers, W. Moorcroft, and C.P. Pollak. 2003. The Role of Actigraphy in the Study of Sleep and Circadian Rhythms. *Sleep* 26(3):342-392.

Andreassi, J. 2007. Psychophysiology: Human Behavior and Physiological Response. 5th Edition. Mahwah, NJ: Lawrence Erlbaum Associates, Inc.

Annett, J. 2002. Subjective Rating Scales: Science or Art? *Ergonomics* 45(14):966-987.

Aston-Jones, G. 2005. Brain Structures and Receptors Involved in Alertness. *Sleep Medicine* 6(Suppl 1):S3-S7.

Automotive Fleet. 2009. 2010 Mercedes-Benz E-Class Addresses Drowsy Driving. Automotive Fleet: The Car and Truck Fleet and Leasing Management Magazine.

Aviation, Space, and Environmental Medicine. 2004. Proceedings of the Fatigue and Performance Modeling Workshop, June 13-14 2002, Seattle, WA 75(3):Sector II, March. (3 Supplement): A1-199. [No authors listed].

Axelsson, J., G. Kecklund, T. Åkerstedt, P. Donofio, M. Lekander, and M. Ingre. 2008. Sleepiness and Performance in Response to Repeated Sleep Restriction and Subsequent Recovery During Semi-Laboratory Conditions. *Chronobiology International* 25(2/3):297-308.

Baker, K., J. Olson, and D. Morriseau. 1994. Work Practices, Fatigue, and Nuclear Power Plant Safety Performance. *Human Factors* 36(2):244-257.

Baker, Stephanie, Darrelle Bowman, Jeff Hickman, Akiko Nakata, and Rich Hanowski. 2007. Focus Groups in Support of an Operator Drowsiness Monitoring System. Final Report. U.S. Department of Transportation University Transportation Center and ITS Implementation Center Research Project.

Baker, T., S. Campbell, K. Linder, and M. Moore-Ede. 1990. Control-*Room Operator Alertness and Performance in Nuclear Power Plants*. EPRI-6748.

Balkin, T.J., W.J. Horrey, R.C. Graeber, C.A. Czeisler, and D.F. Dinges. 2011. The Challenges and Opportunities of Technological Approaches to Fatigue Management. *Accident Analysis and Prevention* 43(2):565-572.

Balkin T., P. Bliese, G. Belenky, H. Sing, D. Thorne, D., and M. Thomas. 2004. Comparative Utility of Instruments for Monitoring Sleepiness-related Performance Decrements in the Operational Environment. *Journal of Sleep Research* 13:219-27.

Banks, S., and D. Dinges. 2011. Chronic Sleep Deprivation. In *Principles and Practice of Sleep Medicine,* 5th Edition. M. Kryger, T. Roth, and W. Dement, Editors. St. Louis, MO: W.B. Elsevier Saunders Company. Pp. 67-75.

Banks, S., H. Van Dongen, G. Maislin, and D. Dinges. 2010. Neurobehavioral Dynamics Following Chronic Sleep Restriction: Dose-Response Effects of One Night for Recovery. *Sleep* 33(8):1013-1026.

Banks, S., and D. Dinges. 2007. Behavioral and Physiological Consequences of Sleep Restriction. *Journal of Clinical Sleep Medicine* 3(5):519-528.

Barnes, Christopher M., and David T. Wagner. 2009. Changing to Daylight Saving Time Cuts into Sleep and Increases Workplace Injuries. *Journal of Applied Psychology* 94(5):1305-1317.

Barr, L., S. Popkin, and H. Howarth. 2009. *An Evaluation of Emerging Driver Fatigue Detection Measures and Technologies.* Cambridge, MA: Volpe National Transportation Systems Center.

Bartley, S. and E. Chute. 1947. *Fatigue and Impairment in Man.* New York: McGraw-Hill Book Company, Inc.

Barton, J., E. Spelten, P. Totterdell., L. Smith, S. Folkard, and G. Costa. 1995. The Standard Shiftwork Index: A Battery of Questionnaires for Assessing Shiftwork-Related Problems. *Work and Stress* 9(1):4-30.

Basner, M., D. Mallicone, and D.F. Dinges. 2011. Validity and Sensitivity of a Brief Psychomotor Vigilance Test (PVT-B) to Total and Partial Sleep Deprivation. *Acta Astronautica* 69:949-959.

Basner, M., K. Fomberstein, F. Razavi, S. Banks, J. William, R. Rosa, and D. Dinges. 2007. American Time Use Survey: Sleep Time and Its Relationship to Waking Activities. *Sleep* 30(9):1085-1095.

Baulk, S.D., S.N. Biggs, K.J. Reid, C.J. van den Heuvel, and D. Dawson. 2008. Chasing the Silver Bullet: Measuring Driver Fatigue Using Simple and Complex Tasks. *Accident Analysis and Prevention* 40:396-402.

Beersma, D., and M. Gordijin, M. 2007. Circadian Control of the Sleep-Wake Cycle. *Physiology and Behavior* 90:190-195.

Beersma, D. 2005. Why and How Do We Model Circadian Rhythms *Journal of Biological Rhythms* 20:304-313.

Beersma, D. 1998. Models of Human Sleep Regulation. *Sleep Medicine Review* 2:31-43.

Behn, C., E. Brown, T. Scammell, and N. Kopell. 2007. A Mathematical Model of Network Dynamics Governing Sleep-Wake Patterns in Mice. *Journal of Neurophysiology* 97(6):3828-3840.

Belenky, G., N. Wesensten, D. Thorne, M. Thomas, H. Sing, D. Redmond, M. Russo, and T. Balkin. 2003. Patterns of Performance Degradation and Restoration During Sleep Restriction and Subsequent Recovery: A Sleep Dose-Response Study. *Journal of Sleep Medicine* 12:1-12.

Bes, F., M. Jobert, and H. Schulz. 2009. Modeling Napping, Post-Lunch Dip, and Other Variations in Human Sleep Propensity. *Sleep* 32(3):392-398.

Beshany, R. 2009. *Analysis of Navy Flight Scheduling Methods Using Flyawake.* Master's Thesis. Naval Postgraduate School. Monterey, CA.

Blatter, K., and C. Cajochen. 2007. Circadian Rhythms in Cognitive Performance: Methodological Constraints, Protocols, Theoretical Underpinnings. *Physiology and Behavior* 90:196-208.

Bojkowski, C., J. Arendt, M. Shih, and S. Markey. 1987. Melatonin Secretion in Humans Assessed by Measuring Its Metabolite, 6-sulfatoxymelatonin. *Clinical Chemistry* 33:1343-1348.

Bonnefond, A., A. Muzet, A. Winter-Dill, C. Bailloeuil, F. Bitouze, and A. Bonneau. 2001. Innovative Working Schedule: Introducing One Short Nap during the Night Shift. *Ergonomics* 44(10):937-945.

Borbély, A. 1982. A Two Process Model of Sleep Regulation. *Human Neurobiology* 1:195-204.

Borbély, A., and R. Achermann, R. 2003. Mathematical Models of Sleep Regulation. *Frontiers of Bioscience* 8(Supplement):S683-S693.

Borbély, A., and R. Achermann. 1999. Sleep Homeostasis and Models of Sleep Regulation. *Journal of Biological Rhythms* 14(6):557-568.

Burian, D., V. White, M. Jenkins, D. Kupfer, D. Harville, R. DeLorenzo and J. McQuade. 2009. Gene Expression Changes in Response to 36 Hours Sleeplessness. *2009 International Conference on Fatigue Management in Transportation Operations*. Boston, MA.

Cabon, P., R. Mollard, F. Debouck, L. Chaudron, J. Grau, and S. Deharvengt. 2009. From Flight Time Limitations to Fatigue Risk Management Systems. Accessed 11.2010 at http://www.resilience-engineering.org/RE3/papers.

Cairns, H., B. Hendry, A. Leather, and J. Moxham. 2008. Outcomes of the European Working Time Directive. *British Medical Journal* 337(3):a942 - a942.

Caldwell, J. 2009. Pharmacological Management of Fatigue. 2009 International Conference on Fatigue Management in Transportation Operations. Boston, MA.

Caldwell, J., and J. Caldwell. 2005. Fatigue in Military Aviation: An Overview of Military-Approved Pharmacological Countermeasures. *Aviation, Space, and Environmental Medicine* 76(7 Supplement):C39-51.

Caldwell, J., J. Caldwell, and R. Schmidt. 2008. Alertness Management Strategies for Operational Contexts. *Sleep Medicine Reviews* 12:257-273.

Caldwell, J., M. Mallis, J. Caldwell, M. Paul, J. Miller, and D. Neri. 2009. Fatigue Countermeasures in Aviation. *Aviation, Space, and Environmental Medicine* 80(1):29-59.

Carskadon, M., and W. Dement. 1982. Nocturnal Determinants of Daytime Sleepiness. *Sleep* 9:519-524.

Caterpillar Corporation. 2008. *Operator Fatigue: Detection Technology Review.* http://www.fmig.org/CATFatigue%20Technology%20report%202008.pdf.

Chalder, T., G. Berelowitz, T. Pawlikowska, L. Watts, S. Wessely, D. Wright, and E. Wallace. 1993. Development of a Fatigue Scale. *Journal of Psychosomatic Research* 37(2):147-153.

Charlton, S., and M. Ashton. 1998. Review of Fatigue Management Strategies in the Transport Industry. Land Transport Safety Authority. Wellington, New Zealand. Technical Report 97-91.

Cirelli, C. 2005. A Molecular Window on Sleep: Changes in Gene Expression between Sleep and Wakefulness. *Neuroscientist* 11(1):63-74.

Cirelli, C., and G. Tononi. 2009. Sleep and Sleep States: Gene Expression. *Encyclopedia of Neuroscience.* Pp. 903-909.

Cirelli, C., and G. Tononi. 2008. Is Sleep Essential? *PLoS Biology* 6(8):1605-1611.

Cirelli, C., and G. Tononi. 2000. The Search for the Molecular Correlates of Sleep and Wakefulness. *Sleep Medicine Reviews* 5(5):399-410.

Cirelli, C., and G. Tononi. 1998. Differences in Gene Expression Between Sleep and Waking as Revealed by mRNA Differential Display. *Molecular Brain Research* 56:293-305.

Cohen, J., B. Galef, B. Saha, T. Keans, and R. Craft. 2009. Analyses of Fatigue-related Large Truck Crashes, the Assignment of Critical Reason, and Other Variables Using the Large Truck Crash Causation Study. *2009 International Conference on Fatigue Management in Transportation Operations.* Boston, MA.

Comer, D. 2004. A Case Against Workplace Drug Testing. 1994. *Organization Science* 5(2):259-267.

Comte, J., M. Schatzman, P. Ravassard, P. Luppi, and P. Salin. 2006. A Three States Sleep-Waking Model. *Chaos, Solitons, and Fractals* 29:808-915.

Copen, M., and D. Sussman. 2000. Fatigue and Alertness in the United States Railroad Industry Part II: Fatigue Research in the Office of Research and Development at the Federal Railroad Administration. *Transportation Research Part F*:221-228.

Craig, A., Y. Tran, N. Wijesuriya, and P. Boord. 2006. A Controlled Investigation into the Psychological Determinants of Fatigue. *Biological Psychology* 72:78-87.

Czeisler, C., and D. Dijk. 2001. Human Circadian Physiology and Sleep-Wake Regulation. In *Handbook of Behavioral Neurobiology: Circadian Clocks.* J. Takahashi, F. Turek, and R. Moore, Editors. New York: Kluwer Academic/Plenum Publishing Co. Pp. 531–561.

Czeisler, C., C. George, C. Guilleminault, S. Ancoli-Israel, and A. Pack. 2009. Addressing Obstructive Sleep Apnea in Commercial Drivers. *2009 International Conference on Fatigue Management in Transportation Operations.* Boston, MA.

Czeisler, C., and J. Gooley. 2007. Sleep and Circadian Rhythms in Humans. *Cold Spring Harbor Symposium on Quantitative Biology 2007* 72:579-597.

Daan, S., D. Beersma, and A. Borbély. 1984. Timing of Human Sleep: Recovery Process Gated by a Circadian Pacemaker. *American Journal of Physiology – Regulatory, Integrative, and Comparative Physiology* 246:161-183.

Darlington, K., L. Palacio, T. Dowler, and F. LeDuc. 2006. Situational Awareness, Crew Resource Management, and Operational Performance in Fatigued Two-Man Crews Using Three Stimulant Countermeasures. Fort Rucker, AL: U.S. Army Aeromedical Research Laboratory.

Datta, S., and R. MacLean. 2007. Neurobiological Mechanisms for the Regulation of Mammalian Sleep-Wake Behavior: Reinterpretation of Historical Evidence and Inclusion of Contemporary Cellular and Molecular Evidence. *Neuroscience and Biobehavioral Reviews* 31:775-824.

Davis, D., D. Popovic, R. Johnson, C. Berka, and M. Mitrovic. 2009. Building Dependable EEG Classifiers for the Real World – It's Not Just About the Hardware. In *Augmented Cognition, HCII.* D. Schmorrow et al., Editors. Berlin: Springer-Verlag. Pp. 355-364.

Dawson, D., Y. Noy, M. Härmä, T. Åkerstedt, and G. Belenky. 2011. Modeling Fatigue and the Use of Fatigue Models in Work Settings. *Accident Analysis and Prevention* 43:549-564.

Dawson, D. 2009. A Critical Review of Current Fatigue and Performance Models and Their Use. *2009 International Conference on Fatigue Management in Transportation Operations.* Boston, MA.

Dawson, D., and D. Holland. 2006. Union Pacific Railroad Fatigue Risk Management Safety Case. Accessed 08.2009 at http://files.me.com/drew.dawson/wb98eo.

Dawson, D., and K. McCulloch. 2005a. Managing Fatigue: It's About Sleep. *Sleep Medicine Reviews* 9:365-380.

Dawson, D., and P. Zee. 2005b. Work Hours and Reducing Fatigue-related Risk: Good Research vs Good Policy. *Journal of the American Medical Association* 294:1104-1106.

Dawson, D., and K. Reig. 1997. Fatigue, Alcohol, and Performance Impairment. *Nature* 388(17 July):235.

Dawson, D., and S. Armstrong. 1996. Chronobiotics – Drugs That Shift Rhythms. *Pharmacology and Therapeutics* 69(1):15-36.

Dawson, D., Gibbon, S. and Singh, P. 1996. The Hypothermic Effect of Melatonin on Core Body Temperature: Is More Better? *Journal of Pineal Reseach* 20:192–197.

de Croon, E., J. Sluiter, and M. Frings-Dresen. 2006. Psychometric Properties of the Need for Recovery After Work Scale: Test-Retest Reliability and Sensitivity to Detect Change. *Occupational and Environmental Medicine* 63:202-206.

Dickinson, D., and S. Drummond. 2008. The Effects of Total Sleep Deprivation on Bayesian Updating. *Judgment and Decision Making* 3(2):181-190.

Dijk, D., and S. Archer. 2010. PERIOD3, Circadian Phenotypes, and Sleep Homeostasis *Sleep Medicine Review* 14:151-160.

Dijk, D., and M. von Schantz. 2005. Timing and Consolidation of Human Sleep, Wakefulness, and Performance by a Symphony of Oscillators. *Journal of Biological Rhythms* 20(4):279-290.

Dijk, D., and C. Cajochen. 1997. Melatonin and the Circadian Regulation of Sleep Initiation, Consolidation, Structure, and the Sleep EEG. *Journal of Biological Rhythms* 12:627-635.

Dijk, D., and J. Duffy. 1997. Melatonin and the Circadian Regulation of Sleep Initiation, Consolidation, Structure, and the Sleep EEG. *Journal of Biological Rhythms* 12:627-635.

Dijk, D., T. Shanahan, J. Duffy, J. Ronda, and C. Czeisler. 1997. Variation of Electroencephalographic Activity during Non-Rapid Eye Movement and Rapid Eye Movement Sleep with Phase of Circadian Melatonin Rhythm in Humans. *Journal of Physiology* 505(Pt 3):851-858.

Dijk, D., and C. Czeisler. 1995. Contribution of the Circadian Pacemaker and the Sleep Homeostat to Sleep Propensity, Sleep Structure, Electroencephalographic Slow Waves, and Sleep Spindle Activity in Humans. *The Journal of Neuroscience* 15(5):3526-3538.

Di Milia, L., M. Smolensky, G. Costa, H. Howarth, M. Ohayon, and P. Philip. 2011. Demographic Factors, Fatigue, and Driving Accidents: An Examination of the Published Literature. *Accident Analysis and Prevention* 43:516-532.

Di Milia, L., P.A. Smith, and S. Folkard. 2005. A Validation of the Revised Circadian Type Inventory in a Working Sample. *Personality and Individual Differences* 39:1293-1305.

Dinges, D., and M. Maislin. 2009. Truck Driver Fatigue Management Survey. Paper presented at the *2009 International Conference on Fatigue Management in Transportation Operations.* Boston, MA. March 24-26.

Dinges, D., G. Maislin, G. Krueger, R. Brewster, and R. Carroll. 2005a. *Pilot Test of Fatigue Management Technologies.* U.S. Department of Transportation, Federal Motor Carrier Safety Administration. Report No. FMCSA-RT-05-002. Accessed 03.2009 at http://www.fmcsa.dot.gov/facts-research/research-technology/publications/pilot-test/pilottest-fmt-finalreport.htm?printer=true.

Dinges, D., R. Rider, J. Dorrian, E. McGlinchey, N. Rogers, Z. Cizman, S. Goldenstein, C. Vogler, S. Venkataraman, and D. Metaxas. 2005b. Optical Computer Recognition of Facial Expressions Associated with Stress Induced by Performance Demands. *Aviation, Space, and Environmental Medicine* 76(6) Section II:B172-B182.

Dinges D., N. Rogers, and M. Baynard. 2005c. Chronic Sleep Deprivation. In *Principles and Practice of Sleep Medicine,* 4th Edition. M. Kryger, T. Roth, and W. Dement, Editors. St. Louis, MO: W.B. Elsevier Saunders Company.

Dinges, D. 2004. Critical Research Issues in Development of Biomathematical Models of Fatigue and Performance. *Aviation, Space, and Environmental Medicine* 75(3, Supplement):A181-191.

Dinges, D., M. Mallis, S. Mejdal, and T. Nguyen. 2004. Summary of the Key Features of Seven Biomathematical Models of Human Fatigue and Performance. *Aviation, Space, and Environmental Medicine* 75(3):Section II.

Dinges, D. 1998. *PERCLOS: A Valid Psychophysiological Measure of Alertness As Assessed by Psychomotor Vigilance.* Department of Transportation Tech Brief: Federal Highway Administration, Office of Motor Carriers.

Dinges, D., and M. Mallis. 1998. Managing Fatigue by Drowsiness Detection: Can Technological Promises Be Realized? Proceedings of the 3rd Fatigue in Transportation Conference: Managing Fatigue in Transportation. Fremantle, Western Australia. (Reprinted in *Managing Fatigue in Transportation.* Hartley, L., editor. Bingley, UK: Emerald Group Publishing Ltd. Pp. 209-230).

Dinges, D., R. Graeber, M. Rosekind, A. Samel, and H. Wegmann. 1996. *Principles and Guidelines for Duty and Rest Scheduling in Commercial Aviation.* Ames Research Center, CA: National Aeronautics and Space Administration.

Dorn, L., Editor. 2010. *Driver Behaviour and Training, Vol IV.* Surrey: Ashgate.

Dorrian, J., G. Roach, A. Fletcher, and D. Dawson. 2007. Simulated Train Driving: Fatigue, Self-Awareness and Cognitive Disengagement. *Applied Ergonomics* 38:155-166.

Dorrian, J., N. Lamond, and D. Dawson. 2000. The Ability to Self-Monitor Performance When Fatigued. *Journal of Sleep Research* 9:137-144.

Dufour, R., N. Edenborough, D. Edwards, D. Sommer, B. Sirois, T. Dawson, A. Aguirre, and U. Trutxchel. 2009. Fatigue Monitoring Technology Evaluation: Detection of Unsafe Driving Using Delphi's DSM System. *2009 International Conference on Fatigue Management in Transportation Operations.* Boston, MA.

Durmer, J., and D. Dinges. 2005. Neurocognitive Consequences of Sleep Deprivation. *Seminars in Neurology* 25:117-129.

Eddy, D., and S. Hursh. 2001. Fatigue Avoidance Scheduling Tool (FAST). Brooks Air Force Base, TX: U.S. Air Force Research Laboratory.

Edery, I. 2000. Circadian Rhythms in a Nutshell. *Physiological Genomics* 3:59-74.

Edwards, D., B. Sirois, T. Dawson, A. Aguirre, B. Davis, U. Trutschel. 2007. Evaluation of Fatigue Management Technologies Using Weighted Feature Matrix Method. *Proceedings of the 4th International Driving Symposium on Human Factors in Driver Assessment, Training, and Vehicle Design.* Stevenson, WA.

Eek, F., A. Garde, A. Hansen, R. Persson, P. Orbaek, and B. Karlson, 2006. The Cortisol Awakening Response – An Exploration of Intraindividual Stability and Negative Responses. *Scandinavian Journal of Work, Environment and Health* (Suppl. 2):15-21.

Eye-Com Corporation Website. The Future in the Blink of an Eye... Description of Eye-Com EC7T™ Eye Tracking System. Accessed 11.2010 at http://eyecomcorp.com/eyecom-technology/ec7t and http://eyecomcorp.com/about-eye-com.

Falleti, M., P. Maruff, A. Collie, D. Darby, and M. McStephen. 2003. Qualitative Similarities in Cognitive Impairment Associated with 24 H of Sustained Wakefulness and a Blood Alcohol Concentration of 0.05%. *Journal of Sleep Research* 12:265–274.

Federal Aviation Administration (FAA). 2010. Notice of Proposed Rulemaking. Flightcrew Member Duty and Rest Requirements. September 3, 2010. Accessed 08.2011 at www.faa.gov/regulations_policies/rulemaking/.../FAA_2010_22626.pdf.

Federal Aviation Administration (FAA). 2006. Advisory Circular No. 120-92. Introduction to Safety Management Systems for Air Operators. Accessed 08.2009 at http://www.airweb.faa.gov/Regulatory_and_Guidance_Library/rgAdvisoryCircular.nsf/0/6485143D5EC81AAE8625719B0055C9E5?OpenDocument&Highlight=120-92.

Ferrara, M. and L. De Gennaro. 2001. How Much Sleep Do We Need? *Sleep Medicine Reviews* 5(2):155-179.

Folkard, S. 1997. Black Times. *Accident Analysis and Prevention* 29(4):417-430.

Folkard, S., and T. Åkerstedt. 2004. Trends in the Risk of Accidents and Injuries and Their Implications for Models of Fatigue and Performance. *Aviation, Space, and Environmental Medicine* 75 (3):A161-A167.

Folkard, S., and D. Lombardi. 2004. Toward a "Risk Index" to Assess Work Schedules. *Chronobiology International* 21(6):1063-1072.

Folkard, S., D. Lombardi, and M. Spender. 2006. Estimating the Circadian Rhythm in the Risk of Occupational Injuries and Accidents. *Chronobiology International* 23(6):1181-1192.

Folkard, S., T. Monk, and M. Lobban. 1979. Towards a Predictive Test of Adjustment to Shift Work. *Ergonomics* 22:79-91.

Folkard, S., E. Spelten, P. Totterdell, J. Barton, and L. Smith. 1995. The Use of Survey Measures to Assess Circadian Variation in Alertness. *Sleep* 18:355-361.

Fourie, C., A. Holmes, S. Bourgeois-Bougrine, C. Hilditch, and P. Jackson. 2010. Fatigue Risk Management Systems: A Review of the Literature. *Road Safety Research Report No. 10.* London Department for Transport.

Franken, P., and D. Dijk. 2009. Circadian Clock Genes and Sleep Homeostasis. *European Journal of Neuroscience* 29(9):1820-1829.

Fridlund, A., and J. Cacioppo. 1986. Guidelines for Human Electromyographic Research. *Psychophysiology* 23(5):567-589.

Friedl, K., M. Mallis, S. Ahlers, S. Popkin, and W. Larkin. 2004. Research Requirements for Operational Decision-Making Using Models of Fatigue and Performance. *Aviation, Space, and Environmental Medicine* 75 (3):A192-A199.

Gander, P., L. Hartley, D. Powell, P. Cabon, E. Hichcock, A. Mills, and S. Popkin. 2011. Fatigue Risk Management: Organizational Factors at the Regulatory and Industry/Company Level. *Accident Analysis and Prevention* 43:573-590.

Gartner, J., S. Popkin, W. Leitner, S. Wahl, T. Åkerstedt, and S. Folkard. 2004. Analyzing Irregular Working Hours: Lessons Learned in the Development of RAS 1.0-The Representation and Analysis Software. *Chronobiology International* 21(6):1025-1035.

Gerashchenko, D., J. Wison, D. Burns, R. Reh, P. Shiromani, T. Sakurai, H. de la Iglesia, and T. Kilduff. 2008. Identification of a Population of Sleep-Active Cerebral Cortex Neurons. *Proceedings of the National Academy of Science* 105(29):10227-10232.

Gerson, B., R. Barnett, and D. Holland. 2009. Screening for and Confirmation of Excessive Daytime Sleepiness (EDS) and Obstructive Sleep Apnea (OSA) in Railroad Workers. *2009 International Conference on Fatigue Management in Transportation Operations.* Boston, MA.

Gertler, J., F. Miller, and T. Raslear. 2009 The Fatigue Status of the U.S. Railroad Industry: A Preliminary Analysis. Paper presented at the *2009 International Conference on Fatigue Management in Transportation Operations.* Boston, MA. March 24-26.

Gilliland, K., and R. Schlegel. 1995. Readiness-to-Perform Testing and the Worker. *Ergonomics in Design" The Quarterly of Human Factors Applications* 3(1):14-19.

Gilliland, K., and R. Schlegel. 1993. *Readiness to Perform Testing: A Critical Analysis of the Concept and Current Practices.* Technical Report DOT/FAA/AM-93/13. Norman, OK: University of Oklahoma.

Goel, N., S. Banks, E. Mignot, and D. Dinges. 2009a. PER3 Polymorphism Predicts Cumulative Sleep Homeostatic But Not Neurobehavioral Changes to Chronic Partial Sleep Deprivation. *PLoS ONE* www.plosone.org 4(6):e5874.

Goel, N., H. Rao, J. Durmer, and D. Dinges. 2009b. Neurocognitive Consequences of Sleep Deprivation. *Seminars in Neurology* 29(4):320-339.

Goldwater, B. 1972. Psychological Significance of Pupillary Movements. *Psychological Bulletin* 77(5):340-355.

Golz, M., D. Sommer, M. Holzbrecher, and T. Schnupp. 2007. Detection and Prediction of Driver's Microsleep Events. *Proceedings of the 14th International Conference on Road Safety on Four Continents.* Bangkok, Thailand.

Grace, R. 2006. Drowsy Driver Monitor and Warning System. Pittsburgh, PA: Robotics Institute, Carnegie Mellon University. Accessed 12.2008 at www.attentiontechnology.com/docs/DrowsyDriverMonitor.pdf.

Graeber, R. 1985. Proceedings of the Flight Safety Foundation 38th International Air Safety Seminar. Arlington, VA: Flight Safety Foundation.

Greeley, H., J. Berg, E. Friets, J. Wilson, G. Greenough, J. Picone, J. Whitmore, and T. Nesthus. 2007. Fatigue Estimation Using Voice Analysis. *Behavior Research Methods* 39(3):610-619.

Groeger, J., A. Viola, J. Lo, M. Schantz, S. Archer, and D. Dijk. 2008. Early Morning Executive Functioning During Sleep Deprivation Is Compromised by a PERIOD3 Polymorphism. *Sleep* 31(8):1159-1167.

Gundel, A., K. Marsalek, and C. ten Thoren. 2007. A Critical Review of Existing Mathematical Models for Alertness. *Somnologie* 11:148-156.

Gundel, A., K. Marsalek, and C. ten Thoren. 2005. Support of Mission and Work Scheduling by a Biomedical Fatigue Model. In *Strategies to Maintain Combat Readiness during Extended Deployments – A Human Systems Approach.* Meeting Proceedings, Neuilly-sur-Seine. Pp. 28-1-28-12. Accessed 08.2009 at http://www.rto.nato.int/abstracts.asp.

Guilleminault, C., and S. Brooks. 2001. Excessive Daytime Sleepiness: A Challenge for the Practising Neurologist. *Brain* 124(8):1482-1491.

Gunzelmann, G., and K. Gluck. 2008. Approaches to Modeling the Effects of Fatigue on Cognitive Performance. In *Proceedings of the Seventh Conference on Behavior Representation in Modeling and Simulation.* J. Hansberger, Editor. Pp. 136-145. Orlando, FL: Simulation Interoperability Standards Organization.

Gunzelmann, G., J. Gross, K. Gluck, and D. Dinges. 2009. Sleep Deprivation and Sustained Attention Performance: Integrating Mathematical and Cognitive Modeling. *Cognitive Science* 33:880-910.

Gunzelmann, G., L. Moore, K. Gluck, H. Van Dongen, and D. Dinges. 2008. Individual Differences in Sustained Vigilant Attention: Insights from Computational Cognitive Modeling. In *Proceedings of the Thirtieth Annual Meeting of the Cognitive Science Society*. B.C. Love, K. McRae, and V. M. Sloutsky, Editors. Austin, TX: Cognitive Science Society. Pp. 2017-2022.

Hachol, A., W. Szczepanowska, H. Kasprzak, I. Zawojska, A. Dudzinski, R. Kinasz, and D. Wygledowska-Promienska. 2007. Measurement of Pupil Reactivity Using Fast Pupillometry. *Physiological Measurement* 28:61-72.

Harrison, R., S. Chaiken, D. Harville, J. Fischer, D. Fisher, and J. Whitmore. 2008. The Identification of Fatigue Resistant and Fatigue Susceptible Individuals. Brooks City, TX: Air Force Research Laboratory Human Effectiveness Directorate Biosciences and Protection Division.

Harrison, Y., and J. Horne. 2000. The Impact of Sleep Deprivation on Decision Making: A Review. *Journal of Experimental Psychology: Applied* 6(3):236-249.

Hartley, L., T. Horberry, N. Mabbott, and G. Krueger. 2000. *Review of Fatigue Detection and Prediction Technologies*. National Road Transport Commission.

Hefner, R., D. Edwards, C. Heinze, D. Summer, M. Golz, B. Sirois, and U. Trutschel. 2009. Operator Fatigue Estimation Using Heart Rate Measures. *Proceedings of the International Conference on Driving Assessment*. Big Sky, MT. Pp.110-117.

Heitmann, A., H. Bowles, K. Hansen, M. Holzbrecher-Morys, T. Langley, and D. Schmipke. 2009. Seeking a New Way to Detect Human Impairment in the Workplace. *2009 International Conference on Fatigue Management in Transportation Operations*. Boston, MA.

Heitmann, A., R. Guttkuhn, A. Aguirre, U. Trutschel, and M. Moore-Ede. 2001. Technologies for the Monitoring and Prevention of Driver Fatigue. Proceedings of the First International Driving Symposium on Human Factors in Driver Assessment, Training and Vehicle Design.

Hennessy, D., and D. Wiesenthal. 1999. Traffic Congestion, Driver Stress, and Driver Aggression. *Aggressive Behaviour* 25:409-423.

Hersman, D. 2009. Fatigue Management in Transportation Operations. Keynote Address to the 2009 International Conference on Fatigue Management in Transportation Operations. Boston, MA. March 24-26.

Hill, S., G. Tononi, and M. Ghilardi. 2008. Sleep Improves the Variability of Motor Performance. *Brain Research Bulletin* 76:605-611.

Hitchcock, E., and G. Matthews. 2005. Multidimensional Assessment of Fatigue: A Review and Recommendations. *Proceedings of the International Conference on Fatigue Management in Transportation Operations*. Seattle, WA. September 2005.

Hobbs, A., and A. Williamson. 2003. Associations between Errors and Contributing Factors in Aircraft Maintenance. *Human Factors* 45:186-201.

Hobbs, A., and A. Williamson. 2000. Aircraft Maintenance Safety Survey: Results. Canberra: Australian Transport Safety Bureau.

Hobson, J., and E. Pace-Schott. 2002. The Cognitive Neuroscience of Sleep: Neuronal Systems, Consciousness, and Learning. *Nature Reviews/Neuroscience* 3:670-693.

Hoddes, E., W. Dement, and V. Zarcone. 1972. The Development and Use of the Stanford Sleepiness Scale. *Psychophysiology* 9:150.

Hofer-Tinguely, G., P. Achermann, H. Landlolt, S. Regel, J. Rétey, R. Dürr, A. Borbély, and J. Gottselig. 2005. Sleep Inertia: Performance Changes after Sleep, Rest, and Active Waking. *Cognitive Brain Research* 22:323-331.

Holland, D. 2008. Union Pacific Railroad's Fatigue Risk Management System (FRMS). *International Railway Safety Conference.* Denver, CO.

Horne, J. 2011. The End of Sleep: "Sleep Debt" versus Biological Adaptation of Human Sleep to Waking Needs. *Biological Psychology*.

Horne, J. 1988. Why We Sleep: The Functions of Sleep in Humans and Other Mammals. Oxford: Oxford University Press.

Horne, J., and S. Baulk. 2004. Awareness of Sleepiness When Driving. *Psychophysiology* 41:161-165.

Horne, J., and O. Ostberg. 1976. A Self-assessment Questionnaire to Determine Morningness–eveningness in Human Circadian Systems. *International Journal of Chronobiology* 4(2):97-110.

Horne, J., and L. Reyner. 1995. Sleep Related Vehicle Accidents *BMJ* 310:565-567.

Human Systems Information Analysis Center (HSIAC). 2005. Draft Fatigue Measurement and Prediction Technology. Prepared for the U.S. Nuclear Regulatory Commission. No Final Report Prepared.

Hursh, S. 2009a. Modeling Sleep and Performance within the Integrated Unit Simulation System (IUSS). Nadick, MA: United States Army Soldier Systems Command; Natick Research, Development, and Engineering Center.

Hursh, S. 2009b. Promise and Limitations of Fatigue and Performance Modeling as a Tool for Fatigue Risk Management in Transportation. Paper presented at the *2009 International Conference on Fatigue Management in Transportation Operations.* Boston, MA. March 24-26.

Hursh, S., and T. Balkin. 2004. Response to Commentary on Fatigue Models for Applied Research in Warfighting. *Aviation, Space, and Environmental Medicine* 74(3, Suppl.):A57-60.

Hursh, S., D. Redmond, M. Johnson, D. Thorne, G. Belenky, T. Balkin, W. Storm, J. Miller, and D. Eddy. 2004. Fatigue Models for Applied Research in Warfighting. *Aviation, Space, and Environmental Medicine* 74(3, Suppl.):A44-53.

Hursh, S., T. Raslear, A. Kaye, and J. Fanzone. 2008. *Validation and Calibration of a Fatigue Assessment Tool for Railroad Work Schedules*, Final Report. (DOT/FRA.ORD-08/04). Washington, DC: Federal Railroad Administration.

Hossain, J., L. Reinish, R. Heselegrave, G. Hall, L. Kayumov, S. Chung, P. Bhuiya, D. Jovaniovic, N. Huterer, J. Volkov, and C. Shapiro. 2004. Subjective and Objective Evaluation of Sleep and Performance in Daytime Versus Nighttime Sleep in Extended-Hours Shift-Workers at an Underground Mine. *Journal of Occupational and Environmental Medicine* 46(3):212-226.

Hussain, A., B. Bais, S. Samad, and S. Hendi. 2008. Novel Data Fusion Approach for Drowsiness Detection. *Information Technology Journal* 7(1):48-55.

Ingre, M., T. Åkerstedt, B. Peters, A. Anund, and G. Kecklund. 2006a. Subjective Sleepiness, Simulated Driving Performance and Blink Duration: Examining Individual Differences. *Journal of Sleep Research* 15:47-53.

Ingre, M., T. Åkerstedt, B. Peters, A. Anund, G. Kecklund, and A. Pickles. 2006b. Subjective Sleepiness and Accident Risk Avoiding the Ecological Fallacy. *Journal of Sleep Research* 15:142-148.

Institute of Medicine. 2008. see Ulmer et al. 2008.

Jackson, P., A. Holmes, and C. Fourie. 2009. A Review of Fatigue Risk Management Systems and their Potential for Managing Fatigue within the UK Road Transport Industry. *2009 International Conference on Fatigue Management in Transportation Operations*. Boston, MA.

Ji, Q., P. Lan, and C. Looney. 2006. A Probabilistic Framework for Modeling and Real-Time Monitoring Human Fatigue. *IEEE Transaction on Systems, Man, and Cybernetics* 36(5):862-875.

Ji, Q, Z. Zhu, and P. Lan. 2004. Real-time Nonintrusive Monitoring and Prediction of Driver Fatigue. *IEEE Transportation Vehicle Technology* 53(4):1052-68.

Johns, M. 2002. Sleep Propensity Varies with the Behaviour and the Situation in Which it is Measured: The Concept of Somnificity. *Journal of Sleep Research* 11:61-67.

Johns, M. 1998. Rethinking the Assessment of Sleepiness. *Sleep Medicine Reviews* 2(1):3-15.

Johns, M. 1991. A New Method for Measuring Daytime Sleepiness: The Epworth Sleepiness Scale. *Sleep* 14:540-545.

Johns, M., R. Chapman, K. Crowley, and A. Tucker. 2008. A New Method for Assessing the Risks of Drowsiness while Driving. *Somnologie* 12:66-74.

Johns, M., A. Tucker, R. Chapman, K. Crowley, and N. Michael. 2007. Monitoring Eye and Eyelid Movements by Infrared Reflectance Oculography to Measure Drowsiness in Drivers. *Somnologie* 11(4):234-242.

Johns, M., A. Tucker, R. Chapman, N. Michael, and C. Beale. 2006. A New Scale of Drowsiness Based on Multiple Characteristics of Blinks: The Johns Drowsiness Scale. *Sleep* 29:A365.

Jones, C., J. Dorrian, S. Jay, N. Lamond, S. Ferguson, and D. Dawson. 2006. Self-Awareness of Impairment and the Decision to Drive After an Extended Period of Wakefulness. *Chronobiology International* 23(6):1253-1263.

Jovanis, P., K. Chen, and C. Chen. 2009. Modeling the Association of Hours-of-Service to Motor Carrier Crash Risk. *2009 International Conference on Fatigue Management in Transportation Operations*. Boston, MA.

Kaida, K., M. Takahashi, T. Åkerstedt, A. Nakata, Y. Otsuka, T. Haratani, and K. Fukasawa. 2006. Validation of the Karolinska Sleepiness Scale Against Performance and EEG Variables. *Clinical Neurophysiology* 117:1574-1581.

Kalia, Madhu. 2006. Neurobiology of Sleep. *Metabolism: Clinical and Experimental* 55(Supplement 2):S2-S6.

Kamdar, B., K. Kaplan, E. Kezirian, and W. Dement. 2004. The Impact of Extended Sleep on Daytime Alertness, Vigilance, and Mood. *Sleep Medicine* 5:441-448.

Karasek, R. 1979. Job Demands, Job Decision Latitude, and Mental Strain: Implications for Job Redesign. *Administrative Science Quarterly* 24:285-308.

Karlen, W., C. Mattiussi, and D.I. Floreano. 2007. Adaptive Sleep/Wake Classification Based on Cardiorespiratory Signals for Wearable Devices. *Biomedical Circuits and Systems Conference Proceedings*. BioCAS-IEEE. 27-30 November, Montreal.

Karrar, M., E. Zilberg, M. Xu, D. Burton, and S. Lal. 2009. Detection of Driver Drowsiness using EEG Alpha Wave Bursts: Comparing Accuracy of Morphological and Spectral Algorithms. *2009 International Conference on Fatigue Management in Transportation Operations*. Boston, MA.

Kay, G. 1994. Phase C Cognitive Function Test Development: Final Report. Submitted to the Federal Aviation Administration, Civil Aeronautical Medical Institute, Oklahoma City, Oklahoma. Contract No. DTFA-02-90-C-90118.

Kecklund, G., and T. Åkerstedt. 1997. Objective Components of Individual Differences in Subjective Sleep Quality. *Journal of Sleep Research* 6:217-220.

Kecklund, G., C. Eriksen, and T. Åkersted. 2008. Police Officers' Attitudes to Different Shift Systems: Association with Age, Present Shift Schedule, Health, and Sleep/Wake Complaints. *Applied Ergonomics* 39:565-571.

Kecklund, G., M. Ingre, T. Åkerstedt, A. Anund, M. Wahde, and D. Sandberg. 2009. Sleepiness and Driving Performance: a Simulator Study of the Effects of Sleep Loss and Time of Day. *2009 International Conference on Fatigue Management in Transportation Operations*. Boston, MA.

Killgore, W., D. Killgore, G. Ganesan, A. Krugler, and G. Kamimori. 2006. Trait-Anger Enhances Effects of Caffeine on Psychomotor Vigilance Performance. *Perceptual and Motor Skills* 103:883-886.

Killgore, W., J. Richards, D. Killgore, G. Kamimori, and T. Balkin. 2007. The Trait of Introversion-Extraversion Predicts Vulnerability to Sleep Deprivation. *Journal of Sleep Research* 16:354-363.

King, A., G. Belenky, and H. Van Dongen. 2009. Performance Impairment Consequent to Sleep Loss: Determinants of Resistance and Susceptibility. *Current Opinion in Pulmonary Medicine* 15(6):559-564.

King, R. 2005. A Novel Approach to Encouraging Proper Fatigue Management in British Army Aviation Training and Operations. *Aeronautical Journal* 109(1096):293-296.

Kirby, A., M. Clayton, P. Rivera, and C. Comperatore. 2007. Melatonin and the Reduction or Alleviation of Stress. *Journal of Pineal Research* 27(2):78-85. Published Online: 30 Jan 2007. Accessed 08.2009 at http://www3.interscience.wiley.com/journal/11907603/abstract#c1#c1.

Klemets, T. and E. Romig. 2009. Flight Crew Scheduling Based on Fatigue Risk Guidance. *2009 International Conference on Fatigue Management in Transportation Operations*. Boston, MA.

Krajewski, J., U. Trutschel, M. Golz, D. Sommer, and D. Edwards. 2009. Estimating Fatigue from Predetermined Speech Samples Transmitted by Operator. *Proceedings of the International Conference on Driving Assessment*. Big Sky, MT. Pp.468-474.

Kristjansson, S., J. Stern, T. Brown, and J. Rohrbaugh. 2009. Detecting Phasic Lapses in Alertness Using Pupillometric Measures. *Applied Ergonomics* 40:978-986.

Kronauer,R. and B. Stone. 2004. Commentary on Fatigue Models for Applied Research in Warfighting. *Aviation, Space, and Environmental Medicine* 74(3, Suppl.):A54-56.

Krueger, G. 2004. Technologies and Methods for Monitoring Driver Alertness and Detecting Driver Fatigue: A Review Applicable to Long-haul Truck Driving. Technical report for American Transportation Research Institute and Federal Motor Carrier Safety Administration. Washington, DC.

Krueger, G., and H. Leaman. 2009. Stimulants, Hypnotics, Nutritional Aids, Medications, and Other Chemical Substances. *2009 International Conference on Fatigue Management in Transportation Operations*. Boston, MA.

Kryger, M., T. Roth, and W. Dement. 2000. *Principles and Practice of Sleep Medicine*. Philadelphia: W.B. Saunders.

Lac, G., and A. Chamoux. 2003. Elevated Salivary Cortisol Levels as a Result of Sleep Deprivation in a Shift Worker. *Occupational Medicine* 53:143-145.

Lal, S., and A. Craig. 2002. Driver Fatigue: Electroencephalography and Psychological Assessment. *Psychophysiology* 39(3):313-321.

Lal, S., and A. Craig. 2001. Electroencephalography Activity Associated with Driver Fatigue: Implications for a Fatigue Countermeasure Device. *Journal of Psychophysiolog* 15(3):183-189.

Lal, S., A. Craig, P. Boord, L. Kirkup, and H. Nguyen. 2003. Development of an Algorithm for an EEG-based Driver Fatigue Countermeasure. *Journal of Safety Research* 34:321-328.

Lammers-van der Holst, H., H. Van Dongen, and G. Kerkhof. 2006. Are Individuals' Nighttime Sleep Characteristics Prior to Shift-Work Exposure Predictive for Parameters of Daytime Sleep After Commencing Shift Work? *Chronobiology International* 23(6):1217-1227.

Lamond, N., and D. Dawson. 1999. Quantifying the Performance Impairment Associated with Fatigue. *Journal of Sleep Research* 8:255-262.

Lamond, N., D. Dawson, and G. Roach. 2005. Fatigue Assessment in the Field: Validation of a Hand-Held Electronic Psychomotor Vigilance Task. *Aviation, Space, and Environmental Medicine* 76(5):486-489.

Lamond, N., S. Jay, J. Dorrian, S. Ferguson, C. Jones, and D. Dawson. 2007. The Dynamics of Neurobehavioural Recovery Following Sleep Loss. *Journal of Sleep Research* 16:33-41.

Lamond, N., S. Jay, J. Dorrian, S. Ferguson, G. Roach, and D. Dawson. 2008. The Sensitivity of a Palm-based Psychomotor Vigilance Task to Severe Sleep Loss. *Behavior Research Methods* 40(1):347-352.

Lamond, N., G. Roach, and D. Dawson. 2002. Is There an Alternative to the 10-minute PVT for Field Studies? In *Susila IGN*. Kumpulan Makalah Ergonomi. Denpasar: Udayana University Press. Pp. 512–21.

Landrigan, C., J. Rothschile, and J. Cronin. 2004. Effect of Reducing Interns' Work Hours on Serious Medical Errors in Intensive Care Units. *New England Journal of Medicine* 52(4):445-453.

Larue, G., A. Rakotonirainy, and A. Pettitt. 2010. Predicting Driver's Hypovigilance on Monotonous Roads: Literature Review. Presented at the Conference on Driver Distraction and Inattention. Queensland, Australia.

Lavie, P. 2001. Sleep-Wake As a Biological Rhythm. *Annual Review of Psychology* 52:277-303.

Lawton, C., D. Miller, and J. Campbell. 2005. Human Performance Modeling for System of Systems Analytics: Soldier Fatigue. Albuquerque, NM: Sandia National Laboratories.

Leaman, H., and G. Krueger. 2009. Medical Provider Practices regarding Drugs and Commercial Driver Certification. *2009 International Conference on Fatigue Management in Transportation Operations*. Boston, MA.

Lee, M., B. Swanson, and H. de la Iglesia. 2009. Circadian Timing of REM Sleep is Coupled to an Oscillator within the Dorsomedial Suprachiasmatic Nucleus. *Current Biology* 19(10):848-852.

Lehrer, A. 2009. Stress Resistance, Wellness, and 24/7 Performance: A Predictive, Integrative Model. Paper presented at the International Conference on Fatigue in Transportation Operations. Boston, MA. March 24-26.

Lehrer, A. 2005. *Shiftwork Stress Resistance, Health, and Performance: A Predictive, Integrative Model.* PhD. Dissertation. Austin, TX: The University of Texas at Austin.

Leproult, R., E. Colecchia, A. Berardi, R. Stickgold, S. Kosslyn, and E. Cauter. 2003. Individual Differences in Subjective and Objective Alertness during Sleep Deprivation Are Stable and Unrelated. *American Journal of Physiology – Regulatory, Integrative, and Comparative Physiology* 284:R280-290.

Lewis, P. 1985. Recommendations for NRC Policy on Shift Scheduling and Overtime Policy at Nuclear Power Plants. NUREG/CR-4248.

Lin, C., R. Wu, T. Jung, S. Liang, and T. Huang. 2005. Estimating Driving Performance Based on EEG Spectrum Analysis. *EURASIP Journal on Applied Signal Processing* 19:3165-3174.

Lin, Y. 2009. EEG-based Model for Real-time Driver Drowsiness Recognition and Prevention. *2009 International Conference on Fatigue Management in Transportation Operations*. Boston, MA.

Littner, M., C. Kushida, W.M. Anderson, D. Bailey, R. Berry et al. 2003. Practice Parameters for the Role of Actigraphy in the Study of Sleep and Circadian Rhythms: An Update for 2002. Standards of Practice Committee of the American Academy of Sleep Medicine. *Sleep* 26(3):337-341.

Lockley, S., S. Rajaratnam, C. O'Brien, M. Moritz, S. Qadri, L. Barger, and C. Czeisler. 2009. Operation Healthy Sleep: An Occupational Screening and Treatment Program for Obstructive Sleep Apnea in a City Police Department. *2009 International Conference on Fatigue Management in Transportation Operations*. Boston, MA.

Lowenstein, O., R. Feinberg, and I. Loewenfeld. 1963. Pupillary Movements during Acute and Chronic Fatigue. *Investigative Ophthalmology and Visual Science* 2(2):138–157.

Lu, J., D. Sherman, M. Devor, and C. Saper. 2006. A Putative Flip-Flop Switch for Control of REM Sleep. *Nature* 441:589-594.

Lüdtke, H., B. Wilhelm, M. Adler, F. Schaeffel, and H. Wilhelm. 1998. Mathematical Procedures in Data Recording and Processing of Pupillary Fatigue Waves. *Vision Research* 38(19):2889-2896.

Luna, T., J. French, and J. Mitcha. 1997. A Study of USAF Air Traffic Controller Shiftwork: Sleep, Fatigue, Activity, and Mood Analyses. *Aviation, Space, and Environmental Medicine* 68:18-23.

Mallis, M., S. Medjal, T. Nguyen, and D. Dinges. 2004. Summary of the Key Features of Seven Biomathematical Models of Human Fatigue and Performance. *Aviation, Space, and Environmental Medicine* 75 (3):A4-A14.

Maruff, P., M. Falleti, A. Collie, D. Darby, and M. McStephen. 2005. Fatigue-related Impairment in the Speed, Accuracy and Variability of Psychomotor Performance: Comparison With Blood Alcohol Levels. *Journal of Sleep Research* 14(1):21-27.

Marx, D., and R. Graeber. 1994. Human Error in Aircraft Maintenance. In *Aviation Psychology in Practice*. Johnston, N., McDonald, N., and Fuller, R., editors. Pp. 87-104. Aldershot: Avebury.

Mason, D. 2009. *A Comparative Analysis between the Navy Standard Workweek and the Work/Rest Patterns of Sailors Aboard U.S. Navy Cruisers.* Master's Thesis. Naval Postgraduate School. Monterey, CA.

Matthews, G., E. Hitchcock, D. Saxby, and L. Langheim. 2009. Development of a Multi-Dimensional Scale for Driver Fatigue. *2009 International Conference on Fatigue Management in Transportation Operations.* Boston, MA.

Matthews, G., P.A. Desmond, L. Joyner, B. Carcary, and K. Gilliland. 1996. Validation of the Driver Stress Inventory and the Driver Coping Questionnaire. Unpublished report. Also presented at the International Conference on Traffic and Transport Psychology, Valencia, Spain, 1996.

Matthews, G., P.A. Desmond, L. Joyner, B. Carcary, and K. Gilliland. 1997. A Comprehensive Questionnaire Measure of Driver Stress and Affect. In Traffic and Transport Psychology: Theory and Application. T. Rothengatter, and E. C. Vaya, editors. Pp. 317-324. Amsterdam: Pergamon.

McCallum, M., T. Sanquist, M. Mitler, and G. Krueger. 2003. *Commercial Transportation Operator Fatigue Management Reference.* U.S. Department of Transportation. Accessed 02.2009 at http://www.fra.dot.gov/downloads/research/fatigue_management.pdf.

McCallum, M., M. Raby, and A. Rothblum. 1996. *Procedures for Investigating and Reporting Human Factors and Fatigue Contributions to Marine Casualties* (Final Report No. CG-D-97). Washington, DC: United States Coast Guard.

McCarley, R., and C. Sinton. 2008. Neurobiology of Sleep and Wakefulness. *Scholarpedia* 3(4):3313.

McCarley, R. 2007. Neurobiology of REM and NREM Sleep. *Sleep Medicine* 8:302-330.

McCauley, P., L. Kalachev, D. Mollicone, G. Belenky, D. Dinges, and H. Van Dongen. 2009a. Modeling Fatigue in Split Sleep Schedules with a New Biomathematical Model for the

Homeostatic Effects of Sleep Loss. Paper presented at the *2009 International Conference on Fatigue Management in Transportation Operations*. Boston, MA. March 24-26.

McCauley, P., L. Kalachev, A. Smith, G. Belenky, D. Dinges, and H. Van Dongen. 2009b. A New Mathematical Model for the Homeostatic Effects of Sleep Loss on Neurobehavioral Performance. *Journal of Theoretical Biology* 256:227-239.

McClaren, J., P. Hauri, S. Lin, and C. Harris. 2002. Pupillometry in Clinically Sleepy Patients. *Sleep Medicine* 3:347-352.

McColgan, J., and D. Nash. 2009. Findings from International Railway Roster Studies – Moving Beyond Fatigue Management Limitations of Current Schedule Design. *2009 International Conference on Fatigue Management in Transportation Operations*. Boston, MA.

McKenna, B., D. Dickinson, J. Orff, and S. Drummond. 2007. The Effects of One Night of Sleep Deprivation on Known-Risk and Ambiguous-Risk Decisions. *Journal of Sleep Research* 16:245-252.

Michael, R., and R. Meuter. 2009. Differential Effects of Monotony versus Fatigue on Driving Performance According to Multiple Psycho-Physiological and Behavioral Measures: Evidence for Independent Constructs. Paper presented at the *2009 International Conference on Fatigue Management in Transportation Operations*. Boston, MA. March 24-26.

Miller, J. 1996. Fit for Duty? *Ergonomics in Design* 4(2):11-17.

Miller, N., P. Matsangas, and L. Shattuck. 2007. Fatigue and Its Effect on Performance in Military Environments. Chapter 12. *Performance Under Stress*. Ashgate Publications.

Mistlberger, R. 2005. Circadian Regulation of Sleep in Mammals: Role of the Suprachiasmatic Nucleus. *Brain Research Review* 49:429-454.

Mitler, M., M. Carskadon, C. Czeisler, W. Dement, D. Dinges, and R. Graeber. 1988. Catastrophes, Sleep, and Public Policy: Consensus Report. *Sleep* 11(1):100-109.

Mitler, M., M. Carskadon, and M. Hirshkowitz. 2000. Evaluating Sleepiness. In *Principles and Practices of Sleep Medicine*. Kryger, M., Roth, T., and Dement, W., editors. Pp. 1251-1257. Philadelphia: W.B. Saunders.

Mitler, M., K. Gujavarty, and C. Browman. 1982. Maintenance of Wakefulness Test: A Polysomnographic Technique for Evaluating Treatment Efficacy in Patients with Excessive Somnolence. *Electroencephalogy and Clinical Neurophysiology* 53:658-661.

Mollicone, D., H. Van Dongen, N. Rogers, S. Banks, and D. Dinges. 2010. Time of Day Effects on Neurobehavioral Performance during Chronic Sleep Restriction. *Aviation, Space, and Environmental Medicine* 81(8):735-744.

Monticelli, F., E. Tutsch-Bauer, W. Hitzl, and T. Keller. 2009. Pupil Function as a Parameter for Assessing Impairment of the Central Nervous System from a Traffic-Medicine Perspective. *Legal Medicine*11:5331-5332.

Moore, R. 2007. Suprachiasmatic Nucleus in Sleep-Wake Regulation. *Sleep Medicine* 8:S27-S33.

Moore-Ede, M., U. Trutschel, R. Guttkuhn, A. Aguirre , and A. Heitmann. 2009. Fatigue Countermeasure Rebound: Temporary Alertness Gain from Caffeinated Chewing Gum Repaid as Excessive Sleepiness after Countermeasure Cessation. Paper presented at the *2009 International Conference on Fatigue Management in Transportation Operations. Boston*, MA. March 24-26.

Moore-Ede, M., A. Heitmann, R. Guttkuhn, U. Trutschel, A. Aguirre, and D. Croke. 2004. Circadian Alertness Simulator for Fatigue Risk Assessment in Transportation: Application to Reduce Frequency and Severity of Truck Accidents. *Aviation, Space, and Environmental Medicine* 75(3 Supplement):A107-A118.

Moore-Ede, M. 1993. We Have Ways of Keeping You Alert. *New Scientist* 140:30-35.

Morad, Y., B. Azaria, I. Avni, Y. Barkana, D. Zodak, R. Kohen-Raz, and E. Barenboim. 2007. Posturography as an Indicator of Fatigue Due to Sleep Deprivation. *Aviation, Space, and Environmental Medicine* 78(9):859-863.

Morad, T., Y. Barkana, D. Zadok, M. Hartstein, E. Pras, and Y. Bar-Dayan. 2009. Ocular Parameters as an Objective Tool for the Assessment of Truck Drivers' Fatigue. *Accident Analysis and Prevention* 41:856-860.

Morrow, S., B. Walsh, M. Badiee, T. Stentz, D. Nash, V. Clark, J. Barnes-Ferrell, and J. Impara. 2010. Work Schedule Manager Gap Analysis: Assessing the Future Training Needs of Work Schedule Managers Using a Strategic Job Analysis Approach. Prepared by the Volpe National Transportation Systems Center. Washington, DC: U.S. Department of Transportation, Federal Railroad Administration.

Mott, C., H. Van Dongen, K. Kan, J. Huang, and D. Mallicone. 2009. Improving Individual Performance Predictions with Bayesian Estimation using Traits Learned from Reference Data. *2009 International Conference on Fatigue Management in Transportation Operations. Boston*, MA.

Murphy, M. 2010. Monitoring Driver Inattention, Distraction, and Drowsiness. *AutomotiveWorld.Com*. April 13. Accessed 11.2010 at http://www.automotiveworld.com/news/components/81626-monitoring-driver-inattention-distraction-and-drowsiness.

Muscio, B. 1921. Is a Fatigue Test Possible? *British Journal of Psychology* 12:31-46.

Myers, L., J. Caldwell, and J. Downs, III. 2009. Novel Identification of Optimal Physiological Indices for Monitoring Cognitive Fatigue. *2009 International Conference on Fatigue Management in Transportation Operations. Boston*, MA.

Nakao, M., A. Karashima, and N. Katayama. 2007. Mathematical Models of Regulatory Mechanisms of Sleep-Wake Rhythms. *Cellular and Molecular Life Sciences* 64:1236-1243.

National Center on Sleep Disorders Research (of the U.S. Department of Health and Human Services). 2003. *2003 National Sleep Disorders Research Plan*. Washington, DC.

National Transportation Safety Board. 1990. *Fatigue, Alcohol, Other Drugs, and Medical Factors in Fatal-to-the Driver Heavy Truck Crashes* (Safety Study 1990, NTSB/SS-90/01). Washington, DC.

Naval Safety Center. 2005. Drinking and Driving, Fatigued-Driving, and Prevention: A Focus Group Study. June. Accessed 09.2009 at http://safetycenter.navy.mil/bestpractices/ashore/downloads/GW_focus_group_synopsis.doc

Neri, D. 2004. Preface: Fatigue and Performance Modeling Workshop, June 13-14, 2002, Seattle, WA. *Aviation, Space, and Environmental Medicine,* 75(3):Sector II, March 2004.

Nilsson, T., T. Nelson, and D. Carlson. 1997. Development of Fatigue Symptoms During Simulated Driving. *Accident Analysis and Prevention* 29(4):479-488.

Nocera, A., and D. Kursandi. 1998. Doctors' Working Hours: Can the Medical Profession Afford to Let the Courts Decide What Is Reasonable? *Medical Journal of Australia* 168:616-618.

Noy, Y., W. Horrey, S. Popkin, S. Folkard, H. Howarth, and T. Courtney. 2008. Future Directions in Fatigue and Safety Research. *Report on the 4th Hopkinton Conference on Accident Analysis and Prevention.*

Nuclear Energy Institute. 2008. NEI 06-11 Rev1. Managing Personnel Fatigue at Nuclear Power Reactor Sites. Nuclear Energy Institute, Washington, DC. October.

Ogilvie, R. 2001. The Process of Falling Asleep. *Sleep Medicine Reviews* 5(3):247-270.

Ogilvie, R., R. Wilkinson, and S. Allison. 1989. The Detection of Sleep Onset: Behavioural, Physiological, and Subjective Convergence. *Sleep* 12:458-474.

O'Hanlon, J., and G. Kelley. 1977. Comparison of Performance and Physiological Changes between Drivers Who Perform Well and Poorly during Prolonged Vehicular Operation. In *Vigilance: Theory, Operational Performance, and Physiological Correlates*. R. Mackie, Editor. New York: Plenum Press. Pp. 97-109.

Oken, B., M. Salinsky, and S. Elsas. 2006. Vigilance, Alertness, or Sustained Attention: Physiological Basis and Measurement. *Clinical Neurophysiology* 117:1885-1901.

Olson, Rebecca. 2006. Assessment of Drowsy-Related Critical Incidents and the 2004 Revised Hours-of-Service Regulations. Master of Science Thesis in Industrial and Systems Engineering. Virginia Polytechnic Institute and State University.

Olson, R., M. Blanco, and R. Hanowski. 2009. Restart Period and Sleep for Commercial Motor Vehicle Drivers. *2009 International Conference on Fatigue Management in Transportation Operations. Boston*, MA.

Oman, C., and A. Liu. 2007. *Locomotive In-Cab Alerter Technology Assessment, Man Vehicle Laboratory.* Department of Aeronautics and Astronautics, Massachusetts Institute of Technology, Cambridge, MA 02139, Report #07.30. Accessed 09.2009 at http://mvl.mit.edu/MVLreports.php?reportnum=07.30.

Oman, C., A. Liu, S. Popkin, J. Pollard, H. Howarth, and A. Aboukhalil. 2009. Locomotive Alerter Technology Assessment. *2009 International Conference on Fatigue Management in Transportation Operations. Boston*, MA.

Oron-Gilad, T., D. Shinar. 2000. Driver Fatigue Among Military Truck Drivers. *Transportation Research Part F: Traffic Psychology and Behavior* 3(4):195-209.

Öz, Bahar, Türker Özkan, and Timo Lajunen. 2010. Professional and Non-Professional Drivers' Stress Reactions and Risky Driving. *Transportation Research Part F 13*(2010):32-40.

Pace-Schott, E., and J. Hobson. 2002. Basic Mechanisms of Sleep: New Evidence on the Neuroanatomy and Neuromodulation of the NREM-REM Cycle. In *Neuropsychopharmacology: The Fifth Generation of Progress.* Davis, K., Charney, D., Coyle, J., and Nemeroff, C., editors. American College of Neuropsychooharnacology. Philadelphia: Lippincott Williams and Wilkins. Pp.1859-1877.

Paim, S., M. Pires, L. Bitencourt, R. Silva, R. Santos, A. Esteves, A. Barreto, S. Tufik, and M. de Mello. 2008. Sleep Complaints and Polysomnographic Findings: A Study of Nuclear Power Plant Shift Workers. *Chronobiology International* 25(2/3) 321-331.

Pandi-Perumal, S., A. Moscovitch, V. Srinivasan, D. Spence, D. Cardinali, and G. Brown. 2009. Bidirectional Communication between Sleep and Circadian Rhythms and Its Implications for Depression: Lessons from Agomelatine. *Progress in Neurobiology* 88:264-271.

Philip, P., and T. Åkerstedt. 2006. Transport and Industrial Safety, How are They Affected by Sleepiness and Sleep Restriction? *Sleep Medicine Reviews* 10:347-356.

Phillips, A., and P. Robinson. 2008. Sleep Deprivation in a Quantitative Physiologically Based Model of the Ascending Arousal System. *Journal of Theoretical Biology* 255:413-423.

Ptitsyn, A., S. Zvonic, and J. Gimble. 2007. Digital Signal Processing Reveals Circadian Baseline Oscillation in Majority of Mammalian Genes. PLos 3(6):1108-1114.

Price, J. 2009. Incidence and Predictors of Fatigue-related Aviation Accidents. 2009 International Conference on Fatigue Management in Transportation Operations. Boston, MA.

Purnell, M., A. Feyer, and G. Herbison. 2002. The Impact of a Nap Opportunity During the Night Shift on the Performance and Alertness of 12-h Shift Workers. Journal of Sleep Research 11(3):219-227.

QinetiQ Centre for Human Sciences and Simon Folkard Associates Ltd. 2006a. *The Development of a Fatigue/Risk Index for Shiftworkers*. Prepared for the Health and Safety Executive. Norwich, UK: HSE Books.

QinetiQ Centre for Human Sciences and Simon Folkard Associates Ltd. 2006b. *Fatigue and Risk Index Calculator*. Prepared for the Health and Safety Executive. Norwich, UK: HSE Books.

Queensland Transport. 2005. 2003 *Road Traffic Crashes in Queensland: A Report on the Road Toll*. Accessed 04.2009 at http://www.transport.qld.gov.au/resources/file/ebbb01022986d5b/Road_traffic_crash_report_2003_V2.pdf.

Rajaraman, S., A. Gribok, N. Wesensten, T. Balkin, and J. Reifman. 2009. Individualized Biomathematical Models for Performance Prediction of Sleep-Deprived Individuals. *2009 International Conference on Fatigue Management in Transportation Operations*. Boston, MA.

Rajaraman, S., A. Gribok, N. Wesensten, T. Balkin, and J. Reifman. 2007. Individualized Performance Prediction of Sleep-Deprived Individuals With the Two-Process Model. *Journal of Applied Physiology* 104:459-468.

Raslear, T., and M. Coplen. 2004. Fatigue Models as Practical Tools: Diagnostic Accuracy and Decision Thresholds. *Aviation, Space, and Environmental Medicine* 75(3):A168-A172.

Rempe, M., J. Best, and D. Terman. 2009. A Mathematical Model of the Sleep/Wake Cycle. *Journal of Mathematical Biology* (June).

Reynolds, A., and S. Banks. 2010 Total Sleep Deprivation, Chronic Sleep Restriction, and Sleep Disruption. Progress in Brain Research 185:91-103.

Roach, G., D. Dawson, and N. Lamond. 2006. Can a Shorter Psychomotor Vigilance Task Be Used as a Reasonable Substitute for the Ten Minute Psychomotor Vigilance Task? *Chronobiology International* 23:1379-1387.

Roach, G., D. Dawson, and N. Lamond. 2005. Can a Shorter PVT Be Used as a Reasonable Substitute for the 10-minute PVT? *Shiftwork International Newsletter* 22:129.

Roach, G., J. Dorrian, A. Fletcher, and D. Dawson. 2001. Comparing the Effects of Fatigue and Alcohol Consumption on Locomotive Engineers' Performance in a Rail Simulator. *Journal of Human Ergology* 30(1-2):125-130.

Roach, G., N. Lamond, J. Dorrian, H. Burgess, A. Holmes, A. Fletcher, K. McCulloch, and D. Dawson. 2005. Changes in the Concentration of Urinary 6-sulphatoxymelatonin during a Week of Simulated Night Work. *Industrial Health* 43(193-196).

Roehrs, T., E. Burduvalie, A. Bonahoom, C. Drake, and T. Roth. 2003. Ethanol and Sleep Loss: A Dose Comparison of Impairing Effects. *Sleep* 26(8):981-985.

Roehrs, T., and T. Roth. 2007. The Physiology of Sleep. *Neurology and Clinical Neuroscience*. 180-184.

Roehrs, T., and T. Roth. 2001. Sleep, Sleepiness, and Alcohol Use. *Alcohol Research and Health* 25(2):101-109.

Rosekind, M. 2009. Use of Biomathematical Models of Fatigue: Safety, Liability, Confidentiality, Adherence, and Consequences. *2009 International Conference on Fatigue Management in Transportation Operations*. Boston, MA.

Rosekind, M. 2004. Preventing Pilot Fatigue. *Air Line Pilot* November:22.

Rosekind, M., K.B. Gregory, and M. Mallis. 2006. Alertness Management in Aviation Operations: Enhancing Performance and Sleep. *Aviation, Space, and Environmental Medicine* 77(12):1256-1266.

Rosekind, M., P. Gander, K. Gregory, R. Smith, D. Miller, R. Oyung, L. Webbon, and J. Johnson. 1996a. Managing Fatigue in Operational Settings 1: Physiological Considerations and Countermeasures. National Aeronautics and Space Administration. Ames, California. Printed in *Behavioral Medicine* 21:157-165.

Rosekind, M., P. Gander, K. Gregory, R. Smith, D. Miller, R. Oyung, L. Webbon, and J. Johnson. 1996b. Managing Fatigue in Operational Settings 2: An Integrated Approach. National Aeronautics and Space Administration. Ames, California.

Rosenberg, E., and Y. Caine. 2001. Survey of Israeli Air Force Line Commander Support for Fatigue Prevention Initiatives. *Aviation, Space, and Environmental Medicine* 72(4):352-356.

Rüger, M., M. Gordijn, D. Beers, B. De Vries, and S. Daan. 2005. Weak Relationships Between Suppression of Melatonin and Suppression of Sleepiness/Fatigue in Response to Light Exposure. *Journal of Sleep Research* 14:221-227.

Ruppert, Barb. 2009. In the Blink of an Eye. *Military Medical/CBRN Technology* 13(8):10-11.

Rupp, T., N. Wesensten, and T. Balkin. 2009. Sleep History Affects Performance During Subsequent Sleep Restriction and Recovery. *2009 International Conference on Fatigue Management in Transportation Operations*. Boston, MA.

Rupp, T., N. Wesensten, P. Bliese, and T. Balkin. 2009. Banking Sleep: Realization of Benefits During Subsequent Sleep Restriction and Recovery. *Sleep 32(3):311-321.*

Russo, M., A. Vo, R. Labutta, I. Black, W. Campbell, J. Greene, J. McGhee, and D. Redmond. 2005. Human Biovibrations: Assessment of Human Life Signs, Motor Activity, and Cognitive Performance using Wrist-Mounted Actigraphy. *Aviation, Space, and Environmental Medicine* 76(7, Suppl):C64-C74.

Russo, M., M. Thomas, D. Thorne, H. Sing, D. Redmond, L. Rowland, D. Johnson, S. Hall, J. Krichmar, and T. Balkin. 2003. Oculomotor Impairment during Chronic Partial Sleep Deprivation. *Clinical Neurophysiology* 114:723-736.

Sack, R., D. Auckley, R. Auger, M. Carskadon, K. Wright, M. Vitiello, and I. Zhdanova. 2007. Circadian Rhythm Sleep Disorders: Part 1, Basic Principles, Shift Work and Jet Lag Disorders. *Sleep* 30(11):1460-1483.

Sakuri, T. 2007. The Neural Circuit of Orexin (Hypocretin): Maintaining Sleep and Wakefulness. *National Review of Neuroscience* 8(3):171-181.

Samn, S., and L. Perelli. 1982. *Estimating Aircrew Fatigue: A Technique with Implications to Airlift Operations*. Technical Report. SAM-TR-82-21. Brooks Air Force Base. San Antonio, TX: USAF School of Aerospace Medicine.

Saper, C., G. Cano, and T. Scammell. 2005a. Homeostatic, Circadian, and Emotional Regulation of Sleep. *The Journal of Comparative Neurology* 493:92-98.

Saper, C., J. Lu, T. Chou, and J. Gooley. 2005b. The Hypothalamic Integrator for Circadian Rhythms. *TRENDS in Neurosciences* 28(3):152-157.

Saper, C., T. Scammell, and Jan Lu. 2005c. Hypothalamic Regulation of Sleep and Circadian Rhythms. *Nature* 437(7063):1257-1263.

Saper, C., T. Chou, and T. Scammell. 2001. The Sleep Switch: Hypothalamic Control of Sleep and Wakefulness. *TRENDS in Neurosciences* 24(12):726-731.

Saxby, D., G. Matthews, E. Hitchcock, and J. Warm. 2007. Development of Active and Passive Fatigue Manipulations using a Driving Simulator. In *Proceedings of the Human Factors and Ergonomics Society 51st Annual Meeting.* Pp. 1237-1241. Santa Monica, CA: Human Factors and Ergonomics Society.

Saxby, D., G. Matthews, E. Hitchcock, and J. Warm. 2007. Fatigue States are Multidimensional: Evidence from Studies of Simulated Driving. In *Proceedings of the Driving Simulator Conference, North America.* Iowa City. 12-14 September.

Schleicher. R., N. Galley, S. Breist, and L. Galley. 2008. Blinks and Saccades as Indicators of Fatigue in Sleepiness Warnings: Looking Tired? *Ergonomics* 51(7):982-1010.

Schmidt, C., F. Collette, and Y. Leclercq, et al. 2009a. Homeostatic Sleep Pressure and Responses to Sustained Attention in the Suprachiasmatic Area. *Science* 324(5926):516-519.

Schmidt, E., M. Schrauf, M. Simon, M. Fritzsche, A. Buchner, and W. Kincses. 2009b. Drivers' Misjudgement of Vigilance State during Prolonged Monotonous Daytime Driving. *Accident Analysis and Prevention* 41:1087-1093.

Schweitzer, P., A. Randazzo, K. Stone, M. Erman, and J. Walsh. 2006. Laboratory and Field Studies of Naps and Caffeine as Practical Countermeasures for Sleep-wake Problems Associated with Night Work. *Sleep* 29(1):39-50.

Shahidi, P., S. Southward, and M. Ahmadian. 2009. A Holistic Approach to Estimating Crew Alertness from Continuous Speech. *2009 International Conference on Fatigue Management in Transportation Operations.* Boston, MA.

Shapiro, C., C. Auch, M. Reimer, L. Kayumov, R. Heslegrave, N. Huterer, H. Driver, and G. Devins. 2006. A New Approach to the Construct of Alertness. *Journal of Psychosomatic Research* 60:595-603.

Shen, J., J. Barbera, and C. Shapiro. 2006. Distinguishing Sleepiness and Fatigue: Focus on Definition and Measurement. *Sleep Medicine Reviews* 10:63-76.

Shinkai, S., S. Watanabe, Y. Kurokawa, and J. Torii. 1993. Salivary Cortisol for Monitoring Circadian Rhythm Variation in Adrenal Activity During Shiftwork. *International Archives of Occupational and Environmental Health* 64(7):499-502.

Siegel, J. 2004. Brain Mechanisms that Control Sleep and Waking. *Naturwissenschaften* 91(8):355-365.

Signal, T., and P. Gander. 2007. Rapid Counterclockwise Shift Rotation in Air Traffic Control: Effects on Sleep and Night Work. *Aviation, Space, and Environmental Medicine* 78(9):878-885.

Simoens, S., and J. Hurst. 2006. The Supply of Physician Services in OECD Countries. OECD Health Working Papers. No. 21.

Sirois, W., T. Dawson, and M. Moore-Ede. 2007. Assessing Driver Fatigue as a Factor in Road Accidents. *Proceedings of the Fourth International Driving Symposium on Human Factors in Driver Assessment, Training, and Vehicle Design.* Iowa City, IA: University of Iowa.

Smith, S., M. Carrington, and J. Trinder. 2005. Subjective and Predicted Sleepiness While Driving in Young Adults. *Accident Analysis and Prevention* 37:1066-1073.

Smith, S., M. Horswill, and G. Parker. 2009. Objective Sleepiness Predicts Performance on a Hazard Perception Simulator Task. *2009 International Conference on Fatigue Management in Transportation Operations.* Boston, MA.

Smolensky, M., L. Di Milia, M. Ohayon, and P. Philip. 2011. Sleep Disorders, Medical Conditions, and Road Accident Risk. *Accident Analysis and Prevention* 43:533-548.

Somers, V., M. Dyken, A. Mark, and F. Abboud. 1993. Sympathetic-Nerve Activity During Sleep in Normal Subjects. *The New England Journal of Medicine* 328:303-307.

Sommer, D., M. Golz, and J. Krajewski. 2009. Consecutive Detection of Extreme Central Fatigue. In *4th European Conference of the International Federation for Medical and Biological Engineering.* J. Sloten, P. Verdonck, M. Nyssen, and J. Haueisen, Editors. Berlin: Springer. Pp. 243-246.

Spencer, M., and K. Robertson. 2007. The Application of an Alertness Model to Ultra-Long-Range Civil Air Operations. *Somnologie* 11:159-166.

Spencer M. and K. Robertson. 2007. Aircrew Fatigue: A Review of Research Undertaken on Behalf of the UK Civil Aviation Authority. CAA Paper 2005/04. Accessed 10.2010 at http://www.caa.co.uk/docs/33/CAAPaper2005_04.pdf.

Spencer M., K. Robertson, and S. Folkard. 2006. *The Development of a Fatigue/Risk Index for Shiftworkers.* Prepared for the Health and Safety Executive, U.K. QinetiQ Center for Human Sciences and Simon Folkard Associates Limited.

Spiegel, K., R. Leproult, and E. van Couter. 1999. Impact of Sleep Debt on Metabolic and Endocrine Function. *The Lancet* 354:1435-1439.

Stentz, T., D. Nash, S. Morrow, B. Walsh, V. Clark, M. Badiee, J. Barnes-Farrell, M. Coplen, and S. Popkin. 2009. Transient Risk Factor Models for Fatigue and Human Factors Rail Accidents. *2009 International Conference on Fatigue Management in Transportation Operations.* Boston, MA.

Stephan, K., S. Hosking, M. Regan, A. Verdoorn, K. Young, and N. Haworth. 2006. *The Relationship Between Driving Performance and the Johns Drowsiness Scale as Measured by the OPTALERT System.* Victoria, Australia: Monash University Accident Research Centre. Report No. 252.

Stern, J., and T. Brown. 2005. *Detection of Human Fatigue.* Final Technical Report, Air Force Office of Scientific Research, AFRL-SR-AR-TR-05-0351.

Stewart, S., D. Brown, C. Turner, S. Bond, and A. Fletcher. 2009. Fatigue Risk Management Integrated within an Airline Management System. Presented at the *2009 International Conference on Fatigue Management in Transportation Operations.* Boston, MA.

Storm, W. 2008. *A Fatigue Management System for Sustained Military Operations.* Fort Detrick, MD: U.S. Army Medical Research and Materiel Command.

Sullinen, M., M. Härmä, R. Akila, A. Holm, R. Luukkonen, H. Mikola, K. Müller, and J. Virkkala. 2004. The Effects of Sleep Debt and Monotonous Work on Sleepiness and Performance during a 12-h Dayshift. *Journal of Sleep Research* 13:285-294.

Tassinary, L., J. Cacioppo, and E. Vanman. 2007. The Skeletomotor System: Surface Electromyography. In *Handbook of Psychophysiology, 3rd Edition.* J. Cacioppo, L. Tassinary, and G. Berntson, Editors. Cambridge: Cambridge University Press.

Teikari, P. 2007. Automated Pupillometry. Helsinki University of Technology. Accessed 10.2009 at http://users.tkk.fi/~jteikari/Teikari_AutomatedPupillometry.pdf.

Thorne, D., D. Johnson, M. Kautz, and D. Redmond. 2005a. Further Comparisons of a PDA-based PVT with the Industry Standard, Using Shortened Test Durations to Improve Compliance in Field Studies. *Sleep* 28(Suppl.):A145.

Thorne, D., D. Johnson, E. Redmond, H. Sing, and G. Belenky. 2005b. The Walter Reed Palm-Held Psychomotor Vigilance Test. *Behavior Research Methods* 35(1):111-118.

Tobler, I., and P. Achermann. 2007. Sleep Homeostasis. *Scholarpedia* 2(10):2432. Accessed 08.2011 at http://www.scholarpedia.org/article/Sleep_homeostasis.

Tomasi, D., R. Wang, F. Teland, V. Boronikolas, M. Jayne, G. Wang, J. Fowler, and N. Volkow. 2009. Impairment of Attentional Networks After 1 Night of Sleep Deprivation *Cerebral Cortex* 19:233-240.

Tononi, G. 2005. The Neuro-Biomolecular Basis of Alertness in Sleep Disorders. *Sleep Medicine* 6(Suppl 1):S8-S12.

Tononi, G., and C. Cirelli. 2006. Sleep Function and Synaptic Homeostasis. *Sleep Medicine Review* 10:49-62.

Toquam, J. and A. Bittner, Jr. 1996. Fitness-for-Duty Testing in the Transit Workplace. Battelle Human Affairs Research Centers. Seattle, WA.

Toquam, J., and A. Bittner, Jr. 1994. Performance-Based Testing for Fitness-for-Duty (FFD): Ready for Industrial Applications or Not?. In *Advances in Industrial Ergonomics*. F. Aghazadeh, Editor. London: Taylor and Francis. Pp. 11-18.

Torgovitsky, R., W. Wang, V. DeGruttola, and E. Klerman. 2009. Subject-Specific Evaluation of Perforamnce Based on Forced Desynchrongy Data. *2009 International Conference on Fatigue Management in Transportation Operations*. Boston, MA.

Tregear, S., J. Williams, M. Tiller, and J. Reston. 2009. Obstructive Sleep Apnea and Motor Vehicle Crashes: A Systematic Review and Meta-Analysis. *2009 International Conference on Fatigue Management in Transportation Operations*. Boston, MA.

Trutschel, U., B. Sirois, A. Aguirre, T. Dawson, M. Moore-Ede, D. Dommer, and M. Golz. 2009. Shiftwork Adaptation Testing System (SATS). *2009 International Conference on Fatigue Management in Transportation Operations*. Boston, MA.

Trutschel, U., R. Guttkuhn, A. Heitmann, A. Aguirre, and M. Moore-Ede. 2003. Expert System for Simulating and Predicting Sleep and Alertness Patterns. In *Knowledge-Based Intelligent Information and Engineering Systems*. V. Palade, R. Howlett, and L. Jain, Editors. Berlin: Springer-Verlag. Pp. 104-110.

Tucker, A., D. Dinges, and H. Van Dongen. 2007. Trait Interindividual Differences in the Sleep Physiology of Healthy Young Adults. *Journal of Sleep Research* 16:170-180.

Ulmer, C., D. Wolman, and M. Johns, Editors. 2008. *Resident Duty Hours: Enhancing Sleep, Supervision, and Safety*. Institute of Medicine. Washington, DC: National Academies Press.

U.S. Congress. 2008. Public Law 110-233, The Genetic Information Nondiscrimination Act of 2008. 122 Statute 881.

U.S. Department of Transportation, Federal Aviation Administration. 2010. Advisory Circular. No. 120-103. Fatigue Risk Management Systems for Aviation Safety.

U.S. Department of Transportation, Federal Aviation Administration. 2008. *Proceedings of the Aviation Fatigue Management Symposium: Partnerships for Solutions*. Vienna, Virginia. 17-19 June.

U.S. Department of Transportation, Federal Motor Carriers Safety Administration. 2000. Hours of Service of Drivers; Driver Rest and Sleep for Safe Operations; Proposed Rule (49 CFR Parts 350, et al.). *Federal Register* 65(85):25540-25611. May 2.

U.S. Nuclear Regulatory Commission (NRC). 2009. Fatigue Management for Nuclear Power Plant Personnel. *Regulatory Guide 5.73*. Accessed 10.20009 via http://www.nrc.gov/reactors/operating/ops-experience/fitness-for-duty-programs/reg-guide-ifmp.html at pbadupws.nrc.gov/docs/ML0834/ML083450028.pdf.

U.S. Nuclear Regulatory Commission (NRC). 2008. 10 CFR Part 26 Fitness for Duty Programs; Final Rule. *Federal Register* 73(62):16966-17235.

U.S. Nuclear Regulatory Commission (NRC). 2005. Federal Register Notice of Proposed Rule. Fitness For Duty Programs. *Federal Register* 70(165):5049. August 26. Accessed 02.2009 at http://www.nrc.gov/reading-rm/doc-collections/commission/secys/2005/secy2005-0074/2005-0074scy.html.

U.S. Nuclear Regulatory Commission (NRC). 2003. Security Order EA-03-038. Order for Compensatory Measures Related to Fitness-for-Duty Enhancements Applicable to Nuclear Facility Security Force Personnel. April 29. Accessed 10.2010 at http://www.nrc.gov/reading-rm/doc-collections/enforcement/security/2003/ml030940198-ffd-04-29-03.pdf.

U.S. Nuclear Regulatory Commission (NRC). 2001. Fatigue of Workers at Nuclear Power Plants. *SECY-01-0113*. June 22. Attachments 1, 2, 3. Accessed 02.2009 at http://www.nrc.gov/reading-rm/doc-collections/commission/secys/2001/secy2001-0113/2001-0113scy.pdf#pagemode=bookmarks.

U.S. Nuclear Regulatory Commission (NRC). 1989. 10 CFR Part 26 Fitness for Duty Programs; Final Rule. *Federal Register* 54:24468-24494.

U.S. Nuclear Regulatory Commission (NRC). 1986. Letter Stating NRC Policy on Need for FFD Programs at Nuclear Power Reactors.

U.S. Nuclear Regulatory Commission (NRC). 1985. Recommendations for NRC Policy on Shift Scheduling and Overtime at Nuclear Power Plants. *NUREG/CR-4248*.

U.S. Nuclear Regulatory Commission (NRC). 1982. Policy on Factors Causing Fatigue of Operating Personnel at Nuclear Reactors. *Generic Letter (GL) 82-12: Nuclear Power Plant Staff Working Hours*.

U.S. Nuclear Regulatory Commission (NRC). 1980a. *Inspection and Enforcement (IE) Circular* (No. 80-02).

U.S. Nuclear Regulatory Commission (NRC). 1980b. Clarification of TMI Action Plan Requirements. Washington, DC. *NUREG-0737* Item I.A1.3.

Vandewalle, G., P. Maquet, and D. Dijk. 2009. Light as a Modulator of Cognitive Brain Function. *TRENDS in Cognitive Science* 13(10):429-438.

Van Dongen, H. 2009. Biomathematical Modeling of Fatigue: Basic Theoretical, Mathematical and Scientific Concepts. *2009 International Conference on Fatigue Management in Transportation Operations.* Boston, MA.

Van Dongen, H. 2004. Comparison of Mathematical Model Predictions to Experimental Data of Fatigue and Performance. *Aviation, Space, and Environmental Medicine* 75(3):A15-A36.

Van Dongen, H., M. Baynard, G. Maislin, and D. Dinges. 2004. Systematic Interindividual Differences in Neurobehavioral Impairment from Sleep Loss: Evidence of Trait-like Differential Vulnerability. *Sleep* 27(3):423-433.

Van Dongen, H., M. Baynard, G. Nosker, and D. Dinges. 2002. Repeated Exposure to Total Sleep Deprivation: Substantial Trait Differences in Performance Impairment Among Subjects. *Sleep* 25(3):121-123.

Van Dongen, H., G. Maislin, J. Mullington, and D. Dinges. 2003. The Cumulative Cost of Additional Wakefulness: Dose-Response Effects on Neurobehavioral Functions and Sleep Physiology From Chronic Sleep Restriction and Total Sleep Deprivation. *Sleep* 26(2):117-126.

Van Dongen, H., C. Mott, J. Huang, D. Mollicone, F. McKenzie, and D. Dinges. 2007. Optimization of Biomathematical Model Predictions for Cognitive Performance Impairment in Individuals: Accounting for Unknown Traits and Uncertain States in Homeostatic and Circadian Processes. *Sleep* 30(9):1129-1143.

Van Dongen, H., K. Vitellaro, and D. Dinges. 2005. Individual Differences in Adult Human Sleep and Wakefulness: Leitmotif for a Research Agenda. *Sleep* 28(4):479-496.

Vassalli, A., and D. Dijk. 2009. Sleep Function: Current Questions and New Approaches. *European Journal of Neuroscience* 29(9):1830-1841.

Venkatraman V., Y. Chuah, and S. Huettel. 2007. Sleep Deprivation Elevates Expectation of Gains and Attenuates Response to Losses Following Risky Decisions. *Sleep* 30, 603-9.

Verwey, W., and D. Zaidel. 2000. Predicting Drowsiness Accidents from Personal Attributes, Eye Blinks, and Ongoing Driving Behaviour. *Personality and Individual Differences* 28:123-142.

Vila, B. 2006. Impact of Long Work Hours on Police Officers and the Communities They Serve. *American Journal of Industrial Medicine* 49:972-980.

Viola, A. 2009. Genetic Predictors of Physiological and Behavioural Responses to Sleep Loss. *2009 International Conference on Fatigue Management in Transportation Operations.* Boston, MA.

Viola, A., L. James, S. Archer, and S. Dijk. 2008. PER3 Polymorphism and Cardiac Autonomic Control: Effects of Sleep Debt and Circadian Phase. *American Journal of Physiology – Heart and Circulatory Physiology* 295(5):H2156-H2163.

Viola, A., L. James, L. Schlangen, and D. Dijk. 2008. Blue-enriched White Light in the Workplace Improves Self-Reported Alertness, Performance, and Sleep Quality. *Scandanian Journal of Work and Environmental Health* 34(4):297-306.

Viola, A., S. Archer, L. James, J. Groeger, J. Lo, D. Skene, M. von Schantz, and D. Dijk. 2007. PER3 Polymorphism Predicts Sleep Structure and Waking Performance. *Current Biology* 17(7):613-618.

Von Thaden, T., S. Spain, and S. Woo. 2009. Safety Culture and Self-Reported Fatigue in Commercial Aviation Operations. Paper presented at the 2009 International Conference on Fatigue Management in Transportation Operations. Boston, MA. March 24-26.

Wehr, T., D. Aeschbach, and W. Duncan, Jr. 2001. Evidence for a Biological Dawn and Dusk in the Human Circadian Timing System. *Journal of Physiology* 535:937-951.

Wehr, T., D. Moul, and G. Barbato. 1993. Conservation of Photoperiod-Responsive Mechanisms in Humans. *American Journal of Physiology* 265:R846-R857.

Wierwille, W., and L. Ellsworth. 1994. Evaluation of Driver Drowsiness by Trained Raters. *Accident Analysis and Prevention* 26(5):571-581.

Wierwille, W., M. Lewin, and R. Fairbanks, III. 1996. *Final Reports: Research on Vehicle-Based Driver Status/Performance Monitoring, Part 1.* DOT HS 808 638. U.S. Department of Transportation.

Wijesuriya, N., Y. Tran, and A. Craig. 2007. The Psychophysiological Determinants of Fatigue. *International Journal of Psychophysiology* 63:77-86.

Williamson, A., and A. Feyer. 2000. Moderate Sleep Deprivation Produces Impairments in Cognitive and Motor Performance Equivalent to Legally Prescribed Levels of Alcohol Intoxication. *Occupational and Environmental Medicine* 57:649-655.

Williamson, A., and R. Friswell. 2009. Disentangling the Relative Effects of Time of Day and Sleep Deprivation on Fatigue and Performance. *2009 International Conference on Fatigue Management in Transportation Operations. Boston*, MA.

Williamson, A., S. Lombardi, S. Folkard, J. Stutts, and T. Courtney. 2011. The Link Between Fatigue and Safety. *Accident Analysis and Prevention* 43:498-515.

Yamaguchi, M., M. Deguchi, J. Wakasugi, S. Ono, N. Takai, T. Higashi, and Y. Mizuno. 2006. Hand-held Monitor of Sympathetic Nervous System Using Salivary Amylase Activity and Its Validation by Driver Fatigue Assessment. *Biosensors and Bioelectronics* 21(7):1007-1014.

Yin, W. 2007. A Mathematical Model of the Sleep-Wake Cycle. Masters Thesis. Georgia Institute of Technology. School of Bioengineering.

Zeitzer, J., A. Morales-Villagran, N. Maidment, E. Behnke, L. Ackerson, and F. Lopez-Rodriguez et al. 2006. Extracellular Adenosine in the Human Brain During Sleep and Sleep Deprivation: An in Vivo Microdialysis Study. *Sleep* 29:455-461.

Zhang, C., C. Zheng, X. Yu, and Y. Ouyang. 2008. Estimating VDT Mental Fatigue Using Multichannel Linear Descriptors and KPCA-HMM. *EURASIP Journal on Advances in Signal Processing* 2008:1-11.

3.8 Alphabetic List of Technologies for Chapter 3: Fatigue Management

24/7 Lifestyle Planner tools (Family Planner and Personal Pocket Planner)

Advisory System for Tired Drivers (ASTID™ – Pernix LTD

Art90 – Act-React-Test system 90

Artificial Neural Network – George Washington University's Center for Intelligent System Research

AutoVue – Iteris Inc.

BLT impairment test – Bowles-Langley Technology, Inc.

Circadian Alertness Simulator (CAS) Model – Circadian Technologies, Inc. (biomathematical fatigue model)

Cogscreen™

CoPilot® – Attention Technologies

Critical Tracking Task (CIT) – (see Factor 1000)

Delta-WP – Essex Corporation

Driver Alert Support (DAS) – Volvo Technology

Driver Drowsiness Monitoring System (DDMS) – prototype – Virginia Tech

Driver Fatigue Monitor – Attention Technologies

Driver Fatigue Questionnaire (DFQ) and Driver Stress Inventory (DSI)

Driver State Monitor (DSM) – Delphi

Driver State Sensor (DSS) – Seeing Machines

Drowsy Driver Detection System (DDDS) – John Hopkins University Applied Physics Laboratory

Engine Drive/Driver Vigilance Telemetric Control System (EDVTCS) – Neurocom

EyeCheck – MCJ, Inc.

Eye-Com Biosensor-Communicator-Controller (EC-6) – Eye-Com Corp

Eye Dynamics – EyeDynamics.com

Eyegaze Analysis System –LC Technologies, Inc

Excessive Daytime Sleepiness Questionnaire

FaceLab – Seeing Machines

Fatigue Accident Causation Testing System (FACTS) – Circadian Technologies, Inc.

Fatigue Audit InterDyne (FAID) – Center for Sleep Research, University of South Australia (biomathematical fatigue model)

Fatigue Avoidance Safety Tool (FAST) – Archinoetics

Fatigue Avoidance Scheduling Tool (FAST) – Archinoetics

Fatigue Index Risk Module (FIRM) and Fatigue Risk Index (FRI)

Fitness Impairment Tester (FIT 2000)

HaulCheck – Acumine

Interactive Neurobiobehavioral model (biomathematical fatigue model) – Megan Jewett

Johns Drowsiness Score (JDS) – Optalert system

Maintenance of Wakefulness Test (MWT) (biomathematical fatigue model)

MIcroNod Detection System (MINDS™) – Advanced Safety Concepts Inc.

MobileEye – Vision/Radar Sensor

NapZapper – Safety Products Unlimited

NOVAlert – Atlas Ltd

NovaScan – NTI, Inc.

Optalert™ – Sleep Diagnostic Ltd.

OSPAT – OSPAT PTY Ltd

Palm-PVT – Walter Reed Army Research Institute

QinetiQ Alertness Model – QinetiQ, Inc.

Queensland Hazard Perception test –

Ready Shift™ – Evaluations Systems Inc

Retrospective Alertness Inventory –

SAFE – System for Aircrew Fatigue Evaluation (biomathematical fatigue model)

SAFTE – Sleep, Activity, Fatigue, and Task Effectiveness model (biomathematical fatigue model)

SafeTRAC® – Applied Perception and AssistWare Technology, Inc

SENSATION – advanced sensor development for attention, stress, vigilance, and sleep/wakefulness monitoring

Shiftwork Adaptation Testing System (SATS) – Circadian Technologies, Inc.

Shiftwork Survey Index
Sleep Band – Archinoetics
Sleep Control Helmut System – Security Electronic Systems
Sleepiness Detection System (SDSTM) – Biocognisafe Canada Inc.
Sleep/Wake Predictor (SWP) – (biomathematical fatigue model)
SleepWatch® Actigraph – Precision Control Design (Walter Reed Army Institute of Research)

Swedish Occupational Fatigue Inventory (SOFI)
Three Process Model of Alertness (TPMA) – associated with Åkerstedt, Karolinska Institute and National Institute for Psychosocial Medicine
Truck Operator Proficiency System
Vehicle Driver's Anti-Dozing Aid (VDADA) – BRTRC Technology Research Cor

3.9   Glossary

Acute Fatigue: Acute fatigue is defined by the NRC as fatigue resulting from causes occurring within the past 24 hours; however, definitions vary, with some using a longer time interval (e.g., days).

Alertness: The ability to remain awake and sustain attention. The state of physical and mental readiness that maximizes individual performance. Alertness is a dynamic state and may vary from second to second.

Biological Clock: A common term referring to a bundle of nerves in the brain that regulates the timing of biological functions associated with the circadian sleep/wake cycle, as well as fluctuations in alertness while awake and stages of sleep; also referred to as the internal clock or circadian clock.

Circadian Disruption: A disturbance of the circadian rhythm resulting from deviations in expected external cues. These deviations could result from such factors as flying across multiple time zones, working irregular schedules, working at night, or exposure to light at irregular times of day.

Circadian Pacemaker: A cluster of neurons residing in the pineal gland, the section of the brain known as the suprachiasmatic nucleus (SCN) that exerts gentle control over the sleep-wake cycle; the biological clock.

Circadian Rhythms: Circadian rhythms are biological rhythms of physiology and behavior that have a period of approximately 1 day (24 hours) (*circa* translates as "around or about" and *dies* translates as "day").

Cumulative Fatigue: The increase in fatigue over consecutive sleep-wake periods resulting from inadequate rest.

Electroencephalogram Spectra: The range and distribution of wavelengths obtained by an electroencephalograph.

Electromyography (EMG): A technique for evaluating and recording the activation signal of muscles using an instrument that detects the electrical potential generated by muscle cells when these cells are both mechanically active and at rest to produce a record called an electromyogram.

Fatigue: The degradation in an individual's cognitive and motor functioning resulting from inadequate rest.

Fatigue Management: Activities designed to identify, assess, and address factors that contribute to fatigue.

Hertz: A unit of frequency, a hertz is the number of complete cycles per second.

Homeostasis: The regulatory mechanisms that maintain the constancy of the physiology of organisms; sleep homeostasis is the mechanisms by which a sleep deficit elicits a compensatory increase in the intensity and duration of sleep and excessive sleep reduces sleep propensity (Tobler and Achermann 2007).

Hypovigilance: Intermediate status between waking and sleeping, or status as when under the influence of narcotics, in which the body's powers of observation and analysis are very small.

Limbo Time: Time on duty but not on actual work duty.

Melatonin: A hormone produced by the pineal gland in the brain that is a biological marker for the onset of sleep.

Ocular Motor: Relating to or causing movements of the eyeball.

Pupillometry: Measurement of the pupil of the eye and its reaction to stimuli.

Saccade: A small rapid jerky movement of the eye especially as it jumps from fixation on one point to another.

Sleep Debt: A condition that occurs when a person gets less sleep than his/her biological sleep need; sleep debt can accumulate over time.

Sleep Homeostasis: The sleep-wake-dependent aspect of physiological sleep regulation that counteracts deviations from an average "reference level" of sleep, augmenting sleep propensity when sleep is curtailed or absent and reducing sleep propensity in response to sleep. Sleep timing depends upon a balance between homeostatic sleep propensity, the need for sleep as a function of the amount of time elapsed since the last adequate sleep episode, and circadian rhythms that determine the ideal timing of a correctly structured and restorative sleep episode.

Sleep Hygiene: Conditions and practices that promote good sleep. These include regular sleep timing; limiting food, alcohol and caffeine before sleep; controlling noise, light and temperature; good-quality bedding; and use of relaxation or sleep-inducing techniques, as needed.

Sleep Inertia: A period of drowsiness and impaired performance immediately after the transition from sleep to wakefulness.

Sleep Intensity: The intensity component of sleep is slow-wave activity, whose level correlates positively with the threshold to arouse subjects or animals. Slow-wave activity is defined as spectral power of the electroencephalogram (EEG) in the frequency range of approximately 0.5 – 4.0, or 4.5 Hz.

Sleep Latency: The length of time it takes to fall asleep.

APPENDIX A: APPLIED FATIGUE RELATED TECHNOLOGIES

A.1 Physical/Physiological Monitoring Systems

A.1.1 Computer Vision Systems: Dashboard Mounted and Eyewear

Driver State Sensor (DSS) – from Seeing Machines

This system uses an in-cab camera that does not use infra-red (IR) reflections. As a result, it is able to operate during daytime as well as night and when prescription lenses or sunglasses are worn. The key data recorded include: head position, eyelid closure levels, and eye gaze relative to direction of the vehicle. It measures immediate acute sleepiness and has a long-term drowsiness metric based on eyelid movement (degradation of alertness before sleep or microsleeps occur). The technology can detect the initial stages of drowsiness before the onset of microsleeps and can warn the driver. There are also two-way dispatcher/driver warnings that allow for timely driver intervention that could significantly reduce drowsy driving and increase safety.

Status: Seeing Machines is a developer of advanced computer vision systems, including non-contact head and eye-tracking systems. In 2005 it joined with the Australian National Laboratory for Information and Communications Technology (NICTA) to develop a system that tracked and combined measurements of the driver (such as head movement, blink rate) with measurements of steering wheel movement and lane departure. The resulting Driver State Sensor (DSS) platform analyzes this information to detect and help the driver counter drowsiness and distraction. In 2010, the DSS system was in operation in a number of U.S. mining company fleets (source: SM- News, May 2010).

FaceLAB™ 5 – from Seeing Machines

The following information is from the evaluation of emerging driver fatigue detection measures and technologies by the Volpe Center (Barr, et al. 2009:16-17). FaceLAB™ 5 is identified by Seeing Machines as a research rather than a user tool. It provides complete face and eye tracking and generates data on eye movement, head position and rotation, eyelid aperture, lip and eyebrow movement, and pupil size. It includes a suite of fully-integrated analysis tools called EyeWorks™. Eyeworks™ can provide real-time blink analysis and PERCLOS. The camera and illuminator are installed on the dashboard as in the image above. The system operates through a range of lighting and movement conditions and recovers quickly if the subject leaves the field of view, and can accommodate subjects wearing sunglasses by relying on analysis of head pose, eyes, and mouth. Specifications of the measures and analyses are available at the Seeing Machine website.

Status: FaceLAB™ is installed in several driving simulators at universities and has been used in test track experiments by NHTSA. A single camera system is installed on a Volvo Safety Truck demonstration vehicle. The road system configuration employs two cameras and an infrared light illuminator. A prototype in-vehicle system for Volkswagen that will include a warning alert feature is being developed.

Sleepiness Detection System (SDS™) – developed by Biocognisafe Canada, Inc.

Detects and warns operators when hypovigilances and micro-sleeps are occurring. The technology has incorporated improvements on past limitations of ocularmotor monitoring devices, mainly limitations with lighting conditions and eye-tracking with movement or types of eyewear.

Status: Prototype; not commercially available. In 2009, the company announced that its patent had been successfully registered in France and Germany.

CoPilot® – manufactured by Attention Technologies of Pittsburgh, PA.

CoPilot uses percentage of slow eyelid closure (PERCLOS). The infrared measurement system is located to the right of the steering wheel, and the alertness feedback is located to the left of the infrared unit. The feedback unit provides a digital numeric readout with values from 0 – 99, with 0 indicating maximum eyelid closure and 99 indicating maximum eyelid opening. The CoPilot technology is shown in the accompanying picture.

Pilot Test Results: This technology was reviewed by Grace (2001) and pilot tested in a study sponsored by FMCSA (see results in Dinges et al. 2005). Drivers in the study did not perceive CoPilot to be helpful and very few indicated they would want one in their trucks or recommend it.

Status: Prototype; piloted tested.

Eye-Com™ Biosensor-Communicator-Controller (EC-9) – developed by Eye-Com Corp.

EC-6 is a wearable, wireless eyeframe that can accommodate any lens (sunglass or prescription lens). It has been tested in a car/truck simulator study that was sponsored by DOT and DoD and has been certified as airworthy for Black Hawk helicopter use by the U.S. Army. It uses PERCLOS, eye blink duration/frequency, eye gaze, pupil size and speed of dilation as indices (a total of 20 eye measures) of driver impairment. It includes an arousal alarm, which can also be sent to a remote source. It uses a nontrackable wideband signal to ensure privacy. Eye-Com Corp. was awarded a Congressional Initiative grant in late 2009 to develop the next version (EC-9). This version will have a more streamlined design and improved capabilities, and will be tested it in different simulated operational scenarios (such as underwater).

Status: Being adapted to specific in-field applications and is in use with the U.S military.

Driver Fatigue Monitor (DFM) DD850 – developed by Attention Technology, Inc.

The following information is from the evaluation of emerging driver fatigue detection measures and technologies by the Volpe Center (Barr, et al. 2009:12-13). The DD850 DFM, designed for nighttime driving by commercial truck drivers, is mounted on a vehicle's dashboard to provide a continuous, real-time measurement of eye position and eyelid closure. It includes a camera module mounted on a rotating base, which allows adjustment of the camera angle by the driver. Its display has a visual gauge that indicates the driver's drowsiness level and includes a three-stage warning signal capability that activates at preset drowsiness thresholds. The DFM estimates PERCLOS and a "bright pupil" method to track eye movement. Reviews have indicated that sunlight can interfere with its functioning. The device has been field tested on the road in a project co-sponsored by FMCSA and NHTSA. The field test involved 37 vehicles and 102 truck drivers, each driving for 17 weeks.

Eyegaze Analysis System – developed by LC Technologies, Inc.

Much of the following information and the image are from the evaluation of emerging driver fatigue detection measures and technologies by the Volpe Center (Barr, et al. 2009:22). The Eyegaze Analysis System is a device for measuring, recording, playing back, and analyzing an individual's eye gaze. The system includes basic video equipment, computer hardware, and Eyegaze software that is used to develop and run eye tracking applications. The remote video camera is mounted below a computer monitor. The Eyegaze Analysis System tracks the subject's gaze point on the screen automatically and in real time at 60 Hz. The system determines gaze direction using the pupil center corneal reflection (PCCR) method. The system includes a small, low-power infrared illuminator that is located at the center of the camera lens to illuminate the eye and provide a direct reflection off the cornea of the eye. It also includes a warning system. The goal of the system is to monitor the driver's eye point-of-regard, saccadic and fixation activity, and percentage eyelid closure reliably, in real time, and accurately under all anticipated environmental conditions.

Status: No specific applications have been developed, although work has continued on the development of the camera/sensor instrument and data integration systems.

Optalert™ – developed by Sleep Diagnostic, Ltd.

Optalert is an eyewear technology, developed primarily for the transportation sector, that is based on infrared reflectance oculography. The system continually monitors the user's drowsiness level with brief pulses of infrared light (500/sec) that are directed to the wearer's left eyelid. The reflected light is detected by a phototransmitter mounted on the eyeglass frame, which also detects the level of environmental infrared light immediately prior to each light pulse. A microprocessor housed in the arm of the glasses controls the timing of the light pulses and digitizes the analog output from the sensors. The power supply and serial output from the glasses are provided via a cable, which connects to a USB port mounted alongside the subject's seat and then to the processing unit that is permanently installed in the vehicle. The processing unit provides a variety of analyses, including a calculation of drowsiness based on the Johns Drowsiness Scale. It issues a visual and audible warning when predefined scores are reached. The developer of this technology, M.W. Johns, works at the Epworth Sleep Center in Australia and has published a number of evaluations of the methodology in peer reviewed journals.

Information obtained primarily from the Optalert website.

Status: Optalert has been field tested and is in commercial use.

Driver State Monitor (DSM) – developed by Delphi, Inc.

Much of the following information is from the evaluation of emerging driver fatigue detection measures and technologies by the Volpe Center (Barr, et al. 2009:12-13). Caterpillar and Delphi worked together to evaluate this technology for use in large mining equipment such as Caterpillar's large off-highway haul trucks. The device is designed to address both drowsiness and distraction by drivers. The device employs a single camera and two infrared illumination sources to track the driver's facial features, eye closures, and head pose and a single high-fidelity imaging sensor that uses an algorithm to predict AVCLOS. AVCLOS, a binary measure indicating whether the eye is open or closed, is the main fatigue parameter of the DSM system. This allows use of a less powerful data processor than is required to analyze PERCLOS. Validation testing of the DSM by Delphi showed overall strong correlations between AVCLOS and PERCLOS and with driving performance as measured by Variation of Lane Deviation (VLD).

Status: According to the Volpe review, Delphi is developing an automotive-grade system.

Driver Drowsiness Monitoring System (DDMS)/Drowsy Driver Warning System (DDWS) – developed by Virginia Polytechnic Institute and State University

The Virginia Tech Transportation Institute conducted a field operational test of the DDWS to collect data on the safety benefits and operational capabilities and limitations of a system that detected indications of driver drowsiness and provided feedback to drivers. The study was funded by the U.S. DOT as a research project and generated over 12 terrabytes of data. The system used an instrumented heavy vehicle and collected video and sensor data from the 103 driver participants, some of whom drove the vehicle on their regular routes for periods up to 6 weeks. The system included four cameras and an actigraph. The system used cameras to capture images of the driver's face and eyes as well as the forward roadway and adjacent lanes. Robust multi-dimensional monitoring system that combines metrics, particularly PERCLOS and lane position. Several students prepared theses analyzing data from the operational field test (see Wierwille et al. 1996; Olson 2006; Baker 2007).

Status: Prototype for operational field test; not commercially available.

Drowsy Driver Detection System (DDDS) – developed by John Hopkins University

The following information is from the evaluation of emerging driver fatigue detection measures and technologies by the Volpe Center (Barr, et al. 2009:23) and the John Hopkins Applied Physics Laboratory website. The DDDS is a small sensor system that measures and analyzes general activity level, speed, frequency, and duration of eyelid closure, heart rate, and respiration to alert drivers when they are becoming drowsy. The device collects information on the speed, frequency, and duration of eyelid closure, rate of heartbeat and respiration, and pulse rate via a Doppler radar system combined with a transceiver similar to those used in garage door openers. The system is small (approximately 3x3x2 inches for the transceiver and 1x2x3 inches for the support electronics. A limited test showed good correlation between the system measurements and those taken with a validated PERCLOS methodology.

Status: Prototype available for licensing; not commercially available. No updated information about applications of this technology past 2004 was found.

Smart Eye Pro 5.4 – developed by Smart Eye AB

The following information is from the evaluation of emerging driver fatigue detection measures and technologies by the Volpe Center (Barr, et al. 2009:18-19) and the Smart Eye website. Smart Eye provides hardware/software systems that produce high accuracy in tracking head, eyelid, and gaze. Smart Eye specializes in computer vision software that enables computers and machines to sense and make use of human face and eye movements. Smart Eye Pro™ 5.0 and 5.4 providea system of up to five high speed 60-Hz cameras with flexible mounts to enable non-intrusive measurements of head pose and eye gaze in real time and in locations that may pose challenges for camera location. The system tracks an individual's facial features relative to a 3D head model. The combination of multiple cameras and the 3D modeling allows precise identification of gaze and can track a subject's movements within a large "head box" area. It employs a fast face detection procedure to relocate the subject's face and resume tracking. Measurement data may be synchronized with external time sources and the system can be set up for remote control functionality for the automation of experiments. Video recordings from the Smart Eye cameras can be analyzed offline. Third-party products such as E-prime™, Gaze Tracker™, and Net Station™ can be integrated with the system.

SmartEye AntiSleep™ 2.0 is designed specifically for the automotive industry. The single camera and IR illuminator is typically mounted in an integrated unit, although they can be placed separately, if necessary. The IR illuminators and filters are tuned to frequencies designed to perform well in both daylight and nighttime conditions. The device measures the driver's head position and orientation, gaze direction, and eyelid opening (at a rate of 60 Hz). The system detects generic and person-specific facial features and maps them onto a generic 3D head model. A patented illumination technique is used to eliminate reflections from eyeglasses, and the system has been tested and calibrated for male and female drivers from a wide range of ages and ethnic groups. The measurement data output include confidence values based on the estimated quality of the measurements. However, according to the Barr assessment, it does not include an algorithm to monitor drowsiness. Technical details available from Smart Eye website.

Status: Volvo, Volkswagen, BMW, and all European truck manufacturers are testing a Smart Eye Pro single camera system with a PC-based processor. Initial indications are that the on-road field tests have gone well and that it has demonstrated the ability to work in all illuminations and with individuals of varying eye traits and with and without eyeglasses.

InSight™ – developed by SensoMotoric Instruments, GMBH

SensoMotoric Instruments specializes in gaze and eye-tracking systems that are being applied to a widening range of psychological and usability research and other specialized applications. The following information is from the evaluation of emerging driver fatigue detection measures and technologies by the Volpe Center (Barr, et al. 2009:19-20) and the company website. Another advanced, non-invasive computer vision-based operator monitoring system, InSight™ measures head position and orientation, gaze direction, eyelid opening, and pupil position and diameter. The camera's sampling rate is 120 Hz for head pose and gaze measurement and for eyelid closure and blink measurement, and 60 Hz for combined gaze, head pose, and eyelid measurement. The system calculates PERCLOS to determine the operator's state of alertness. The eye closure measurement is reported to have an accuracy of 1 mm. It employs an automatic and robust tracking algorithm that the developers affirm operate under all lighting conditions.

Status: Extensive studies have been conducted with truck drivers and passenger car drivers and the system has been used to compare simulator and on-road performance (detecting drowsiness and microsleeps). In 2005, SensoMotoric Instruments was working with Volkswagen and BMW to incorporate the system in the non-test driving environment. SensoMotoric Instruments indicates that more than 4,000 of their systems are in operation worldwide. In 2010, automotive fatigue monitoring was not highlighted on their website, and no updated information was found.

A.2 Biological/Neurological Monitoring Systems

B-Alert™ – developed by Advanced Brain Monitoring, Inc.

Patented wireless sensor headset system that acquires high quality electroencephalography (EEG) and electrocardiography (ECG or EKG) signals via a sensor headset that is light-weight (less than 4 ounces) and designed to fit under helmets or other headgear. Its patented EEG sensor headset eliminates the need for hair or scalp preparation. The analog circuit combined with EEG amplification close to the sensors and online impedance monitoring enhances high-quality EEG data. The wireless EEG allows the user freedom of movement without generating artifacts obtained with wired systems. Versions that provide quantitative information about head movement and position, and pulse rate, are also available. Price for EEG 6-channel sensor headset – B-Alert configuration; 4 wireless sensor caps, and external syncing unit (ESU), approximately $20,000. The company has developed an "Alertness and Memory Profiler) that uses a test battery of the 3-choice vigilance test, image recognition, image recognition with interference, verbal/number-image paired associate learning, and the Sternberg verbal memory scan to assess physiological and neurocognitive factors. According to Inc magazine, the company has received over $14 million in grants from the National Institutes of Health and the U.S. Department of Defense (see http://www.inc.com/inc5000/profile/advanced-brain-monitoring).

Status: Being marketed for fatigue and alertness monitoring and sleep apnea evaluation.

Sleep Band/ReadiBand – developed by Archinoetics, LLC and its subsidiary/Fatigue Science

Most companies sell the software that outputs the basic sleep/wake calculation. These technologies tend to cost between $1000-1500. Archinoetics has a different strategy. Their actigraphs are less expensive because they sell their sleep band without the analysis software. The sleep band is commercially available but the software to calculate the results is intentionally not sold to the customer. Rather, there is a method for downloading the data from the sleep band and to send the downloaded data via the Internet back to Archinoetics for analysis. This approach allows the system to cost less and permits more sophisticated analyses to be conducted. Archinoetics takes the sleep patterns and integrates them into a fatigue risk analysis. In 2010, the ReadiBand Report uses data from the ReadiBand to provide statistics on sleep efficiency, sleep duration, and time to sleep onset, as well as a calculation of "fatigue risk levels." According to the Fatigue Science website, the ReadiBand is endorsed by the Federal Aviation Administration, U.S. Department of Defense, and the Federal Railway Administration

Sleep Band/Readiband is part of a larger set of fatigue risk analysis products. Archinoetics/Fatigue Science has four basic products:

- Fatigue Avoidance Scheduling Tool (FAST) – a tool to manage and evaluate fatigue as part of the scheduling process, with a version designed specifically for aviation;
- Sleep Band (in 2010, this product was called ReadiBand) – this provides the method of assessing where interventions or changes resulted in reduced fatigue. The data from the sleep bands can produce individual and group level results;
- XSRiskPro – a schedule optimizer that ranks different schedule designs on fatigue risk (in 2010, this was a "real-time" assessment of ReadiBand data to provide a fatigue risk level indicator);

- Fatigue Risk Management System technical support.

Status: All the products are commercially available; a few field applications.

SleepWatch® Actigraph – produced by Precision Control Design, Inc.

The SleepWatch, a product of Integrated Safety Support, combines actigraphy with a biomathematical fatigue model called the *Sleep Management System*, developed by the Walter Reed Army Research Institute.

FMCSA sponsored a pilot study both in the United States and Canada that included this technology (see results of this study in Dinges et al. 2005). The SleepWatch is based on a similar concept and technology as the ReadiBand, described above. In comparisons conducted on relatively early versions of these two products, the SleepWatch received somewhat lower ratings. The SleepWatch is being used by a variety of customers, including Action Rescue and EMQ Helicopter Rescue in Australia, and the U.S. military. The 2007 versions can monitor/report environmental conditions (temperature, humidity, and solar radiation), and life measures (to address time when the watch is not worn).

Status: Four versions of the SleepWatch are commercially available.

SENSATION – a European Sleep and Research Society

SENSATION focuses on advanced sensor development for use in attention, stress, vigilance, and sleep/wakefulness monitoring. This system is currently under development. It is exploring a wide range of micro- and nano-sensor technologies, with the aim of developing unobtrusive, cost-effective, real-time monitoring, detection, and prediction of the human physiological state in relation to wakefulness, fatigue , and stress anytime, everywhere, and for everybody. They are working on 17 micro sensors and two nano sensors for brain monitoring, that will include wearable, eye-related posture and motility and autonomic functions sensors that are wirelessly integrated through a body/local/wide area network. Applications include medical diagnosis and treatment as well as monitors to detect and predict hypovigilance in operators in a variety of industrial and environments settings.

Status: Unknown. Although the website is current, little new information has been posted since 2005.

A.3 Performance Testing Technologies

Factor 1000 – developed by Systems Technology, Inc

Factor 1000 is a version of the critical tracking test to measure hand-eye motor skills. The computer-based test requires the test-taker to control the random movements of a cursor between two markers on a computer screen. The test was developed by Systems Technology Inc. of Hawthorne, California, which licensed it to Performance Factor Inc. It was developed in part to provide an alternative to drug and alcohol testing.

Status: Factor 1000 was commercially available between 1995 and 2000, but no current references to its use were found.

Palm-performance vigilance test (PVT) – developed by the Walter Reed Army Research Institute (Thorne et al. 2005b)

This hand-held device collects PVT data in a 5-minute period. The original PVT was delivered on a microcomputer that was made available commercially by Ambulatory Monitoring Inc., in Model 192. It required a 10-minute data collection period on a device that measured 21 x 11 x 6 cm and weighed 658 g. Feedback from field studies suggested that a briefer and more portable version of the PVT would be advantageous. The Palm-PVT described below achieved both objectives. There is ongoing work within NASA to further reduce PVT duration to 3 minutes (Dinges, personal communication; NASA website on PVT self-test).

The device was evaluated in several experiments (Lamond et al. 2005; Lamond et al. 2008), which showed that the 5-minute portable PVT is sensitive to sustained wakefulness, as indicated by reliable increases in reaction time lapses (Thorne et al. 2005b).

Pilot Test Results: The Palm PVT technology was pilot tested in a study sponsored by the FMCSA (see results in Dinges et al. 2005).

Status: No longer commercially available, significant because it served as market leader for subsequent hand-held devices.

Bowles-Langley Technology (BLT) Impairment Test – developed by Bowles-Langley Technology, Inc.

Developed in part with funding from the National Institute for Occupational Health and Safety (NIOSH), the BLT impairment test is a brief, inexpensive, computerized shape recognition test that requires the user to make a Yes/No decision about whether all items in a given screen are the same. After a series of 50 screens, the resulting speed/accuracy-based score is compared to the user's baseline. A stability trial, laboratory sleep deprivation trial, and a workplace feasibility trial with emergency department doctors were used in the process of test development. It has been included in several studies evaluating candidate screening tests. The current test version is sensitive to severe impairment rather than detecting gradual alertness changes – further refinements are need to enhance sensitivity to gradual alertness changes.

A.4 Actual Job Performance Monitoring/Evaluation Technologies

Driver Alert Support (DAS) system – developed by Volvo Technology

The DAS is an elementary lane deviation system that was introduced into the market in 2008.

Pilot Test Results: The Federal Motor Carrier Safety Administration (FMCSA) sponsored a pilot study both in the US and Canada that included this technology. The results were published in 2005 (Dinges et al. 2005). Fewer than half the Canadian drivers and less than 20 percent of the U.S. drivers believed that the lane tracking alertness index was helpful. The DAS illustrates the evolutionary process of technology development in this area. The experience gained from this field test was used as a basis for next-generation technology.

Status: No longer in use.

Artificial Neural Network (ANN) – developed by George Washington University Center for Intelligent System Research

The following information is from the evaluation of emerging driver fatigue detection measures and technologies by the Volpe Center (Barr, et al. 2009: 26). The ANN is designed to detect driver drowsiness by sensing and analyzing the steering angle patterns, classifying them into drowsy- and non-drowsy-driving intervals. The system sensors are focused on vehicle performance rather than driver attributes, and is therefore less invasive. The system used steering wheel movements that prior research has shown are correlated with a driver's state of impairment: when alert, a driver makes small amplitude movements of the steering wheel to maintain the vehicle's position in the lane, when impaired, the steering wheel movements become larger in amplitude and less precise. This causes the vehicle to undergo sharp changes in trajectory. Neural net systems require training, but are capable of dealing with very complex functions. Tests of the ANN system on a simulator after it had been trained confirmed that it was effective (accuracy = 90 percent) in detecting drowsiness.

Status: Prototype, undergoing further testing to further refine and validate the algorithm, conduct additional simulator experiments, and conduct on-road tests. In addition, the plan was to research integration of a warning system with the detection system.

A.5 Fatigue Self-Assessment Technologies

Driver Stress Inventory (DSI), Driver Fatigue Questionnaire (DFQ), and Driver Risk Index (DRI) ™ – developed by DriverMetrics® and distributed in the United Kingdom by Peak Performance

The DSI is a survey instrument that has been used in a number of pilot and research studies (see Hennessy 1999; Matthews et al. 2009; and Öz et al. 2010) in which the objective has been to develop and validate to validate a multi-dimensional scale of fatigue. Research using the inventory has focused on correlating scores on the index with experimental manipulations of drive duration and workload factors. Experiments with a driving simulator have identified seven correlated factors: (1) muscular fatigue, (2) boredom, (3) confusion, (4) performance worries, (5) comfort-seeking, (6) self-arousal, and (7) a single factor representing exhaustion and sleepiness, that are measured in the DSI (Matthews et al. 2009). Performance worries show the largest increase and confusion the smallest, which confirms the importance of using a multi-dimensional scale of fatigue. The DSI is referenced frequently in the fatigue literature.

DFQ is a multi-dimensional fatigue questionnaire developed and tested with drivers in mind but a version of this questionnaire could be tested and validated for other work contexts. Further testing is under way to validate the DFQ. DSI is an earlier version of this subjective fatigue questionnaire, developed by some of the same researchers. The DSI includes a fatigue-proneness scale that was validated in both simulated and real-life driving studies. The DSI fatigue-proneness scale is the strongest predictor of DFQ scores.

The Driver Risk Index (DRI) is targeted to fleet managers, in particular, and draws upon items from the DFQ and DSI. It is based on psychometric principles to "assess the thoughts, feelings and behaviours underlying driver risk." The DSI has been tested for its ability to predict traffic accidents in a number of studies, summarized by Dorn (2010).

Status: DRI is commercially available in the United Kingdom, and the DSI is often used in research.

24/7 Lifestyle Planner Tools (Family Planner and Personal Pocket Planner) – Fatigue Management Solutions Ltd.

Fatigue Management Systems provides both technical assistance and tools that are designed to help employees "manage fatigue risks of their actual work patterns." The tools are built on the platforms of the Activity Rest Cycles (ARC) and Fatigue Index Risk Measurements (FIRM) software systems that provide user-friendly graphics to highlight and illustrate fatigue issues and challenges. FIRM uses biomathematical models to generate indicators of fatigue risk in a planning/scheduling format.

Status: Family Planner and Personal Pocket Planner are commercially available.

Sleepiness Scales (measures subjective sleepiness)

A number of scales have been developed for individuals to use to assess their current (momentary) degree of alertness or sleepiness. The best known of these are the Karolinska Sleepiness Scale, developed by Åkerstedt (1996), the Stanford Sleepiness Scale, developed by Hoddes and colleagues (1972), the Epworth Sleepiness Scale, developed by Johns (1991). In addition, visual analog scales (e.g., a horizontal line with anchors of very sleepy and very alert) are also used. The Karolinska Sleepiness Scale (KSS) is a semi-quantitative standardized 9 or 10-point scale on which individuals rate their sleepiness during the previous 10 minutes. The scale has a known threshold for the occurrence of microsleeps, validated by correlating individual's ratings with EEG and electrooculogram (EOG) signals (that reveal microsleeps). It has also been validated with electroencephalographic, behavioral, and other subjective indicators of sleepiness (for example, the Epworth Sleepiness Scale, the Stanford Sleepiness Scale, and the Samn-Pereli fatigue scale) (Kaida et al. 2006). The process can be implemented as a paper and pencil, hand-held device, or computer-based process.

A.6   Schedule and Roster Management Tools

Fatigue Avoidance Scheduling Tool (FAST) – developed by the U.S. Air Force/Science Applications International Corporation (SAIC)

The U.S. Air Force, using SAIC as a contractor, initiated the development of FAST in 2000 to address safety issues related to scheduling. Consequently, FAST was built with a particular focus on circadian rhythms and the special requirements of scheduling airline crews who may cross many time zones in the course of their shifts. Fatigue Science now markets two versions, FAST® and FAST® Aviation. Because it was intended for use as an actual scheduling tool, FAST was developed as a Windows program. It is designed to be used to analyze existing and past work schedules and past deviations from these schedules (see Gertler et al. 2009). Based on the cumulative experience of this analysis, FAST is also used to inform the design of schedules and to assess schedules after they have been assembled. FAST was derived from the SAFTE simulation, also developed by Dr. Hursh. FAST was validated by Hursh, et al. (2008). Fatigue predictions in FAST are derived from the Sleep, Activity, Fatigue, and Task Effectiveness (SAFTE) simulation model, also developed by Hursh of Johns Hopkins University. It has undergone a number of field tests, including for the military and aviation sectors. The

APPENDIX B: A NATIONAL TRANSPORTATION SAFETY BOARD METHODOLOGY FOR INVESTIGATING OPERATOR FATIGUE IN A TRANSPORTATION ACCIDENT

Initial Screening Questions

If any of the following is true, proceed with the detailed methodology:

- Does the operator's 72-hour history suggest little sleep, or less sleep than usual?
- Did the accident occur during times of reduced alertness (such as 0300 to 0500)?
- Had the operator been awake for a long time at the time of the accident?
- Does the evidence suggest that the accident was a result of inaction or inattention on the part of the operator?

Detailed Methodology

It is important to establish two factors before concluding that operator fatigue contributed to an accident. First, determine whether the operator was susceptible based on sleep lengths, sleep disturbances, circadian factors, time awake, and/or medical issues. Second, if it is determined that the operator was likely experiencing excessive fatigue, evaluate information concerning the operator's performance, behaviors, and appearance at the time of the accident to determine whether they were consistent with the effects of fatigue.

A finding that the operator was susceptible to the development of a fatigued state in the absence of performance or behaviors consistent with fatigue should not be used to support operator fatigue as a probable cause or contributing factor in the accident, but may still be an important safety issue to be addressed in the accident report.

Part 1: Determine whether the operator was susceptible to fatigue.

Sleep Length

Determine whether the operator had acute or chronic sleep loss by documenting sleep/wake patterns for at least 72 hours before accident and learning about the operator's "normal" sleep habits.

- Ask operator:
  - Describe your typical sleep pattern of when you go to bed, awaken, and how much sleep you get during days off.
  - What time did you fall sleep the night before the accident? What time did you wake up? What was the quality of your sleep? (Repeat for two nights before, three nights before, etc.)
  - Did you take any naps? When, where, for how long, and why?
- Interview family members, hotel staff or other witnesses who can help complete the operator's sleep/activity schedule before the accident.
- Use receipts, cell phone records, work schedules, log books, alarm clock setting, or other records to help complete the operator's sleep/activity schedule before the accident.

Fragmented/Disturbed Sleep

Determine if the operator's sleep was fragmented (e.g., multiple sleep episodes per 24-hour period) and/or disturbed (e.g., awakenings during sleep due to internal or environmental factors) in days leading to accident.

- Use sleep/wake information collected in "Sleep Length" to examine the lengths and patterns of sleep episodes for split sleeps or daytime sleep.
- Ask operator (or determine through interviews with family members):
  - Are there factors in your environment (e.g., noise, light, phone calls, etc.) that interfere with your sleep?
  - Was your sleep pattern different or disrupted in the days leading to the accident?

Circadian Factors

Determine if accident happened during a circadian low point. The primary circadian trough is approximately midnight to 0600, especially 0300 to 0500, while a secondary "afternoon lull" occurs at approximately 1500 to 1700. Also, determine if the operator suffered from circadian issues due to recently crossing multiple time zones or to rotating, inverted or variable work/sleep schedules.

Sleep Disorders, Health, and Drug Issues

Determine if sleep disorders or other medical factors (e.g., disease or drug use) were present in the operator's history.

- Ask operator:
  - Do you have difficulty falling asleep or staying asleep?
  - Have you ever told a doctor about how you sleep? If so, why, when, and what was the result?
  - What drugs/medications do you use regularly, and did you take any in days prior to the accident?
  - Do you have any medical concerns that affect sleep (e.g., chronic pain, GERD, etc.)?
- Review operator's toxicological results for substances that may affect sleep or alertness.
  - If applicable, have the operator evaluated by a physician who specializes in sleep medicine.
  - Other evidence sources include the operator's medical or pharmacy records, or any drugs or medicine found within the wreckage.

Time Awake

Determine how long the operator had been awake at the time of the accident, using interviews or records to estimate wake up time from most recent significant sleep before the accident.

Additional Suggestions

- Check work records and records of previous accidents/incidents (including DMV and/or insurance records) for evidence of prior falling asleep during vehicle operation.

- Determine what kind of training the operator had received regarding fatigue management.
- Review operator's environment and tasks for unusual conditions on the accident day that would depress arousability, like low lighting, operational delays, or boredom.
- Determine whether representatives of management of labor union parties have indicated complaints of operator fatigue in the recent past.

Part 2: Determine whether the operator's performance, behaviors, or appearance were consistent with the effects of excessive fatigue, and whether their performance or behaviors contributed to the accident.

Operator Performance

Determine whether the operator's performance was consistent with the effects of fatigue.

- Use available evidence to determine whether the operator's performance was deteriorating prior to the accident. For example:
  - Did the operator overlook or skip tasks or parts of tasks?
  - Was there steering or speed variability?
  - Did operator focus on one task to the exclusion of more important information?
  - Was there evidence of delayed responses to stimuli or unresponsiveness?
  - Was there evidence of impaired decision-making or an inability to adapt behavior to accommodate new information?

Operator Behaviors and Appearance

- Determine whether the person's appearance or behaviors before the accident were suggestive of sleepiness/fatigue, as based on witness interviews, operator report of being tired, audio or video records of the operator's behavior.

NRC FORM 335 (12-2010) NRCMD 3.7	U.S. NUCLEAR REGULATORY COMMISSION	1. REPORT NUMBER (Assigned by NRC, Add Vol., Supp., Rev., and Addendum Numbers, if any.)
	BIBLIOGRAPHIC DATA SHEET *(See instructions on the reverse)*	NUREG/CR-7156 (FINAL) PNNL-19222

2. TITLE AND SUBTITLE	3. DATE REPORT PUBLISHED	
Fitness for Duty in the Nuclear Power Industry: An Update of Technical Issues on Drugs of Abuse Testing and Fatigue Management	MONTH	YEAR
	June	2013
	4. FIN OR GRANT NUMBER	

5. AUTHOR(S)	6. TYPE OF REPORT
Kristi Branch Kathryn Baker Marina Skumanich Nancy Durbin	Technical
	7. PERIOD COVERED (Inclusive Dates)

8. PERFORMING ORGANIZATION - NAME AND ADDRESS (If NRC, provide Division, Office or Region, U. S. Nuclear Regulatory Commission, and mailing address; if contractor, provide name and mailing address.)

Pacific Northwest National Laboratory
P.O. Box 999
Richland, WA 99352

9. SPONSORING ORGANIZATION - NAME AND ADDRESS (If NRC, type "Same as above"; if contractor, provide NRC Division, Office or Region, U. S. Nuclear Regulatory Commission, and mailing address.)

Division of Risk Analysis
Office of Nuclear Regulatory Research
U.S. Nuclear Regulatory Commission
Washington, DC 20555-0001

10. SUPPLEMENTARY NOTES

11. ABSTRACT (200 words or less)

This report is part of a series of updates to technical issues concerning fitness for duty in the nuclear power industry. It discusses technologies relevant to the detection and management of two key elements of a fitness-for-duty program: drug and alcohol testing and fatigue management. On drug and alcohol testing, the report provides an introduction to the pharmacokinetics of drugs of abuse in different bodily fluids and substances (matrices), a review of the technologies used to separate, identify, and quantify drugs in workplace drug testing programs, and a description of emerging research in developing and validating the technology systems capable of testing alternative matrices as well as newly appearing drugs of abuse, both in the laboratory and at the point of collection. On fatigue management, the report reviews recent research on sleep and fatigue, describes efforts under way to develop and deploy technologies to aid fatigue assessment and management, reviews the status of fatigue management in industries and governmental sectors where fatigue is a significant safety concern, and discusses implications for the nuclear power industry. Finally each chapter includes an extensive bibliography of documents to support further, more in-depth reviews.

12. KEY WORDS/DESCRIPTORS (List words or phrases that will assist researchers in locating the report.)

Drug testing
10 CFR Part 26, Fitness for Duty program
Fatigue management
Alternative specimens
Alternative matrices
Alcohol testing
Pharmacokinetics
Point of collection

13. AVAILABILITY STATEMENT
unlimited
14. SECURITY CLASSIFICATION
(This Page)
unclassified
(This Report)
unclassified
15. NUMBER OF PAGES
16. PRICE

Printed
on recycled
paper

Federal Recycling Program

UNITED STATES
NUCLEAR REGULATORY COMMISSION
WASHINGTON, DC 20555-0001

OFFICIAL BUSINESS

NUREG/CR-7156

Fitness for Duty in the Nuclear Power Industry: An Update of Technical Issues on Drugs of Abuse Testing and Fatigue Management

June 2013

www.ingramcontent.com/pod-product-compliance
Lightning Source LLC
Chambersburg PA
CBHW080238180526
45167CB00006B/2331